HERBAL PRODUCT DEVELOPMENT

Formulation and Applications

HERBAL PRODUCT DEVELOPMENT

Formulation and Applications

Edited by
Anil K. Sharma, PhD, MPharm
Raj K. Keservani, MPharm
Surya Prakash Gautam, PhD, MPharm

First edition published 2021

Apple Academic Press Inc.
1265 Goldenrod Circle, NE,
Palm Bay, FL 32905 USA

4164 Lakeshore Road, Burlington,
ON, L7L 1A4 Canada

CRC Press
6000 Broken Sound Parkway NW,
Suite 300, Boca Raton, FL 33487-2742 USA

2 Park Square, Milton Park,
Abingdon, Oxon, OX14 4RN UK

© 2021 Apple Academic Press, Inc.

Apple Academic Press exclusively co-publishes with CRC Press, an imprint of Taylor & Francis Group, LLC

Library and Archives Canada Cataloguing in Publication

Title: Herbal product development : formulation and applications / edited by Anil K. Sharma, PhD, MPharm, Raj K. Keservani, MPharm, Surya Prakash Gautam, PhD, MPharm.

Names: Sharma, Anil K., 1980- editor. | Keservani, Raj K., 1981- editor. | Gautam, Surya Prakash, editor.

Description: Includes bibliographical references and index.

Identifiers: Canadiana (print) 20200292676 | Canadiana (ebook) 20200292692 | ISBN 9781771888776 (hardcover) | ISBN 9781003003182 (PDF)

Subjects: LCSH: Herbs—Therapeutic use. | LCSH: Natural products. | LCSH: Drug development. | LCSH: Materia medica, Vegetable.

Classification: LCC RM666.H33 H47 2021 | DDC 615.3/21—dc23

Library of Congress Cataloging-in-Publication Data

Names: Sharma, Anil K., 1980- editor. | Keservani, Raj K., 1981- editor. | Gautam, Surya Prakash, editor.

Title: Herbal product development : formulation and applications / edited by Anil K. Sharma, Raj K. Keservani, Surya Prakash Gautam.

Description: Burlington ON ; Palm Bay, Florida : Apple Academic Press, [2021] | Includes bibliographical references and index. | Summary: "This new volume, Herbal Product Development: Formulation and Applications, addresses some of the challenges that hinder the path of successful natural products from laboratory to market. Highly skilled, experienced, and renowned scientists and researchers from around the globe offer up-to-date information that describes characteristics of herbs and herbal products, applications, evaluation techniques, and more. There is also a section dedicated to alternative medicinal strategies for the treatment and cure of diverse diseases. Also considered, of course, is the efficacy and safety of herbal products, which are of major concern. This valuable volume will be an important addition to the library of those involved in herbal product development and testing, including researchers, scientists, academicians, industry professionals, and students in this area. Key features: Presents an overview of development and characterization of herbal products Considers the obstacles in the way of herbal product commercialization Recommends a thorough scientific and systemic approach to attempting to ready a viable herbal product for the market Looks at possible side effects of popular plant products Discusses cancer therapeutics through drugs derived from nature along with interactions when co-administered with synthetic counterparts Explores the applications of medicinal mushrooms other than as food Describes the principles and applications of naturopathy"-- Provided by publisher.

Identifiers: LCCN 2020031904 (print) | LCCN 2020031905 (ebook) | ISBN 9781771888776 (hardcover) | ISBN 9781003003182 (ebook)

Subjects: MESH: Plant Preparations | Drug Development | Phytotherapy--methods | Herbal Medicine--methods

Classification: LCC RM301.25 (print) | LCC RM301.25 (ebook) | NLM QV 766 | DDC 615.1/9--dc23

LC record available at https://lccn.loc.gov/2020031904

LC ebook record available at https://lccn.loc.gov/2020031905

ISBN: 978-1-77188-877-6 (hbk)
ISBN: 978-1-00300-318-2 (ebk)

About the Editors

Anil K. Sharma, PhD, MPharm

Anil K. Sharma, PhD, MPharm, is an Assistant Professor (Pharmaceutics) at the School of Medical and Allied Sciences, GD Goenka University, Gurugram, India. He has more than 10 years of experience in academics. He has published 28 peer-reviewed papers in the field of pharmaceutical sciences in both national and international journals as well as 16 book chapters and 13 edited books. His research interests encompass nutraceutical and functional foods, novel drug delivery systems (NDDS), drug delivery, nanotechnology, health science/life science, and biology/cancer biology/neurobiology. He graduated with a degree in pharmacy (BPharm) from the University of Rajasthan, Jaipur, India, and received a Master of Pharmacy (MPharm) from the School of Pharmaceutical Sciences, Rajiv Gandhi Proudyogiki Vishwavidyalaya, Bhopal, India, with a specialization in pharmaceutics. He earned his PhD at the University of Delhi.

Raj K. Keservani, MPharm

Raj K. Keservani, MPharm, is a Faculty of B. Pharmacy, CSM Group of Institutions, Allahabad, India. He has more than 12 years of academic (teaching) experience from various institutes of India in pharmaceutical education. He has published 30 peer-reviewed papers in the field of pharmaceutical sciences in national and international journals, 15 book chapters, 3 co-authored books, and 12 edited books. He is also active as a reviewer for several international scientific journals. Mr. Keservani graduated with a pharmacy degree from the Department of Pharmacy, Kumaun University, Nainital (Uttarakhand), India. He received his Master of Pharmacy (MPharm) (specialization in pharmaceutics) from the School of Pharmaceutical Sciences, Rajiv Gandhi Proudyogiki Vishwavidyalaya, Bhopal, India. His research interests include nutraceutical and functional foods, novel drug delivery systems (NDDS), transdermal drug delivery/drug delivery, health science, cancer biology, and neurobiology.

Surya Prakash Gautam, PhD

Surya Prakash Gautam, PhD, MPharm, LSPER, LIANN, LASR, LIAASSE, DACS, MISART, MINC, MNPhA, MIAGN, is Professor and Head at the CT Institute of Pharmaceutical Sciences, Jalandhar, India. He is doctorate in dendrimers-based drug delivery from SGUV, Jaipur, postgraduate in Pharmaceutics from Rajiv Gandhi Proudyogiki Vishwavidyalaya, Bhopal and graduate from Pune University. He has an experience of 12 years in academics and research. He is the recipient of fellowships from the India Ministry of Human Resource Development (MHRD), New Delhi, India, in 2001 and 2005. He is a lifetime member of SPER, IANN, and ASR, and the IAASSE member of AAPS, ACS, ISART, INC, and NPhA. He has published around 40 research/review publications in peer-reviewed journals, published six books and four book chapters. He has guided 16 MPharm candidates and six PhD students. Currently, he is working on dendrimers-based drug delivery, surface modification of dendrimers, and polyphenol-loaded nano-carriers for cancer targeting.

Contents

Contributors

Lucrece Ahovegbe
Pharmbiotechnology and Traditional Medicine Center of Excellence, School of Medicine, Mbarara University of Science and Technology, P.O. Box 1410, Mbarara, Uganda.
Unite de Formation et de Recherché en Pharmacie, Faculté des Sciences de la Santé, Université d'abomey-Calavi, Cotonou, Benin.

A. H. Auni
Oncological and Radiological Sciences Cluster, Advanced Medical and Dental Institute, Universiti Sains Malaysia, Kepala Batas, Penang 13200, Malaysia.

Sakshi Bajaj
Department of Pharmacognosy and Phytochemistry, Delhi Institute of Pharmaceutical Sciences and Research, Sec-III, Pushp Vihar, M.B. Road, Delhi 110017, India.

Ushmita Gupta Bakshi
Department of Biochemic Medicine, Indian School of Complementary Therapy and Allied Sciences, Ichapur 743144, Kolkata, West Bengal, India.

Himangini Bansal
Department of Pharmaceutical Chemistry, Delhi Institute of Pharmaceutical Sciences and Research, Sec-III, Pushp Vihar, M.B. Road, Delhi-110017, India.

Francisco J. Barba
Nutrition and Food Science Area, Preventive Medicine and Public Health, Food Science, Toxicology and Forensic Medicine Department, Faculty of Pharmacy, Universitat de València, Avda. Vicent Andrés Estellés, s/n, 46100 Burjassot, València, Spain.

Tamirat Bekele
Pharmbiotechnology and Traditional Medicine Center of Excellence, School of Medicine, Mbarara University of Science and Technology, P.O. Box 1410, Mbarara, Uganda.
Department of Pharmacy, Ambo University, P.O. Box 19, Ambo, Ethiopia.

Ibrahim Chikowe
Department of Pharmacy, College of Medicine, University of Malawi, P/Bag 360, Chichiri, Blantyre 3, Malawi.

Rubén Domínguez
Centro Tecnológico de la Carne de Galicia, Rúa Galicia No. 4, Parque Tecnológico de Galicia, San Cibrao das Viñas, 32900 Ourense, Spain.

Elena Movilla Fierro
Complejo Hospitalario Universitario de Ourense, Ourense, Spain.

Surya Prakash Gautam
CT Institute of Pharmaceutical Sciences, Jalandhar, India
E-mail: suryagautam@gmail.com, gautamsuryaprakash@gmail.com

Bey Hing Goh
Institute of Pharmaceutical Sciences, University of Veterinary and animal sciences, Outfall Road, Lahore, Pakistan.

Srijan Goswami
Department of Biochemic Medicine, Indian School of Complementary Therapy and Allied Sciences, Ichapur 743144, Kolkata, West Bengal, India.

Charu Gupta
Amity Institute of Herbal Research and Studies, Amity University, Uttar Pradesh, India.

Raj Keserwani
School of Pharmaceutical Sciences, Rajiv Gandhi Proudyogiki Vishwavidyalaya, Airport Bypass Road, Bhopal 462036, Madhya Pradesh, India.

Tahir Mehmood Khan
School of Pharmacy, Monash University, Bandar Sunway 45700, Selangor, Malaysia.

Danijela Bursać Kovačević
Faculty of Food Technology and Biotechnology, University of Zagreb, Pierottijeva 6, 10000 Zagreb, Croatia.

Fanuel Lampiao
Africa Centre of Excellency in Public health and Herbal Medicine, College of Medicine, University of Malawi, P/Bag 360, Chichiri, Blantyre 3, Malawi.

Learn-Han Lee
School of Pharmacy, Monash University, Bandar Sunway 45700, Selangor, Malaysia.

José M. Lorenzo
Centro Tecnológico de la Carne de Galicia, Rúa Galicia No. 4, Parque Tecnológico de Galicia, San Cibrao das Viñas, 32900 Ourense, Spain.

Annum Malik
Department of Pharmacy, Quaid-i-Azam University, Islamabad, Pakistan.

Adetoun E. Morakinyo
Department of Biochemistry, Adeleke University, Ede, Nigeria.

Andrew G. Mtewa
Department of Chemistry, Malawi Institute of Technology, Malawi University of Science and Technology, P.O. Box 5196, Limbe, Malawi.
Pharmbiotechnology and Traditional Medicine Center of Excellence, School of Medicine, Mbarara University of Science and Technology, P.O. Box 1410, Mbarara, Uganda.

Paulo S. E. Munekata
Centro Tecnológico de la Carne de Galicia, Rúa Galicia No. 4, Parque Tecnológico de Galicia, San Cibrao das Viñas, 32900 Ourense, Spain.

Kennedy Ngwira
School of Chemistry, University of Witwatersrand, P/Bag 3, Wits 2050, Johannesburg, South Africa.

S. R. Nur
Faculty of Applied Sciences, Universiti Teknologi Mara (UiTM), Arau Campus, Arau Perlis 02600, Malaysia.

S. M. N. Nurul
Oncological and Radiological Sciences Cluster, Advanced Medical and Dental Institute, Universiti Sains Malaysia, Kepala Batas, Penang 13200, Malaysia.

Patrick E, Ogwang
Pharmbiotechnology and Traditional Medicine Center of Excellence, School of Medicine, Mbarara University of Science and Technology, P.O. Box 1410, Mbarara, Uganda.

Temitope A. Oyedepo
Department of Biochemistry, Adeleke University, Ede, Nigeria.

Mirian Pateiro
Centro Tecnológico de la Carne de Galicia, Rúa Galicia No. 4, Parque Tecnológico de Galicia,
San Cibrao das Viñas, 32900 Ourense, Spain.

Amrita Poonia
Department of Dairy Science and Food Technology, Institute of Agricultural Sciences,
Banaras Hindu University, Varanasi 221005, Uttar Pradesh, India.

Dhan Prakash
Amity Institute of Herbal Research and Studies, Amity University, Uttar Pradesh, India.

Predrag Putnik
Faculty of Food Technology and Biotechnology, University of Zagreb, Pierottijeva 6,
10000 Zagreb, Croatia.

Bhushan R. Rane
Shri D.D. Vispute College of Pharmacy and Research Center, Panvel, Raigad 410206
(Affiliated to University of Mumbai), Maharashtra, India.

Shahzadi Sidra Saleem
Department of Pharmacy, Quaid-i-Azam University, Islamabad, Pakistan.

Duncan C. Sesaazi
Pharmbiotechnology and Traditional Medicine Center of Excellence, School of Medicine,
Mbarara University of Science and Technology, P.O. Box 1410, Mbarara, Uganda.
Department of Pharmaceutical Sciences, School of Medicine, Mbarara University of Science and
Technology, P.O. Box 1410, Mbarara, Uganda.

Kifayat Ullah Shah
Department of Pharmacy, Quaid-i-Azam University, Islamabad, Pakistan.

H. Shahrul
Oncological and Radiological Sciences Cluster, Advanced Medical and Dental Institute,
Universiti Sains Malaysia, Kepala Batas, Penang 13200, Malaysia.

Anil K. Sharma
School of Medical and Allied Sciences, GD Goenka University, Gurugram, India
E-mail: sharmarahul2004@gmail.com

Sandip A. Tadavi
PSGVPM's College of Pharmacy, Shahada 425409 (Affiliated to Kavayitri Bahinabai Chaudhari North
Maharashtra University), Jalgaon, Maharashtra, India.

M. L. Tan
Healthy Lifestyle Sciences Cluster, Advanced Medical and Dental Institute, Universiti Sains Malaysia,
Kepala Batas, Penang 13200, Malaysia.

Abbreviations

DMH	1,2-dimethyl hydrazine
Ap-1	activator protein
AHCC	active hexose correlated compound
AEs	adverse effects
ALT	alanine aminotransferase
ACE	angiotensin converting enzyme
AST	aspartate aminotransferase
AST	aspartate transaminase
BACs	bioactive compounds
BRMs	biological response modifiers
BD	*Boerhavia diffusa*
BChE	butyrylcholinesterase
CR	caloric limitation
CVDs	cardiovascular diseases
CNS	central nervous system
CHF	chronic congestive heart failure
COX	cyclooxygenase
DM	dry matter
eNOS	endothelial nitric oxide synthase
EOS	essential oils
FDA	Food and Drug Administration
GRAS	generally recognized as safe
HCC	hepatocellular carcinoma
HPA	hypothalamic-pituitary-adrenal
ICAM-1	intercellular adhesion molecule-1
IF	irregular fasting
LVEF	left ventricular ejection fraction
LDL	low-density lipoprotein
MMP-2	matrix metallopeptidase-2
MMP-9	matrix metallopeptidase-9
NK	natural killer
NO	nitric oxide
NDGA	nordihydroguaiaretic acid
PAF	platelet activating factor
PMN	polymorphonuclear

PSP	polysaccharide of P
PSK	polysaccharide-Krestin
Q&S	quality and safety (Q&S)
ROSs	reactive oxygen species
SPG	Schizophyllan
SFE	supercritical liquid extraction
TBA	thiobarbituric acid
TCM	Traditional Chinese Medicine
TNF	tumor necrosis factor
VACM-1	vascular cell adhesion molecule-1

Preface

The eternal association of mankind with nature dates back to the beginning of human civilization. Nature has benevolently blessed us with the things to run our livelihood, which we have exploited in various ways. Several medicines have their origin from natural sources, and applications of the same have witnessed a rise with the evolution of science and technology. Yet, there are some challenges that hinder the path of successful production and supply of natural products from laboratories to the market. For instance, a plant drug may have several phytoconstituents of medicinal significance; nevertheless, it is quite tiresome to isolate and characterize the principal moieties. At the same time, efficacy and safety of herbal products are of a major concern. Often the herbal products are considered to be free from associated ill effects; however, the data from clinical reports have revealed occurrences of toxicity, leading to the withdrawal of such products from the market. Even today there are only a few regulatory guidelines that establishes the authentication of herbal medicines.

The present book strives to provide requisite and relevant information about herbal products to its readers. This book has been written by highly skilled, experienced, and renowned scientists and researchers around the globe with up-to-date information to offer insights into herbal medicines to readers, researchers, academicians, scientists, and industrialists worldwide.

The book *Herbal Product Development* comprises 12 chapters with 3 sections that describe applications of herbs, their characteristics, evaluation techniques, and so on. One particular section is dedicated to alternative medicinal strategies for treatment/cure of diverse diseases.

Section I: Herbal Products and Mankind

Chapter 1, *Medicinal Plants in Natural Health Care as Phytopharmaceuticals,* written by Charu Gupta and Dhan Prakash gives a general introduction of drugs of natural origin. In addition, it enumerates several applications of herbal medicines that have proven prudence to treat ailments.

The details of herbal products that are consumed as food have been presented in Chapter 2, *Herbal Food Product Development and Characteristics,* written by Amrita Poonia. The author has given elementary information regarding uses of herbs as food. The text also mentions obstacles in the way of herbal product commercialization. In conclusion, the author recommends that through a scientific and systemic approach, a viable herbal product can be introduced to the market.

Chapter 3, *Herbal Supplements and Health,* written by Himangini Bansal and Sakshi Bajaj, provides an introduction of herbal supplements. Plants that produce medicinally important substances have been discussed in detail. The side effects of popular plant products have also been provided in a table.

The multidimensional uses of herbal medicines are described in Chapter 4, *Herbal Therapies,* written by Shahrul H and colleagues. This chapter highlights several region-specific herbs and their mechanisms of action. It also mentions some plants that are used among the community but need to be further explored of their potential benefits to human race.

Chapter 5, *Herbs in Cancer Therapy,* written by Annum Malik and associates, focuses on cancer therapeutics through drugs derived from nature. The authors have listed a few herbal drugs having reported interactions when co-administered with synthetic counterparts.

A general introduction to role of plants in curbing the menace of cancer has been provided in Chapter 6, *Herbs in Cancer Therapy: A Preamble,* written by Andrew G. Mtewa and colleagues. The authors recommend in the chapter that if the discovery of drugs against cancers is accorded high priority and funding, more drug-leading molecules can be taken forward, which afterward may be modulated to form better and cheaper anti-cancer drugs. Further, the authors call for the need of collaboration among academia, funders, and the industry so as to make meaningful strides in the advancement of herbal therapy against cancers.

Chapter 7, *Medicinal Mushrooms,* written by Oyedepo and Morakinyo, gives a focused view of applications of mushrooms other than food. The chapter also reviews selected medicinal mushrooms with proven pharmacological activities and the health-promoting benefits of medicinal mushrooms through gut microbiota.

Section II: Herbal Product Development and Evaluation

The multifaceted applications of herbal nutraceuticals have been discussed in Chapter 8, *Herbal Product Development and Characteristics,* written by Mirian Pateiro and colleagues. The authors have strived to compile all nutraceuticals having medicinal value. The authors further press for the need to incorporate herbal products as dietary supplements to achieve the health benefits using a nonfood matrix, for example, pills, capsules, powder, essential oils, and extracts.

Chapter 9, *Brief Overview of Development and Characterization of Herbal Products,* written by Andrew G. Mtewa and associates, has addressed the formulation aspects of herbal dosage forms along with methods to evaluate them. The challenges faced while designing a robust product have also been discussed. The authors have advice for all stakeholders to collaborate for moving ahead from initial screening to full-fledged commercialization.

Section III: Alternative Medicine Approaches

The specialized uses of the science of fragrances to treat diseases are given in Chapter 10, *Aromatic Medicine,* written by Sakshi Bajaj and Himangini Bansal. The chapter begins with a general description of aromatology, followed by its classification and application methods by the clinicians. The chapter takers its readers on a fascinating voyage of role played by essential oils derived from plants.

Chapter 11, *Understanding Classical Naturopathy: The Hippocratic Way of Healing,* written by Srijan Goswami and Ushmita Gupta Bakshi, described the principles and applications of naturopathy. The authors pointed out the postulates given by the Great Hippocrates on the "Art of Healing," which clearly indicates that the system of medicine that he taught people about is the naturopathic way of treatment. The following sections of the chapter explain some of the postulates of the Great Hippocrates, such as how they form the basis of the naturopathic system of medicine and how they are guided by the scientific laws of nature.

Chapter 12, *Naturopathy,* written by Bhushan R. Rane and colleagues described the basics of naturopathy. In this chapter, the authors have also provided a comparison of naturopathy with the allopathic system of medicine. To conclude, the authors have enlisted the techniques and advantages of the holistic system to benefit the ailing mankind.

CHAPTER 1

Medicinal Plants in Natural Health Care as Phytopharmaceuticals

CHARU GUPTA* and DHAN PRAKASH

Amity Institute of Herbal Research and Studies, Amity University, Uttar Pradesh, India

*Corresponding author. E-mail: charumicro@gmail.com.

ABSTRACT

There is a growing awareness in natural plant products as the basis of novel pharmaceuticals and other biologically active compounds. The majority of screening for biological activities of herbal extracts has been done in the hunt for novel anti-cancer, anti-viral and anti-fertility drugs. The advancement of the fast screening trials nowadays used in manufacturing has intended that many other plants can be assessed for an extensive variety of biological activities. There still remains a crucial necessity to develop novel medical medications, and this can be illustrated by the several ailments such as cancer, hypertension, obesity, diabetes and other age-related disorders. Natural goods already have an established track record for various activities, and it is likely that there are more such remedies still to be found from the environment. Regrettably, the outcomes of such trials do not inevitably spread the public domain and are kept in inaccessible business records. This is a timely review of the latest advances and trends in a field which is becoming a commercially significant area of investigation for the pharmaceutical industry.

1.1 INTRODUCTION

The customary structure of medication in which plant derivatives are the chief ingredients has gained worldwide recognition and popularity. The practice of herbal therapy in treating diseases and different disorders for meeting primary health needs is thought to be very effective and safe within prescribed doses. Some recent applications of the early customary scheme of treatment may be cited as Pacific yew (*Taxus* species) that had made quite an impact in the previous decade by providing raw materials to fight against certain types of cancer. Discovery of Ginseng and its therapeutic properties is also noteworthy. Another example is goldenseal (*Hydrastis canadensis*). The most significant development of anti-malarial drug in current ages was based on a traditionally used plant *Artemisia annua* L. Its medicinal properties had been well known in China for more than 2000 years. Medicinal plants as cultivated and collected form the basis of numerous phytomedicines in several countries and are thus a big business indeed. Recently, there has been tremendous upsurge in the interest on plants used by the tribal people or are a part of the ancient systems of medicine such as Ayurvedic, Unani, Chinese, Tibetean and the Aztech (South American). The reasons for such interest are mainly the safety, nominal side effects and trials of centuries. The modern approach to the science of ethnobotany evolved in India, USA, France, South East Asia, China, Jamaica, Combodia, Nepal, Hawai, Turkana, etc. are also engaged in the detailed ethnomedicinal studies of the tribal and aborginal populations of their countries (Gupta and Prakash, 2014; Haidan et al., 2016).

Medicinal and aromatic plants provide various kinds of medicines besides supplying nutrition for the preclusion and management of ailments and constitute an integral part of traditional medical prescriptions. From folk medicine and traditional scheme of drug, medicinal florae were adopted into the contemporary scheme of medication after they have been found effective drugs through chemical and pharmacological screening in the initial stages of growth of contemporary medicine plants. Medicinal plants have been quantified to encompass around 10,000 species and account for roughly 50% of all the higher flowering plant species in India alone. Curative properties of few such plants have been described; however, a large number of plants still used by native folklore are to be unraveled. Ayurveda, Siddha, Unani and Amchi systems of medication offer a good base for scientific investigation of medicinally significant molecules from the environment. Evolving notion of joining Ayurveda with the advanced drug discovery programme is worldwide acceptable. According to the reports of the World Health Organization (WHO), about three quarters of the world's populace presently practise herbs and other customs of customary medicines to treat sicknesses. Customary medicines are broadly used in India and China, and the use of herbal medicines has improved histrionically in the past two decades all over the world (Chikezie and Ojiako, 2015).

Ayurveda (*ayus*—life, *veda*—knowledge, meaning science of life), the Indian customary health care system, is the primogenital medicinal system in the world. It offers a method to prevent and treat diverse ailments by a huge number of therapeutic measures and pharmaceuticals. During the previous few eras, intensive search for plant constituents of potential medicinal importance was being pursued all over the world. The therapeutic value of a plant depends on the occurrence of active principles, which exert impact on the structure or role of the living being. These substances are of varying chemical groups, and several of them are useful in the pharmaceutical industry. This natural system is more accessible and devoid of severe side effects, since the plants are well tested over the centuries. From traditional medication and the traditional system of medicine, medicinal plants were adopted into the modern system of medicine after they have been found effective drugs through chemical and pharmacological selection in the initial stages of development of current medication (Pan et al., 2011).

The activity of medicinal plants depends on phytochemicals present in them, though it is, in general, the outcome of the combined activity of several active composites as well as of inert accompanying substances. These inert components might influence bioavailability, stability, and can minimize side effects or might have additive/synergestic effect. The aim of chemical evaluation of traditional medicines is obviously to ensure their therapeutical efficacy and to establish chemical parameters for the standardization and quality governor of the herbal/Ayurvedic finished/end products. Chemistry of natural produces is a research field with endless potential and is especially important in countries that possess great biodiversity. Nowadays, the customary medication all over the world is revalued by an extensive activity of investigation on diverse herbal species and their beneficial properties. The therapeutic properties of florae have been explored in the light of current technical advances throughout the world, due to their potent pharmacological activities, low toxicity and commercial feasibility (Acharya and Shrivastava, 2008; Bjelakovic et al., 2014).

The Rasayana in ayurveda not only focuses on drug remedy but also includes the practice of revitalization and nutrition, including its movement, circulation and perfusion in the body tissues. In reference to the Rasayana drug therapy, the strong antioxidant activity of any Rasayana has been reported; these compounds were found to be several folds more powerful than synthetic molecules (Bjelakovic et al., 2014; Forman et al., 2016). Phytochemical constituents of a plant play a key role in biological activities of different medicinal plants and are accountable for their effectiveness.

Reactive oxygen species (ROSs) are generated by normal metabolic processes in all oxygen-utilizing organisms. Damage induced by ROSs includes DNA mutation, protein oxidation and lipid peroxidation, contributing to the growth of cancer, diabetes, atherosclerosis, inflammation and premature ageing. Antioxidant properties elicited by plant species have a full range of perspective applications in human healthcare. Natural antioxidants such as carotenoids, tocopherols, ascorbates and polyphenols are generally found in plants, foods, vegetables, fruits, herbs and spices containing many compounds with strong antioxidant activities (Kota et al., 2018).

1.2   FREE RADICALS AND HUMAN HEALTH

In the situation of a disturbed balance between the formation of free radicals and antioxidant defense, in the cell we have oxidative stress and the free radicals that can play a part in the growth of many ailments. The over-production of ROS has been involved in the etiology of neurodegenerative diseases such as cardiovascular, diabetes, cancer, Alzheimer's, retinal degeneration, ishemic dementia and other disorders due to ageing. There are strong indications that antioxidant supplements can ameliorate such conditions (Carocho et al., 2018). They are known to act as antioxidant to defuse toxic free radicals, anti-allergic, anti-inflammatory, immuno-stimulant and anti-hepatotoxic. Specifically, phenols may help to lower the risk of cancer, cardiovascular diseases, age-related vision disorders, asthma and reduce inflammation (Poprac et al., 2017).

1.3   MEDICINAL PLANTS AND ANTIOXIDANT ACTIVITY

The customary medicine all over the world is nowadays enhanced by a wide activity of study on diverse plant species and their beneficial properties. Besides supplying food for the avoidance and management of diseases, plants provide various kinds of medicines. Foods play a main part in the notions of sickness and curing and establish an essential part of customary medicinal treatments. From folk medicine and traditional system of medicine, medicinal plants were adopted into the modern system of medicine after they were found effective in relation to drugs through chemical and pharmacological screening in the initial phases of growth of modern medicine. The use of medicinal plants, herbs and spices has been valued prehistorically, and spices serve not only as a flavour agent but also as a safe food antioxidant and preventing the deterioration of foodstuff. Dietary natural antioxidants strengthen the endogenous antioxidant system by reducing oxidative stress and risk of toxic diseases. Furthermore, the biological, biochemical, physiological, pharmaceutical and medicinal properties of polyphenols have been reviewed (Lobo et al., 2010; Poprac et al., 2017) in regard to their free radical scavenging activity and multiple biological activities, including vasodilatory, antibacterial, immune-stimulating, anti-carcinogenic, anti-inflammatory, anti-allergic, antiviral and estrogenic effects, as well as being inhibitors of specific enzymes.

The biological activity of some medicinal plants, parts used and their phytochemical constituents are mentioned in Table 1.1.

TABLE 1.1 Biological Activity of Some Medicinal Plants, Parts Used and Their Phytochemical Constituents

Botanical Name	Family	Parts Used	Chemical Constituent(s)
Aglaia foveolata	Meliaceae	Bark	Silvestrol
Aglaia roxburghiana	Meliaceae	Bark	Silvestrol
Aglaila sylvestre	Meliaceae	Bark	Silvestrol
Achyranthes aspera	Amaranthaceae	Whole plant	Triterpenoid saponins
Aegle marmelos	Rutaceae	Fruits, stem bark	Imperatorin and related compounds
Abrus precatorius	Fabaceae	Seeds	Protein extract
Acalypha indica	Euphorbiaceae	Whole plant	Beta-sitosterol-beta-d-glucoside
Agave americana	Agavaceae	Leaves	Steroidal saponin, alkaloid, coumarin, isoflavonoid
Agrimonia pilosa	Rosaceae	Whole plant	Agrimonolide, flavonoid, triterpene, tannin and coumarin
Ailanthus altissima	Simaroubaceae	Bark	Triterpene, tannin, saponin and quercetin-3-glucoside
Akebia quinata	Lardizabalaceae	Fruits	Flavonoid and saponin
Albizia lebbeck	Fabaceae	Seeds	Monomeric 5.5-kDa protein
Alpinia calcarata	Zingiberaceae	Leaves, rhizome	Flavonoids and related compounds
Alpinia galanga	Zingiberaceae	Leaves, rhizome	Galangin and related compounds
Alstonia scholaris	Apocynaceae	Leaves, stem bark	Berberine and other compounds
Andrographis paniculata	Acanthaceae	Whole plant	Andrographolide
Anemarrhena asphodeloides	Asparagaceae	Rhizome	Mangiferin
Artemisia vulgaris	Asteraceae	Whole plant	Artimisin
Astragalus membranaceus	Fabaceae	Whole plant	Astragalosides, flavonoids
Azadirachta indica	Meliaceae	Whole plant	Phenols, tannins
Berberis amurensis	Berberidaceae	Dried fruit	Berbamine

TABLE 1.1 *(Continued)*

Botanical Name	Family	Parts Used	Chemical Constituent(s)
Betula alba	Betulaceae	Outer bark	Betulinic acid
Bleekeria vitensis	Apocynaceae	Whole plant	Elliptinium
Bohhervia diffusa	Nyctaginaceae	Whole plant	Phenols, tannins
Broyonia dioica	Cucurbitaceae	Root	Cucurbitacin and glycoside
Bupleurum falcatum	Apiaceae	Whole plant	Isochaihulactone
Calophyllum inophyllum	Calophyllaceae	Stem bark	4-Phenylcoumarins
Calophyllum rubiginosum	Calophyllaceae	Stembark	Calophyllolide
Camellia sinensis	Theaceae	Leaves	Tannins, phenols
Camptotheca acuminate	Cornaceae	Bark, wood and fruit	Topotecan, irinotecan, exatecan, le-sn-38
Cannabis sativa	Cannabinaceae	Leaves	Ns stereo isomers of cannabitriol
Carmona retusa	Boraginaceae	Leaves	Alkaloids
Catharanthus pusillus	Apocynaceae	Whole plant	Ajmalicine, rauwolscine, pusiline
Catharanthus roseus	Apocynaceae	Whole plant	Vinblastine, vincristine, vinorelbine, vindesine, vinflunine
Centella asiatica	Umbelliferae	Whole plant	Asiaticoside
Cephalotaxus hainanensis	Cephalotaxaceae	Leaves	Cephalotaxine and homoharringtonine
Cephalotaxus harrintonia	Cephalotaxaceae	Leaves	Harringtonine, Homoharringtonine
Cephalotaxus qinensis	Cephalotaxaceae	Leaves	Harringtonine, Homoharringtonine
Cerbera odollam	Apocynaceae	Fruits	17Bh-neriifolin, tamoxifen
Chimaphila umbellate	Ericaceae	Whole plant	Ericolin, arbutin, urson and tannin
Cinnamomum camphora	Lauraceae	Leaves , stem bark	Camphor, linalool, 1,8-cineole, nerolidol, safrole, and borneol.
Cinnamomum tamala	Lauraceae	Leaves, stem bark	Beta-caryophyllene, linalool, caryophyllene oxide, eugenol
Cinnamomum verum	Lauraceae	Leaves, stem bark	Cinnamaldehyde, linalool, eugenol, chavicol, phellandrene, tannin, catechin, mannitol, coumarins, mucilage,
Coix lachrymal-jobi	Poaceae	Seeds	Trans-ferulyl stigmasterol
Colchicum autumnale	Colchicaceae	Bulb-like corms	Colchicine
Coleus forskholii	Lamiaceae	Whole plant	Forskolin
Combretum caffrum	Combretaceae	Combretastatins	Combretastatin

TABLE 1.1 *(Continued)*

Botanical Name	Family	Parts Used	Chemical Constituent(s)
Commiphora mukul	Burseraceae	Gum/resin	Myrcene and eugenol, cembrene-a, mukulol 4
Coriandrum sativum	Apiaceae	Fruit/seed	Isoquercitrin, limonene, rhamnetin, rutin
Cucurbita andreana	Cucurbitaceae	Fruits	Cucurbitacin
Cucurbita pepo	Cucurbitaceae	Fruits	Cucurbitacin
Curcuma longa	Zingiberaceae	Rhizome	Curcumin
Cymbopogon citratus	Poaceae	Arial parts	Essential oils
Datura metal	Solanaceae	Aerial parts	Alkaloids
Daucus carota	Apiaceae	Rhixome/root	Tocopherol, butyric-acid, falcarinol, limonene
Digitalis purpuria	Plantaginaceae	Leaves	Steroid glycosides
Dryopteris crassirhizoma	Polypodiaceae	Rhizome	Filicinic and filicic acids, aspidinol and aspidin
Duchesnea indica	Rosaceae	Leaves	Phenolic comounds
Dysoxylum binectariferum	Meliaceae	Leaves or aerial parts	Rohitukine, chromane alkaloids
Echinops setifer	Asteraceae	Whole plant	Echinopsine
Eleutherococcus senticosus	Araliaceae	Leaves, fruits	Eleutherosides, triterpenoid saponins
Equisetum arvense	Equisetaceae	Whole plant	Caffeic-acid, gallic-acid, isoquercitrin, kaempferol, naringenin
Erythronium americanum	Liliaceae	Whole plant	Alpha-methylenebutyrolactone
Erythroxylum pervillei	Erythroxylaceae	Wood or bark	Pervilleines
Foeniculum vulgare	Apiaceae	Seed, fruit	Isoquercitrin, limonene, rutin
Fragaria vesca	Rosaceae	Leaves, fruits	Flavonoid, tannin, borneol and ellagic acid
Fritillaria thunbergii	Liliaceae	Whole plant	Alkaloid and peimine
Galium aparine	Rubiaceae	Whole plant	Iridoid, polyphenolic acid, tannin, anthraquinone and flavanoids
Glechoma hederacea	Lamiaceae	Whole plant	Alpha-terpineol, caffeic-acid, hyperoside, limonene, rutin, tannin, ursolic-acid
Gleditsia sinensis	Fabaceae	Thorn	Saponin
Gloriosa superba	Colchicaceae	Tuber	Colchicine
glycine max	Fabaceae	Seeds	Genistein

TABLE 1.1 *(Continued)*

Botanical Name	Family	Parts Used	Chemical Constituent(s)
Glycyrrhiza glabra	Fabaceae	Whole plant	Glycyrrhizin
Gmelina arborea	Lamiaceae	Root, bark	Flavonoids and other related compounds
Hydrastis canadensis	Ranunculaceae	Whole plant	Isoquinoline alkaloids
Ipomoeca batatas	Convolvulaceae	Leaves	4-Ipomeanol
Iridaceae tea pallasii	Iridaceae	Flower	Irisquinone
Iris kumaonensis	Iridaceae	Rhizomes	Irigenin
Jasminum sambac	Jasmineae	Flowers	Rutin and its derivatives
Junchus effuses	Juncaceae	Whole plant	Tridecanone, effusol, juncanol, phenylpropanoid
Lagerstroemia speciosa	Lythraceae	Bark	Corosolic acid
Lantana camara	Verbenaceae	Whole plant	Alkaloids
Larrea tridentate	Zygophyllaceae	Whole plant	Resin
Leonurus cardiaca	Lamiaceae	Whole plant	Benzaldehyde, hyperoside, isoquercitrin, limonene, rutin, tannin, ursolic-acid
Ligustrum lucidum	Oleaceae	Berries	Tannins
Lonicera japonica	Caprifoliaceae	Whole plant	Tannins, saponins, carotenoids
Myristica fragrans	Myristicaceae	Fruit/nuts	D-pinene, limonene, d-borneol, l-terpineol, geraniol, safrol, myristicin
Nicotiana tabacum	Solanaceae	Leaves/fruits	Beta-carotene, butyric-acid, caffeic-acid, isoquercitrin, kaempferol, lignin, rutin
Ocimum basilicum	Lamiaceae	Leaves/inflorescence	Beta-carotene, caffeic-acid, isoquercitrin, kaempferol, rutin, ursolic-acid, vicenin-2
Origanum vulgare	Lamiaceae	Whole plant	Alpha-terpineol, beta-carotene, caffeic-acid, limonene, naringenin, ursolic-acid, vanillic-acid
Panax quinquefolium	Araliaceae	Root	Ginsenoside, sesquiterpene, limonene
Phyllanthus emblica	Phyllanthaceae	Fruits	Punicafolin, phyllanemblinin, poly phenols
Picrorhiza kurroa	Scrofulariaceae	Whole plant	Picroliv
Piper longum	Piperaceae	Fruits	Piperine and related compounds
Podophyllum emodi	Berberidaceae	Rhizomes	Podophyllotoxin
Podophyllum peltatum	Berberidaceae	Rhizomes	Podophyllotoxin

TABLE 1.1   *(Continued)*

Botanical Name	Family	Parts Used	Chemical Constituent(s)
Potentilla chinensis	Rolsaaceae	Whole plant	Gallic acid and tannin
psoralea corylifolia	Fabaceae	Seeds	Bavachinin, corylfolinin, psoralen
Ribes nigrum	Grossulariaceae	Fruit	Beta-carotene, limonene, rutin
Rosmarinus officinalis	Lamiaceae	Whole plant	Borneol, carnosol, ursolic acid, diterpene, rosmaricine, flavonoids, tannin
Rubia akane	Rubiaceae	Whole plant	Anthraquinone, triterpene
Salvia prionitis	Lamiaceae	Roots	Salvicine
Salvia sclarea	Lamiaceae	Whole plant	Alpha-terpineol, benzaldehyde, butyric-acid, citral, limonene, tannin, ursolic-acid
Saraca asoca	Fabaceae	Flower, bark	Flavonoid fraction
Schisandra chinensis	Schisandraceae	Berries	Dibenzocyclooctadiene lignans
Sida cordifolia	Malvaceae	Whole plant	Cryptonine alkaloid, dihydro benzophenanthridine
Silybum marianum	Asteraceae	Whole plant	Silymarin, silibinin
solanum nigrum	Solanaceae	Fruits	Polyphenols
Soymida febrifuga	Meliaceae.	Bark, leaves	Coumarins
Strychnos nux-vomica	Loganiaceae	Bark, leaves, seeds	Strychnine
Swertia chirayta	Gentianaceae	Whole plant	Mangiferin, amarogenitine
Tabebuia avellanedae	Bignoniaceae	Bark	Betalapachone
Taxus baccata	Taxaceae	Bark	Paclitaxel, docetaxel
Taxus brevifolia	Taxaceae	Bark	Paclitaxel, docetaxel
Terminalia ballarica	Combretaceae	Bark, leaves and fruits	Tannins, phenols
Terminalia chebula	Combretaceae	Bark, leaves and fruits	Tannins, phenols
Thymus vulgaris	Lamiaceae	Whole plant	Tannin, ursolic-acid, vanillic-acid
Tinospora cordifoia	Menispermaceae	Stem	Polyphenols, tannins
Tribulus terrestris	Zygophyllaceae	Fruits, whole plant	Saponins and related compounds
vaccaria segetalis	Caryophyllaceae	Flowers, seeds	Tannins
vicia faba	Fabaceae	Seeds	Lectin
Vitex negundo	Lamiaceae	Leaves	Chrysoplenetin and other flavonoids
Wikstroemia indica	Thymelaeaceae	Roots	Daphnoretin
Withania somnifera	Solanaceae	Berries, leaves	Alkaloids, saponins
Zizyphus mauritiana	Rhamnaceae	Leaves, fruits	Betulinic acid

1.4 PHYTOPHARMACEUTICALS OF SOME THERAPEUTIC FLORAE IN WELL-BEING

1.4.1 *Andrographis paniculata (KALMEGH [KING OF BITTERS]; FAMILY: Acanthaceae)*

Kalmegh is a perennial herb widely cultivated in many states of India. It is hepatoprotective, anti-spasmodic, stomachic, anthelmintic, blood purifier and febrifuge. It acts well on the liver disorders and promotes the secretion of bile. It is used in jaundice, flatulence and diarrhoea of children, colic, strangulation of intestines and splenomegaly, and also for cold and upper respiratory tract infections. This plant is also being used conventionally for the management of array of diseases such as cancer, diabetes, high blood pressure, ulcer, leprosy, bronchitis, skin diseases, flatulence, colic, influenza, dysentery, dyspepsia and malaria for centuries in the Asia, America, and Africa continents. Recent experimental findings indicated that Kalmegh possesses anti-typhoid and antibiotic properties. It has been proved to be a hepato-protective drug (Okhuarobo et al., 2014). The aerial part of the plant, used medicinally, holds a great number of chemical constituents, mainly lactones, diterpenoids, diterpene glycosides, flavonoids and flavonoid glycosides (Jayakumar et al., 2013).

This plant is the most widely used herb in numerous Ayurvedic preparations. The whole plant of *Andrographis paniculata (A. paniculata)* is used widely as an anti-inflammatory and antipyretic drug for the treatment of fever, cold, laryngitis, diarrhoea and inflammation. It has also been used traditionally for sluggish liver as an antidote in the case of colic dysentery and dyspepsia. It is used as a bitter tonic, anti-spasmodic, anti-peristaltic, stomachic and anti-helmintic (Dhiman et al., 2012). Therapeutically active constituent of *A. paniculata* is present in the juice of its fresh leaves and aerial parts, which contains andrographolide. It is a colourless bicyclic diterpene lactone and Kalmeghin (up to 2.5%), with bitter taste. Several active constituents have been isolated from the leaf and rhizome; notable ones are andrographolide, deoxyandrographolide and other diterpenes. The amount of the active constituents present in the plant depends on various factors such as topographical conditions, harvesting time and the processing method.

The leaves of *A. paniculata* grown in the tropical and subtropical areas of China and Southeast Asia contain more than 2% andrographolide before the plant blooms and less than 0.5% afterwards. The content of andrographolide depends on both growing region and the collection time. The stem contains 0.1%–0.4% of andrographolide. Quantitative high-performance liquid chromatography (HPLC) analysis of andrographolide isolated from two different phases of lifespan of the plants also showed a varied array of phytoconstituents (Koteswara Rao et al., 2004). Besides andrographolide, neo-andrographolide and kalmeghnin are the other active principles. The other chemical constituents include andrographanoside, andrographanin, 14-deoxy-12-methoxyandrographolide and deoxyandrographolide (Joselin and Jeeva, 2014).

1.4.2   *Trigonella foenum-graecum* Linn. *(Fenugreek; Family: Fabaceae)*

The genus *Trigonella* family Leguminosae, generally known as 'methi, is spread in the Mediterranean region, Europe, Asia, and South Africa. It is cultured for its therapeutic value that has been known since the time of ancient Greek and in several other customary schemes of medication throughout the world. Over the centuries, it has been established that the high mucilage amount of the seeds makes them a soothing expectorant for bronchitis and chest complaints and that the mashed seeds make an excellent poultice for skin problems and boils—wash will ease mouth ulcers and blistered lips. There are two varieties of methi: the ordinary methi or desi methi or *Trigonella foenum-graecum* and the scented methi also known as kasuri methi *T. corniculata* (Shashikumar et al., 2018).

Fenugreek is fairly tolerant to frostiness and low temperature. It is grown in India in cold weather; however, under irrigation, the crop can be grown throughout the year as potherb. It is cultivated as a flavouring herb, green leafy vegetable and as a source of low-cost natural nutrients. Its dried leaves are refrigent, given internally for vitiated conditions of *pitta* in 'Ayurveda' and also used as condiments to enhance flavours in the Indian cooking system (Shashikumar et al., 2018). Its seeds are bitter, mucilaginous, aromatic and carminative and used as tonic, thermogenic, galactogogue, astringent and aphrodisiac in fever, vomiting, cough, bronchitis and colonitis. The aqueous extract of the seeds possesses antibacterial property and is used externally in the form of poultice in boils, abscess and ulcers. Whole plant has anti-inflammatory, antipyretic, anti-spasmodic, gastroprotective, hypertensive, antineoplastic and central nervous system (CNS) depressant activity. The saponins of leaves flowers and seeds were reported to yield diosgenin, tigogenin, gitogenin neogetogenin, homoorientin, saponaretin, trigonelline triacontane, 22,23-dihydrostigmasterol and major phytosterols such as β-sitosterol and cholesterol. Foliage and seeds of fenugreek were studied for their nutritional composition, antioxidant and free radical scavenging activities (Shashikumar et al., 2018; Snehlata and Payal, 2012).

1.4.3   *Trigonella corniculata Linn.*
(Fenugreek; Family: Fabaceae)

This medicinal plant is a diffuse sub-erect, strongly scented, grows annually up to 30 cm or higher, occurs in Western Himalayas between 1500 and 3600 m, and extends eastwards to Bihar and West Bengal. The plant radiates a spicy odour, which persists on the hands after touching. The whole plant is uprooted and allowed to dry. Its seeds are separated out and further dried. The leaves of this plant are rich in phosphorous. The seeds contain 5.2% fatty oil; from the unsaponifiable fraction of the oil, β-sitosterol is isolated; in the seeds, dihydrostegmasterol is also reported to be present. Two nitrogenous bases, choline and betaine, are present in the alcohlic extract of the seeds. The seeds contain three saponins which on hydrolysis yield diosgenin. The major component of the mixture is yuccagenin. The presence of two new flavone C-glycosides, namely 6-8-β-d-glucopyranosyl acacetin and its monoacetate, in the seeds has also been reported. The saponins from the leaves of the plants yield diosgenin in major and tigogenin in minor quantity. The fruit is bitter, astringent, and styptic (Jain et al., 2013; Roberts, 2011).

The plant has been used in the oldest medical regimes to stimulate lactation, effective in coughs, tuberculosis, bronchitis, fevers, sore throat, neuralgia, sciatica, swollen glands, skin eruptions, wounds tumours, sores, asthma, and emphysema. During the middle ages, fenugreek acquired something of a reputation as an aphrodisiac and as a cure for impotence. Modern chemical analysis has actually detected the presence of a substance diosgenin that acts in a similar way to the body's own sex hormones (Shashikumar et al., 2018). The effect of fenugreek seeds compared to omeprazole was studied on ethanol-induced gastric ulcer. The aqueous extract and a gel fraction isolated from the seed showed significant gastro- and ulcer-protective effecs. The cytoprotective effect of the seeds seemed to be not only due to the anti-secretory action but also due to their effects on mucosal glycoprotiens (Hamden et al., 2010). Anti-inflammatory and antipyretic effects of the *T. foenum-graecum* leaves extract were examined as compared to available anti-inflammatory and antipyretic drugs (steroidal and nonsteroidal); they showed significant anti-inflammatory and antipyretic properties. The aqueous and methanolic extract of *T. foenum-graecum* seeds showed hypoglycaemic effect by delaying glucose absorption and enhancing its utilization. Therefore, its seeds are considered to be potentially useful for glucose control, preventing hyperlipidaemia and atherosclerosis in diabetic subjects (Fatima et al., 2018; Shashikumar et al., 2018).

In a multi-factorial disease like diabetic cataract, compounds that can reverse/inhibit various common pathways in the development of cataract are more likely to succeed. The role of alternative therapeutic approaches has become very popular, and because a single plant may have many pharmaco-logical activities (antidiabetic, antioxidant, and antistress activity) they can be effectively utilized to delay or counter diabetic complications, such as cataract. In this regard, aqueous extract of *T. foenum-graecum* seeds was found to be very effective (Vats et al., 2004).

1.4.4 *Coriandrum sativum L. (Coriander; Family: Umbelliferae)*

This plant is native to Middle East and southern Europe, occasionally found in Britain in fields and waste places and by the sides of rivers but has also been known in Asia. It is found wild in Egypt and the Sudan. It is an aromatic stimulant spice and has been cultivated and used from very ancient times. It was used by Hippocrates and other Greek physicians. Its seed is generally sold dried and available in both forms: whole and grounded. The fresh leaves of the plant are called cilantro and they are used as herbs. The seeds are warm, mild and sweetish. The leaves are used as a culinary herb in many Indian and Mexico dishes; they are always used fresh. They feature in Spanish, Middle Eastern, Indian, Oriental, and South American cookery as well. Both seeds and leaves can be used in salads. The seed is now produced in Russia, India, South America, North Africa especially Morocco and in Holland. Among earliest medics, coriander was known to Hippocratic and Pliny who called it coriandrum because it emits 'buggy' smell, coris being a bug, or possibly because the young seed resembles *Cimex lectularius*, the European bedbug.

In the ripe fruits, the content of essential oil is comparably low (1%), the oil consists mainly of linalool (50%–60%) and about 20% terpenes (pinenes, γ-terpinene, myrcene, camphene, phellandrenes, α-terpinene, limonene, and cymene). In toasted coriander fruits, pyrazines are formed as the main flavour compounds. The taste of the fresh herb is due to an essential oil (0.1%) that is almost entirely made up of aliphatic aldehydes with 10–16 carbon atoms. One finds both saturated (decanal) and α,β-unsaturated (*trans*-2-tridecenal) aldehydes; the same aldehydes appear in the unripe fruits (Matasyoh et al., 2009).

Coriander seed oil exhibits strong antibacterial against several microorganisms. The medicinal property of the coriander usually resides in their seeds, and they have been used as a remedy for indigestion, against worms, rheumatism, and pain in the joints. The seed is an aromatic stimulant, a carminative (remedial in flatulence), appetizer and a digestant stimulating the stomach and intestines. It is generally also beneficial to the nervous system. Volatile components in essential oil, from both seeds and leaves, have been reported to inhibit the growth of a range of microorganisms and inhibition of lipid peroxidation is also reported. In Asia, the herb is used against piles, headache and swellings; the fruit is used in colic, conjunctivitis, rheumatism and neuralgia; and the seeds are used as a paste for mouth ulceration and a poultice for other ulcers (Asgarpanah and Kazemivash, 2012).

Studies have demonstrated the hypoglycaemic action and effects of coriander on carbohydrate metabolism. The effect of coriander seeds on carbohydrate metabolism was studied in rats that were fed with a fat-rich cholesterol diet. The spice exhibited noteworthy hypoglycemic action. There was an increase in the concentration of hepatic glycogen as was evident from the increased activity of glycogen synthase. Activities of glycogen phosphorylase and gluconeogenic enzymes revealed decreased rates of glycogenolysis and gluconeogenesis. The increased activities of glucose-6-phosphate dehydrogenase and glycolytic enzymes suggest the utilization of glucose by the pentose phosphate pathway and glycolysis. These observations clearly indicated that coriander seeds demonstrate good hypoglycemic activity through enhanced glycogenesis, glycolysis and decreased glycogenolysis and gluconeogenesis (Aissaoui et al., 2011).

Epidemiological and experimental studies recommend that dietary factors, particularly saturated fatty acids, play an imperative role in the growth of cancer. Spices are stated to possess carcinogenic and anti-carcinogenic properties. In view of the daily intake of several spices and their positive effects on human beings, the effect of coriander seeds, a commonly used spice in Indian cuisines, was studied on lipid metabolism in 1,2-dimethyl hydrazine (DMH)-induced colon carcinogenesis using rats as experimental animals. Studies showed that in DMH-induced colon carcinoma in rats, there is an increased concentration of lipids and also a decreased excretion of bile acids in the DMH control group compared to the experimental group. The administration of coriander counteracts this effect. The spice also prevents changes in the ratio of cholesterol to phospholipid, thereby maintaining the membrane fluidity, integrity and function. Thus, the inclusion of this spice in the daily diet plays an important role in the protection of colon against chemical carcinogenesis (Milner and Kaefer, 2007).

1.4.5 *Brassica campestris (Mustard; Family: Brassicaceae)*

This plant is distributed in the Mediterranean region, Europe, Asia and South Africa. It is cultivated throughout India as a leafy vegetable, edible oil crop and spice. It has been cultivated and used from very ancient times. Its plant is glabrous with a few bristles at the base up to 1.5 m in height, basal leaves long, broadly ovate, coarsely dentate, persistent middle leaves oblong, fruits siliqua, breaking away from below upward and seeds are attached to the replum. Its seeds contain sinigrin, gluconapin, sinapin base and a volatile isothyocyanate.

Seeds of *Brassica campestris* contain sinigrin, gluconapin, sinapin base and a volatile isothyocyanate. New flavonol glycosides, such as kaempferol 3-O-β-D-(2E-p-coumarylα-D-glucosyl(1-2)-glucopyranoside)7-O-β-D-glucopyranoside, kaempferol 3-O-α-D-(2E-Caffeoyl-β-D-glucosyl-(1→2)glucopyranoside)7-O-β-D-glucopyranoside and quecetin-3-O-β-D-(2Eferuloylβ−D-glucosyl(1→2)-glucopyranoside) 7-O-β-D-glucopyranoside were isolated from leaves. Kaempferol 3-O-β-D-(glucopyranosyl (1→2)-glucopyranoside)7-O-β-D-glucopyranosideandkaempferol-3-O-β-D-(2-sinapoyl-β-D-glucosyl(1→2)glucopyranoside)7-O-β-D-glucopyranoside were also isolated. A phytoalexin, cyclobrasisin sulphoxide was also isolated from leaves. The proximate analysis of crude mucilage and its fractions revealed carbohydrates as the major component (80%–94%) with ash (1.7%–15%) and protein (2.2%–4.4%) as minor constituents. Glucose (22%–35%) was the major monosaccharide present followed by galactose (11%–15%), mannose (6%–6.4%), rhamnose (1.6%–4.0%), arabinose (2.8%–3.2%) and xylose (1.8%–2.0%). Amino acid analyses in seed showed high levels of (g/100 g protein) glutamic acid (20.7), arginine (10.8) and proline (6.5), while histidine (2.9) and tyrosine (2.5) were lowest (Cartea et al., 2010).

The seeds are bitter, acrid, thermogenic, anodyne, digestive, carminative, antihelmintic, aperient sudorific and tonic. They are useful in the vitiated conditions of *vata* (vata dosha is light, dry, mobile, cold, hard, rough, sharp, subtle, flowing, and clear) and *kapha* (kapha dosha is responsible for the structure of the body), dengue fever, intestinal worms, flatulence, anorexia, dyspepsia, inflammations, morbid state of the cerebrospinal system, skin diaeases, splenomegaly and persistent vomiting. It is also used in large doses as an emetic in cases of poisoning, and it will cause hyperdypsia, burning sensation and other disorders of *pitta* (pitta dosha combines the fire and water elements, from which the bile is formed.) (Cavallos-Casals and Cisneros-Zevallos, 2010).

1.4.6 *Coleus forskohlii (COLEUS; FAMILY: LABIATAE)*

Its roots are a natural source of forskolin, the only plant-derived compound presently known to directly stimulate the enzyme adenylate cyclase and subsequently cyclic adenosine monophosphate (cAMP). It inhibits the platelet activation, release of histamine and other allergic compounds, increases force of contraction of the heart muscle and contributes to the relaxation of the arteries and other smooth muscles called vasodilation. It is very high in demand. *Coleus forskohlii* is a source of diterpene forskolin, derived as an active alkaloid from the roots. In traditional Ayurvedic systems of medicine, it is used for treating heart diseases, abdominal colic, respiratory disorder, insomnia, convulsions, asthma, bronchitis, intestinal disorders, burning sensation, constipation, epilepsy and angina (Kavitha et al., 2010). The roots are also used in treatment of worms and to alleviate burning in festering boils. When mixed with mustard oil, the root extract is applied to treat eczema and skin infections. The plant is also used for veterinary purposes. Forskolin is reported to be anti-glaucoma, anti-platelet, broncho-spasmolytic, cardio-tonic, hypo-tensive, anti-ageing, and anti-allergic, smooth muscle and arterial relaxant, and anti-asthmatic (Soni and Singhai, 2012). Coleus also aids in weight loss due to its ability to break down stored fat as well as inhibit the synthesis of adipose tissue. Additionally, it increases thyroid hormone production, thereby increasing metabolism. Forskolin is also used as one of the important ingredients in the preparation of medications for preventing greying of hair and restoring them to their natural colour. Though grouped as a medicinal plant, it also contains essential oil in tubers, which has very attractive and delicate odour with a spicy note. Essential oil has potential uses in the food flavouring industry and can be used as an anti-microbial agent. Forskolin is also used for the treatment of eczema, asthma, psoriasis, cardiovascular disorders and hypertension, where the decreased intracellular cAMP level is believed to be a major factor in the development of the disease process (Kanne et al., 2015). It is being developed as a drug for hypertension, glaucoma, asthma, congestive heart failures and certain types of cancers. Forskolin is in great demand in Japan and European countries for its medicinal use and related research purposes (Kavitha et al., 2010).

1.4.7 Piper longum Linn. (Pippali; Family: Piperaceae)

This plant is commonly found in moist deciduous evergreen forests. It is a slender sub-scandent herb, branchlets erect, straggling or sometimes climbing, hairless, with swollen nodes and those of creeping branches with roots at lower nodes. Its fruits contain resin, alkaloids piperine, piperolactam A, B and C, piperadione, aristolactum A, cepharadione A, cepharanone B, norcepharadione B, guineesine, pluviatilol, methyl pluviatilol, piper longuminine (piplartine), piperlongumine, pipernonaline, piperundecalidine, sesamin, *N*-isobutyl *trans*-2,4-decadienamide, a lignin derivative, a terpenoid and 1% volatile oil (Kumar et al., 2011; Zaveri et al., 2010). It acts as a bioenhancer, antibacterial, antifungal, antitubercular and anti-inflamatory; liquifies thick mucus; stops formation of cough; and is useful in cough, cold, bronchitis, asthma, fever and other respiratory diseases. The roots and fruiting spikes are used in treating diarrhoea, indigestion, jaundice, urticaria, abdominal disorders, hoarseness of voice, asthama, hiccough, cough, piles, malarial fever, flatulence vomiting, thirst, oedema, earache, wheezing, chest congestion, throat infections, worms and sinusitis (Ali et al., 2007; Jin et al., 2009). Its fruits are used for treating respiratory tract diseases (e.g., cough, bronchitis and asthma); they are used as emmenagogue, digestive, appetizer, carminative (in indigestion), general tonic and haematinic (in anaemia and chronic fevers).

1.4.8 Spinacia oleracea L. (Spinach; Family: Chenopodiaceae)

This is a leafy vegetable consumed in most developed countries after the decoction of either fresh or in the frozen form. It is a good source of carotenoids and contain flavonoids such as rutin, quercetin, kaempferol, jaceidin 4'-glucuronide. 5,3',4'-trihydroxy-3-methoxy-6:7-methyle-nedioxyflavone 4'-glucuronide and 5,4'-dihydroxy-3,3'-dimethoxy-6:7-methyle-nedioxyflavone 4'-glucuronide are extracted from its leaves. Some additional constituents including spinatoside (5,7,3',4'-tetrahydroxy-3,6-dimethoxyflavone 4-α-D-glucuronide), patuletin 3-glucosyl-(1→6)[apiosyl (1→2)]-glucoside and patuletin and spinacetin 3-gentiobiosides were also found and biological activities of spinach polyphenols have been reported (Olasupo et al., 2018). Results indicate significant antioxidant activity, principally involving the spinach flavonoids, as these constitute the major water-soluble polyphenols found in it. Patuletin-3-glycopyranosyl-(1→6)-α-gluco-pyranoside Patuletin-3-[α-D-apiofuranosyl-(1→2)]-α-D-glucopyranoside, Patuletin-3-α-

D-glycopyranosyl-(1→6)-[-D-apiofuranosyl-(1→2)]-α-D-glucopyranoside, Spinacetin-3-α-D-glycopyranosyl-(1→6)-α-D-glucopyranoside,40-dihydroxy-3, 30-dimethoxy-6:7-methylenedioxyflavone   40-glucuronide   5,30,40-trihy-droxy-3-methoxy-6:7-methylenedioxyflavone 40-glucuronide 5,7,40-trihy-droxy-3,6,30-trimethoxy-flavone 40-glucuronide 5,7,30,40-tetrahydroxy-3,6-dimethoxyflavone 40-glucuronide were also reported (Metha and Belemkar, 2014).

The β-carotene and polyphenolic flavonoids have been reported with anti-cancer, anti-mutagenic and antioxidant activities. Phytochemicals present like rutin, quercetin, kaempferol posses anti-cariogenic, anti-inflammatory, antibacterial, anti-hypertensive, anti-inflammatory, antimutagenic, anti-acne, anti-ageing and anti-asthmatic activities. Antioxidant activity of spinach water-soluble polyphenolic flavonoids has been reported (Ashok Kumar et al., 2013; Khare, 2007).

1.4.9   *Acacia nilotica* Delile *(Babul; Family: Leguminosae)*

The plant is a medium-sized spiny evergreen tree. Its leaves contain mucilage, tannin, organic acids and traces of asparagin and ash, containing alkaline sulphates, chlorides magnesium phosphate and calcium carbonate. Roots also contain asparagin. Its seeds are used in piles, laxative, expectorant and in gonnorhoea. Leaves are demulcent, locally applied to boils and ulcers. Its decoction is used in toothache and tender gums, given internally for the inflammtion of bladder. Bark has astringent properties and antidiuretic. Experimental studies have proven its anti-diabetic, anti-hypertensive, anti-spasmodic, antibacterial, antifungal, antiplaque, antioxidant, antiplasmodic and antiviral activities; it is catalytic and galactagogue (Kaushal, 2017; Nandkarni, 2010).

1.4.10   *Acorus calamus* Linn. *(Bach; Family: Araceae)*

Acorus calamus is a perennial plant, growing to 1 m (3 ft 3 in) × 1 m (3 ft 3 in) at a medium rate.

It is hardy plant and is not frost tender. It flowers from May to July, and its seeds ripen from July to August. The species is hermaphrodite (has both male and female organs) and is pollinated by insects. The chemical constituents present are calamenol, calamene, calamenine, methyleugenol, eugenol and α-pinene, camphene, palmitic heptylic and butyric acids, asaronal-dehyde, calamol, calamone and azulene, asarone and its β-isomer. Sesquiterpenic ketones such as acorone, calarene, calcone, calacorene, acorenone, acolamone, iso-acolamone, epishyobunone, shyobunone, isoshyo-obunone, acorangermendiol and preisocalamendiol are present. The rhizome is acrid, bitter, thermogenic, intellect, promoting, emetic, laxative, carmina-tive, stomatic, acthelmintic, emmenagogue, diuretic, alexteric, expectorant, anodyne, anti-spasmodic, aphrodisiac, anti-convulsant, resuscitative, anti-inflammatory, sudorific, antipyretic, sialagogue, insecticidal, tranquillizing, nervine tonic, sedative and tonic. It is useful in vitiated conditions of *vata* and *kapha*, stomatopathy, hoarseness, colic flatulence, dyspepsia, helmin-thiasis, amenorrhoea, dysmenorrhoea, nephropathy, calculi, strangury, cough, bronchitis, odontalgia, pectoralgia, hepatodynia, otagia, inflamma-tions, gout, epilepsy, delirium, amentia, convulsions, depression and other mental disorders, tumours, dysentery, hyperdipsia, haemorrhoids, intermit-tent fevers, skin diseases and general debility (Ganjewala and Srivastava, 2011; Rajput et al., 2014).

1.4.11 *Albizia lebbeck (Siris; Family: Mimosaceae)*

The plant is a medium- to large-sized unarmed deciduous tree about 20 cm in height with an umbrella-shaped crown and grey to dark brown rough and irregularly craked bark; leaves are abruptly pinnate, main rachis with a large gland; flowers are white fragrant, in lobose umbellate heads; fruits are long with characteristic pods that are oblong and compressed. The bark yields condensed tannins, namely D-catechin, isomer of lcucocyanadin and melac-acidin, and β-sitosterol. The flowers contain lupeol, and a pigment similar to crocetin. The bark is astringent, bitter, acrid, sweet, expectorant, aphrodisiac, anti-inflammatory and tonic. It is used in the vitiated condition of *pitta* and *kapha*, asthma, skin eruption, leucoderma, sprains, wounds and ulcers. Its seeds are useful in inflammation. The flowers are used in chronic cough and bronchitis (Kajaria, 2015).

1.4.12 *Alpinia galanga (Kulanjan; Family: Zingiberaceae)*

The plant is a perennial herbaceous plant, 1–2 m in height. Its rhizomes are cylindrical, stout, aromatic, covered with scales. Leaves alternate, lanceolate, the upper face shining and glaborous. Inflorescence in terminal dense raceme 20–30-cm long, flowers white, lip veine with red, fruit globose or ovoid. The rhizomes yield an essential oil consisting of cineol, methyl cinnamate and flavones, galangin, alpinin, kaemferide and 3-deoxy 4-methoxy flavone. The rhizomes are antibacterial and a digestive stimulant. It is used in the treatment of dyspepsia, flatulence, vomiting, gastralgia, colic, diarrhea, malaria fever and also applied externally on carious teeth to cure toothache (Chouni and Paul, 2018).

1.4.13   Aloe barbadensis (Ghritkumari; Family: Liliaceae)

It is a coarse perennial with short stem and a shallow root system, leaves fleshy in rosettes, often crowded with horny prickles on the margins, convex below, 45–60-cm long tapering to a blunt point, surface pale green with irregular white blotches. Its flowers are yellow or orange in racemes and fruits loculicidal capsule. The leaves contain barbaloin, chrysophanol glycosid and the aglycone, aloe-emodin. The mucilage of the leaves contains glucose, galactose, mannose and galacturonic acid in addition to an unidentified aldopentose and a protein with 18 amino acids. The plant also contains aloesone and aloesin. The plant is bitter, sweet, cooling, anthelmintic, aperient, carminative, deobstruent, depuretive, diuretic, stomachic emmenagogue, ophthalmic and alexeteric (Gupta et al., 2018).

1.4.14   Amoora rohitika (Amoora rohituka; Family: Meliaceae)

It is a customary plant with potential medicinal uses which is abundantly found in India. *Aphanamixis polystachya* bark is a strong astringent, antimicrobial, used for the treatment of liver and spleen diseases, rheumatism and tumors. Fruits are globular, smooth and yellow when ripe, with seeds scarlet. The seed has rich oil content which is non-edible but a future source for biodiesel. The plant is the large handsome evergreen tree, with a dense spreading crown and a straight cylindrical bole up to 15 m in height and 1.5–1.8 m in width. The chemical constituents present in seeds are palmitic acid and oleic as dominant fatty acids besides linoleic and linolenic acids. It possesses desmosterol, campesterol, α-tocopherol, 1,8-cineole, α-pinene, β-pinene, α-terpinene, α-terpineol, α-terpinyl acetate, terpinene-4-ol, borneol, nerol, nerolidon, nerolidol, geraniol, geranyl acetate, linalool and linalyl acetate. Its seeds are aromatic, acrid, sweet, cooling, stimulant, carminative, digestive, stomachic, diuretic, cardiotonic, abortifacient, alexeteric, expectorant and tonic. It is useful in treatment of asthama, bronchitis, haemorrhoids, strangury, renal and vesical calculi, halitosis, cardiac disorders, anorexia, dyspepsia, gastropathy, hyperdispsia, burning sensation, debility and vitiated conditions of *vata* (Shadid Hossain and Ali, 2016).

1.4.15 *Asparagus racemosus (Satawar; Family: Liliaceae)*

It is most frequently spread species in tropical and subtropical regions of India. The traditional use of Shatavari has been mentioned in customary medication system such as Unani & Siddh in folk and ayurvedic system of Indian Pharmacopeias. This plant is a widely used shrub that is important for its sapogenin content which is the precursor of many pharmacologically active steroids. It is believed that almost all of its parts possess pharmaceutical properties, but its roots, stem and leaves are the most significant parts which are used medicinally. The Shatavari is also used as Rasayanas to enhance the body resistance against infections and improve the immune system. Owing to the presence of many phytochemicals, it is widely used for treatment of various diseases. The remarkable medicinal properties of Shatavari are anti-spasmodic, antioxidant, anti-diabetic, anti-allergic, anti-malarial, hepato-protective, anti-neoplastic activities, immune enhancer, anti-arthritic, anti-inflammatory, anti-periodic, anti-ulcerogenic, immune modulatory, anti-stress, anti-diarrhoeal, anti-depressant, anti-leprotic, anti-abortifacient, antibacterial, antipyretic and analgesic. The plant contains four saponins, namely shatavarin I−VI. It is a glycoside of sarasapogenin having two molecules of rhamnose and one molecule of glucose. It also contains mucilage and starch. Its roots are bitter, sweet, emollient, cool, nervine tonic, constipating, galactogogue, ophthalimic, anodyne, aphrodisiac, rejuvenating, carminative, appetizer, stomachic, anti-spasmodic and tonic. They are useful in treatment of nervine disorders, dyspepsia, diarrhoea, dysentery, tumours, inflammations, vitiated conditions of *vata* and *pitta*, burning sensations, hyperdipsia, strangury, scalding of urine, throat infections, tuberculosis, cough, bronchitis, gleet, gonorrhoea, cough, leucorrhoea, abortion, agalactia and general debility (Singh et al., 2018a).

1.4.16 *Boerhavia diffusa L. (Punarnava; Family: Nyctaginaceae)*

Boerhavia diffusa (BD) Linn is an important medicinal plant used in customary Indian medicine as well as other parts of world, for example, Southern American and African continent. Many of its parts, especially roots, are used for gastrointestinal, hepatoprotective and gynecological indications. BD has been classified as 'Rasayana' herb as it possesses properties such as anti-ageing, re-establishing youth, strengthening life and brain power, and disease prevention. Thus, it increases the resistance of the body against any onslaught and provides hepatoprotection and immunomodulation. In ayurvedic texts, more than 35 formulations of different types contain punarnava as a chief ingredient. Its main chemical constituents are alkaloids, triacontanol hentriacotane, β-sitosterol, ursolic acid, 5,7-dihydroxy-3, 4-dimethoxy-6, 8-dimethyl flavone, glucose, fructose, sucrose and hypoxanthine-9-L-arabinoside, moulding hormone, β-ecdysone. The plant is bitter, astringent, cool, anthelmintic, diuretic, aphrodisiac, cardiac stimulant, diaphoretic, emetic, expectorant, anti-inflammatory, febrifuge and laxative. It is useful in all types of inflammations, strangury, leucorrhoea, ophthalmia, lumbago, myalgia, scabies, cardiac disorders, jaundice, anemia, dyspepsia, constipation, cough and bronchitis (Vasuki et al., 2018).

1.4.17 Bixa orellana L. (Sinduriya; Family: Bixaceaea)

Bixa orellana is a plant native to Brazil but grows in other regions of South and Central America. It is grown in tropical countries such as Peru, Mexico, Ecuador, Indonesia, India, Kenya and East Africa. It is a small evergreen tree; leaves large, cordate acuminate, glaborous on both surface; flowers whire and pink in terminal panicles; fruits reddish brown or green capsules, clothed with soft bristles; seeds trigonous covered with a red pulp. The seeds are considered the plant part of commercial importance, since the pericarp (layer that surrounds the seeds) contains the pigments that have wide industrial application. About 80% of this pigment is the carotenoid known as bixin, which has the dye property and can be extracted with vegetable oils or chemical bases. Depending on the cultivar and climatic conditions of the region, the bixin content can vary from 1% to 6% in the seed aril. The remainder is composed of other dyes and inert substances of minor importance (Raddatz-Mota et al., 2017).

Bixin is the main chemical contituent along with phenolic acids. The roots, seeds and bark are antiperiodic, antipyretic and astringent. They are useful in intermittent fevers and gonorrhoea.

The pulp surrounding the seed is a mosquito repellant and is useful in treating dysentery. A non-toxic dye 'Annato dye' obtained from the pulp is used for coloring edible materials (Raddatz-Mota et al., 2017*)*.

1.4.18   Cassia fistula L. (Amaltas; Family: Caeselpiniaceae)

This is a medium-sized deciduous tree, 8–15 m in height with greenish grey smmoth bark when young and rough and old exfoliating in hard scales; leaves pinnately compound, leaflets 4–8 ovate, acute, glabrous above, paler, main nerves numerous, flowers bright yellow in lax pendulous racemes, fruits cylindric, seeds broadly ovate, horizontally immersed in dark coloured sweetish pulp. Plant contains sennosides A and B, rhein and its glucocides, baraloin, aloin, formic acid, tannins and reducing sugars. Roots are astringrnt, cooling, febrifuge and tonic and are useful in treatment of skin diseases, tuberculous glands, syphilis and burning sensation. The bark is laxative, antihelmintic, emetic, febrifuge, diuretic and is useful in boils pustles, leprosy, ringworm, fever etc. (Ali, 2014). *Cassia fistula* (*C. fistula*) plays an important role in diseases prevention due to the presence of bioactive constituents. They bear important properties such as anti-microbial, antioxidant, anti-cancer through modulation of genetic pathways. Plant parts such as stem, leaf, and flower contain different types of constituents that play a vital role in health care. Previous findings have shown that stem bark of *C. fistula* is a chief source of lupeol, ß-sitosterol and hexacosanol. Fruits and flower of *C. fistula* are a source of important phytochemicals that display a role in health management. A compound such as 1,8-dihydroxy-3-anthraquinone derivative isolated from the fruit pulp and compounds kaempferol, lcucopelargonidin tetramer, rhein, fistulin, and triterpenes are isolated from their flowers. Another finding showed that seeds of *C. fistula* are rich in glycerides with linoleic, oleic, stearic and palmitic acids as chief fatty acids and also contain traces of caprylic and myristic acids (Rahmani, 2015).

1.4.19   Cinnamomum tamala (Tejpatra; Family: Lauraceae)

The plant is found in the Himalayas, in areas of 900–2400-m elevation. It is a medium-sized evergreen tree approximately 7.5 m in height with dark brown or blackish rough bark and pinkish or reddish-brown blaze leaves. The leaves yield an essential oil that resembles cinnmon oil contains D-α-phellendrine and 78% eugenol, cinnamaldehyde, diterpensoids-cinnzeylamin, cinnzeylanol, anhydrocinnzeylanin, anhydrocinzeylanol, cinncassiol-D-4 and its glucoside, coumarin, *trans*-cinnamic acid and protocatechuic acid (Mal et al., 2018). The leaves are bitter, sweet, aromatic, thermogenic, alexeteric, anthelmintic, diuretic, stimulant, carminative and tonic. They are useful in treatment of diabetes, cardiac disorders, inflammations, helminthiasis, dyspepsia, colic, hyperptyalism, ophthalmia, diarrhoea, proctitis, proctalgia, hepatopathy and splenopathy (Hassan et al., 2016).

1.4.20   Curcuma longa L. (Haldi; Family: Zingiberaceae)

It is a perennial herb distributed throughout tropical and sub-tropical regions of the world such as India, Pakistan, Bangladesh and Sri Lanka. Turmeric has long been recognized for its medicinal properties, has received interest from both the medical/scientific world and from culinary enthusiasts, as it is the major source of the polyphenol curcumin. It aids in the management of oxidative and inflammatory conditions, metabolic syndrome, arthritis, anxiety and hyperlipidemia. It may also help in the management of exercise-induced inflammation and muscle soreness, thereby enhancing recovery and performance in active people. In addition, a relatively low dose of the complex can provide health benefits for people that do not have diagnosed health conditions. Majority of these benefits are due to its antioxidant and anti-inflammatory activities. Ingesting curcumin by itself does not lead to the associated health benefits due to its poor bioavailability, which appears to be primarily due to poor absorption, rapid metabolism and rapid elimination. It is an auspicious spice for all religious ceremonies among the 'Hindus'. Its rhizomes contain several polyphenolic compounds (3%–6%) a mixture of curcumin, demethoxycurcumin and bisdemethoxycurcumin, collectively known as curcuminoids. Rhizomes are used as antiseptic, blood purifier, anti-inflammatory, anthelmintic, in dropsy, epilepsy, spleen disorders, rheumatism, bronchitis, cough and cold, in cholera, syphilis, antifungal, anti-inflammatory and antibacterial agent (Krup et al., 2013; Sinoriya et al., 2018).

1.4.21   Euphorbia hirta (Dudhi; Family: Euphorbiaceae)

It is an erect annual or perrenial plant growing to 50 cm, with pointed oval leaves and clusters of small flowers. Pill-bearing spurges contain flavonoids, terpenoids, alkanes, phenolic acids, shikimic acid and choline. The latter two constituents may by partly responsible for the anti-spasmodic action of this plant. *Euphorbia hirta* is used to treat bronchial asthama; pill-bearing spurge relaxes the bronchial tubes and eases breathing. Mildly sedative and expectorant, it is also taken to treat intestinal amoebiasis (Sashini et al., 2018).

1.4.22   Emblica officinalis GAERTN (Aonla; Family: Euphorbiaceae)

It is a medium- to large-sized deciduous tree. The most significant feature of this plant is its branches which are interestingly flat like leaves with flowers on the margins after blossoming. Leaves are simple oblong, blunt, arranged in two ranks. Flowers are greenish yellow in colour in axiallary fascicles. Phyllemblin, phyllemblic acid, gallic acid, emblicol, ellagic acid, pectin, putrnjivain A, two hydrolysable tannins vitamin-C like called emblicanin A and B and not ascorbic acid. Fruits have been used for thousands of years in the traditional Indian medicines for treatment of several diseases such as mutagenesis, clastogenicity, preserves against diseases, antioxidant acitivity, antiviral activity, increases immunodefence, hepatoprotective and have a hypolipidemic effect (Diwan et al., 2018).

1.4.23   Eucalyptus globules Labill (Yucaliptas; Family: Myrtaceae)

It is a lofty tree about 90 cm in height with a clean straight bole and smooth bark; leaves opposite on juvenile shoots, alternate in adult shoots, lanceolate, 20−25 cm long, broad, rather thick and curved; flowers large, white, 1−3 together in axils, fruit a hardened capsule and seeds very small. Eucalyptin, quercetin, quercetrin, quercitol, chrysin, rutin, caffiec, ferulic, gallic, protocatechuic acids are the main chemical constituents. The oil is acrid, bitter, astringent, thermogenic, antiseptic, carminative, antipyretic and carditonic, and is useful in the vitiated condition of *pitta* and *kapha* (Kumar et al., 2018).

1.4.24   Evolvulus alsinoides L. (Shankhapushpi; Family: Convolvulaceae)

Evolvulus alsinoides (*E. alsinoides*) L. is an important plant in Ayurveda known for its therapeutic effects. *E. alsinoides* is commonly known as Shankhapushpi, and it is found throughout India ascending to the 6000 ft in the Himalayas. Its potential therapeutic effect leads to its use in various disorders such as insanity, epilepsy, memory enhancement and nervous debility in the Ayurveda system of medicine. Antioxidant properties of this plant are used to treat low spirits and depression as shown in various in vivo experiments. In various formulations, it is used as an important ingredient which is used for the management of various CNS disorders such as psychosis, epilepsy and other conditions where brain activities are affected. It is widely used as a nervine tonic in various Asian countries as it has potent memory-enhancing activity. It is also included as a Medhya drug in the treatises of Ayurveda like Charaka Samhita, Susruta Samhita and Ashtanga Hridaya. The chemical constituents are Shankhapushpine, evolvine and betain. Fresh plant contains volatile oil, organic acids, fat and saline substances. It is used as tonic in febrifuge, vermifuge, chronic bronchitis, general weakness, fever, dysentery, nervous debility, loss of memory and also in syphilis and scrofula (Gollen and Mehla, 2018).

1.4.25 *Gymnema sylvestre (Gudmar or Madhunasini; Family: Asclepiadaceae)*

The plant has a good medicinal value and market demand; its leaves are used in diabetes, hydrocil and asthama. Its main ingredients are gymnemic acid, a glycoside isolated from its leaves, a destroyer of madhumeha (glycosuria), and other urinary disorders. It is believed that it neutralizes the excess of sugar present in the body in diabetes mellitus (Pothuraju et al., 2014). It is useful in the treatment of dyspepsia, constipation, jaundice, haemorrhoids, renal and vesical calculi, cardiopathy, asthma, bronchitis, amenorrhoea, conjuctivitis and leucoderma. The leaves are also helpful in lowering the serum cholesterol and triglyceride levels. The primary chemical constituents of *Gymnema* include gymnemic acid, gurmarin, stigmasterol, betaine and choline. The water-soluble acidic fractions reportedly provide the hypoglycemic action. Gurmarin and gymnemic acid have been reported to block sweet taste in humans (Bamola et al., 2018; Ramachandran et al., 2010).

1.4.26 **Lawsonia enermis L.** *(Mehndi; Family: Lythraceae)*

It is a glaborous, much branched, deciduous shrub with 4-gonous lateral branches often ending in spines; leaves simple, opposite, entire, lanceolate; flowers white, fruit globose; the principle colouring matter is lawsone, 2-hydroxy-l: 4-naphthaquinone which is present in dried leaves, behenic, arachidic, stearic, palmitic, oleic, etc. The roots are bitter, refrigerant, depurative and diuretic and are useful in treatment of burning sensation, dipsia, leprosy, skin diseases and liver tonic. They are useful in treatment of wounds, ulcers, anti-inflammatory, diarrhoea, dysentery, scabies, boils, etc. Earlier, this plant was used as hair dye and in cosmetics. But nowadays, it is used as a medicinal plant. The leaves of this plant showed anti-inflammatory, antipyretic, analgesic, antifungal and antibacterial activity. Lawsone (2-hydroxy napthaquinone), mucilage, mannite, gallic acid and tannic acid were found to be the main chemical constituents that might be having a role in its medicinal properties (Chakkilam, 2017).

1.4.27   *Myristica fragrans (Jaiphal; Family: Myristaceae)*

It is a lofty tree with branches slender, leaves 7–8 cm, dark green, entire, lanceolate, flowers yellow, dioecious. Fruits are ovoid, sub-globose, 4–5 cm. The plant is widely used as spices as well as for medicinal purpose in the Unani System of Medicine since times immemorial. The plant nutmeg was introduced by a *saydalaneh* (pharmacist in Arabic) for its clinical use. As per Unani nature, nutmeg is hot and dry. It is a very common and prevalent drug used as digestive, carminative, appetizer, exhilarant and mood elevator, anti-tussive, anti-emetic, demulcent, aphrodisiac, stomachic, liver, nervine, cardio and uterine tonic in the Unani System of Medicine. The extensive information for its use is provided in Unani text by Greek physicians so it is used in many compound formulations and useful recipes. Any research fimdings have shown that it possesses good digestive, appetizing, exhilarant properties; it is also used as mood elevator, anti-tussive, anti-emetic, nervine tonic, aphrodisiac, anti-diarrhoeal, liver tonic, cardiotonic and uterine tonic. Recent studies have shown its anti-microbial, hypolipidemic, antioxidant, anti-cancer, sedative, analgesic, and anti-inflammatory activity. Further, the scope of research can also be explored with the help of traditional knowledge of Unani and other customary medication.

Major constituents of nutmeg oil are d-B-pinene, myristicin, linalool, safrole, dipentene and linalyl acetate. The plant is antibacterial, anti-cancer, anti-fertility, anti-gingivitic, antiherpetic, antihistaminic, anti-implantation, anti-inflammatory, anti-leukemic, anti-lymphocytic, anti-mutagenic, antioxidant, culinary (Abourashed and El-Alfy, 2016).

1.4.28   *Santalum album L. (Chandan; Family: Santalaceae)*

Santalum album is a small- to medium-sized ever green, glabrous, semi-parasitic tree with slender drooping branches, opposite leaves, hermaphrodite flowers, axillary or terminal, trichotomous panniculate cymose inflorescence and small globose fruiting berries. The sapwood is white and odourless and heart wood is yellowish in colour and scented. The tree is root parasite; it forms haustoria which establishes contact with the host. It obtains lime and potash directly from soil through roots, and for nitrogen and phosphorus, it partially depends upon its host. Bark is of generally dark grey colour. Leaves are opposite, sometimes alternate, oval, shining and stalks long 0.6–1.2 cm. Flowers are purple brown in colour, in clusters, perianth bell shaped, with four triangular lobes, stamen four. It contains santalol, β-santalic, beta-santalol, epi-β-santalene, *cis*-lanceol, and *cis*-nuciferol. Plant shows antioxidant activity, analgesic, anesthetic, antibacterial, antihemorrhoidal, antiseptic, antisinusitic, antispastic, antiviral, anti-wrinkle and cancer-preventive properties (Moy and Levenson, 2017).

1.4.29   *Sida cordifolia L. (Bala; Family: Malvaceae)*

It is widely distributed along with other species throughout the tropical and sub-tropical plains all over India and Sri Lanka up to an altitude of 1050 m. *Sida cordifolia* is a small, erect, downy shrub. Its main chemical constituents are alkaloids, fatty oil, phytosterol, mucin, resin, acids and potassium nitrate. The leaves, stem and root yield 0.085% and the seeds about 0.3% alkaloid. The main portion of the alkaloid is ephedrine and 0.4% rutin is present in the plant. The plant is reputed for its tonic and aphrodisiac properties. The plant parts are used for the fever, fits, leucorrhoea, micturitin, gonorrhoea, colic, nervous disorders, general debility and heart irregularity. The root juice is used for the healing of the wound and the curing facial paralysis and sciatica and the leaves are used for curing the bloody flux, useful in ophthalmic, rheumatism shivering, fits and improve sexual strength (Singh and Navneet, 2018).

1.4.30   *Saraca indica (Ashok; Family: Leguminosea)*

This plant has an age-old history as it was first recognized as a medicinal tree in the Agnivesa Caraka Samhita which is supposed to have been complied somewhere near 1000 BC. Ashoka is used in various female gynaegological problems includinng urinary tract infections. *Saraca indica* is used in the Ayurvedic system of medicine as hypothermic and diuretic, as a blood purifier, and in stomach ache. It is also used in bleeding piles, bacillary dysentery, and Asokarista, Asokaghrta, Asoka decoction and Asoka pills are the well-known pharmaceutical preparations. The drug Asoka Aristha is traditionally used in India and Sri Lanka to treat menorrhagia.

The bark of *Saraca indica* contains an estrogenic compound called ergo-sterol. The major constituents present in the Saraca asoca stem bark extracts are flavonoids, terpenoid, lignin, cardiac glycosides, phenolic compounds and tannins. The flavonoids and leucoanthocyanidins are present in the bark of plant. These phytoconstituents are responsible for various pharmacological actions, such as antiulcer, anticancer and chemo protective activities. Petroleum ether extract of Saraca indica leaves exhibits potential antitumor and antioxidant activities.

The main chemical constituents are tannins, catechol, sterol, haematoxylene and water-soluble glycoside. The bark is bitter, astringent, sweet, refrigerant, anthelminthic, antioxidant, stytic, stomachic, constipating, febrifuge and demulcent. It is used in dyspepsia, fever, dipsia, burning sensation, visceromegaly, colic, ulcers, menorrhagia, metropathy, leucorrhoea and pimples. The leaves are depurative, and their juice mixed with cumin seeds is used for treating stomachalgia. The flowers are considered to be a uterine tonic and are used in vitiatied conditions of *pitta*, syphilis, cervical adeitis, and hyperdypsia, burning sensation, naemorrhoids, dysentery, scabies in children and inflammation. The dried flowers are used in diabetes and haemorrhagic dysentery, and seeds are used for treating bone fractures, strangury, breast cancer and vesical calculi (Yadav et al., 2015).

1.4.31   *Terminalia chebula (Harad; Family: Combretaceae)*

It is a medium-sized large deciduous tree with a cylindrical bole, rounded crown and spreading branches; leaves ovate; fruits glaborous, shining ellipsoidal, yellow to orange in colour, seeds hard and pale yellow. Main ingredients in seed kernels are palmitic, stearic, arachidic, behnic acid, Fruit contain chebulin, tannic acid, gallic acid, resins, etc. The fruits are astringent, sweet, acrid, bitter, sour, thermogenic and anodynic, anti-inflammatory, alterant, stomachic, laxative, puragative, carmiantive, digestive, antihelmintic, caritonic, aphrodisiac, antiuseptic, diuretic also useful in the vitiated condition tridosha, ulcers, inflammation, gastropathy, anorexia, helmenthiasis, flatulence, etc. (Promila and Madan, 2018).

1.4.32   *Terminalia bellarica (Bahera; Family: Combretaceae)*

It is a large deciduous tree, 20–30 m in height with thick brownish longitudinal fissures; leaves simple, long petioled, flowers pale greenish yellow colour withan offensive odour, fruits ovoid grey drupes. Fruits contain tannins, β-sitosterol, gallic acid, ellagic acid, ethyl gallate and chebulagic acid. The bark is mildly diuretic and useful in anemia and leucoderma. The

Fruits are astringent, acrid, sweet, anthelmintic and antipyretic. It is useful in dropsy, leprosy, skin diseases, flatulence and dysentery. The mature and dry fruit is constipating and is useful in treatment of diarrhoea and dysentery (Kumar and Khurana, 2018).

1.4.33 *Terminlia arjuna (Arjun; Family: Combretaceae)*

It is a large evergreen tree wth buttressed trunk and spreading crown with drooping branches; bark smooth grey outside and fiesh coloured inside; leaves simple ovate, oblong, subopposite; flowers white in panicle of splikes with linear bracteoles; fruits ovoid or oblong with 5−7 wings. Main ingredients are arjunolic acid, tomentosic, ellagic acid, saponin and leucodelphinidin. Bark contains arjunin, a lactone arjunetin, tannin and pyrocatechol. The bark is astringent, aphrodisiac, carditonic and is useful in ulcers, leucorrhoea, diabetes, inflammations, cirrhosis of the liver and hypertension, etc. (Augustine and Sreeraj, 2017).

1.4.34 *Withania somnifera (Ashwagandha; Family: Solanaceae)*

It is also called 'Indian Ginseng' due to its restorative properties. It is a shrub which grows in dry arid soils of subtropical regions. The plant grows throughout the drier parts and in the temperate regions in India, from plain to the height of 1700 m. Alkaloids and withanolides (steroidal lactones) are the major groups of the secondary metabolites of medicinal interest. The other secondary metabolites include flavonol glycosides, glycowithanolides, sterols and phenolics, tropine, pseudotropine, dl-isopelletierine, cuscohygrine, 3-tropyl tigloate anaferine, anahygrine and with asomnine. The major steroidal lactones are withanone and withaferin A. Its roots are generally used as a valuable drug in the Indian customary system of medicine Ayurveda for rheumatic pain, inflammation of joints nervous disorders and epilepsy. It is mainly indicated as aphrodisiac, diuretic, restorative and rejuvenative. Its leaves have also been used as folk remedy for all types of skin lesions, ulcers and boils and in reducing the pus formation and inflammation. It has antibacterial activity, antiviral, antifungal, antitumor, antimitotic, anti-inflammatory, anti-arthritic, anti-stress, antioxidant, anti-microbial, anti-convulsant, immuno-modulatory, nematicidal, insectisidal, antiprotozoal, CNS, adaptogenic, nervine tonic, useful for memory, antipeptic ulcer, rheumatic swelling, aphrodisiac, restorative, consumption debility from old age, emaciation of children, ulcer and insomnia. Steroidal lactones are the major phytoconstituents present in this species. Different pharmacological experiments in a number of in vitro and in vivo models have convincingly demonstrated the ability of *W. somnifera* to exhibit anti-inflammatory, anti-oxidative, anti-microbial, anti-anxiety, aphrodisiac, immunomodulation, anti-diabetic, anti-ulcer, anticancer, CNS depressant and hepatoprotective activities, lending support to the basis behind several of its traditional uses (Azgomi et al., 2018).

Some of the important medicinal plants with their phytochemicals present are indicated in Figure 1.1.

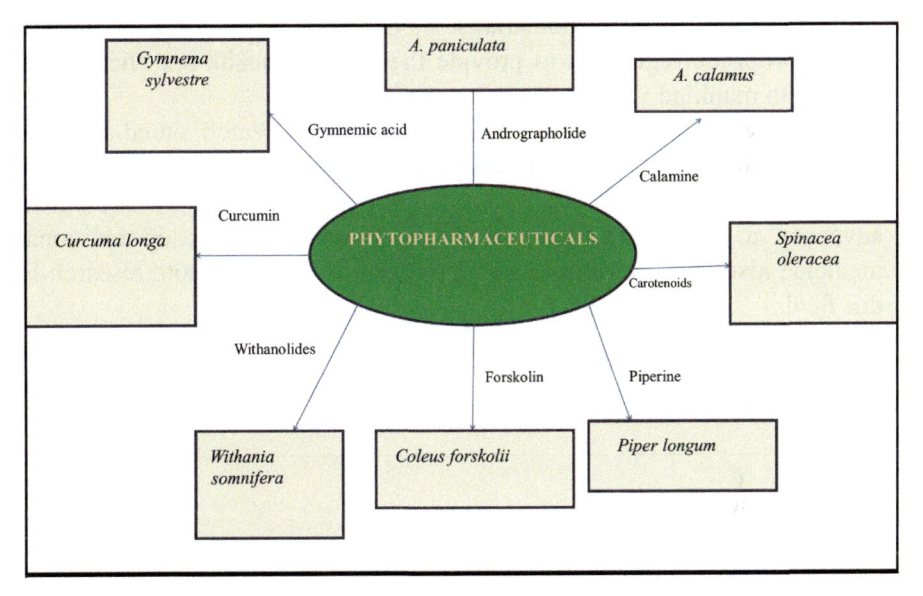

FIGURE 1.1   Important medicinal plants with their phytochemicals.

1.5   CONCLUSION

There is an increasing interest in correlating phytochemical constituents of a plant with its pharmacological activity. Plant-derived phytopharmaceutical is of pronounced significance in the current system of medicine and healthcare. Scientists have even started correlating the botanical properties of plants with their pharmacological activity. The dearth of quality control is a foremost issue and hurdle for the phytophamaceuticals.

The quality of plant material and manufacturing processes used for phytophamaceuticals are regulated by food laws, which lack the specificity required for botanical drugs. This can have grave concerns. Phamaceutical professionals and regulatory bodies should play a key role for safety maintenance and advances of phytophamaceuticals. Future demand of phytophamaceuticals depends upon consumer awareness and the link between diet and disease.

Although phytophamaceuticals and functional foods have a significant role in the promotion and care of human health to prevent diseases, the health professionals, nutritionists, biotechnologists, regulatory toxicologist and phytophamaceuticals industrialist should strategically work together to plan appropriate regulation to provide the ultimate health and therapeutic benefits to mankind with purity, efficacy and safety.

In future, more coordinated multi-dimensional research aimed at correlating botanical and phytochemical properties to specific pharmacological activities is expected. In addition to the proper utilization of technological advances, a logical interpretation of the codified language of traditional medicine also becomes a necessity in order to further promote research in this field.

KEYWORDS

- **phytochemical**
- **medicinal plant**
- **pharmaceutical**
- **anti-microbial**
- **antioxidant**
- **anti-cancer**
- **anti-inflammatory**

REFERENCES

Abourashed, E.A.; El-Alfy, A.T. Chemical diversity and pharmacological significance of the secondary metabolites of nutmeg (*Myristica fragrans* Houtt.). *Phytochem Rev.* 2016, 15(6), 1035–1056.

Acharya, D.; Shrivastava, A. *Indigenous Herbal Medicines: Tribal Formulations and Traditional Herbal Practices.* 1st edn., Jaipur: Aavishkar Publishers, 2008; pp. 440.

Aissaoui, A.; Zizi, S.; Israili, Z.H.; Lyoussi, B. Hypoglycemic and hypolipidemic effects of *Coriandrum sativum* L. in Meriones shawi rats. *J Ethnopharmacol.* 2011, *137*(1), 652–661.

Ali, A.M.; Alam, N.M.; Yeasmin, M.S.; Khan, A.M.; Sayeed, M.A. Antimicrobial screening of different extracts of *Piper longum* Linn. *Res J Agr Bio Sci.* 2007, *3*(6), 852–857.

Ali, M.A. Cassia fistula Linn: A review of phytochemical and pharmacological studies. *Int J Pharm Sci Res.* 2014, 5(6), 2125–2130.

Asgarpanah, J.; Kazemivash, N. Phytochemistry, pharmacology and medicinal properties of *Coriandrum sativum* L. *Afr J Pharm Pharmacol.* 2012, *6*(31), 2340–2345.

Ashok Kumar, C.K.; Divyasree, M.S.; Joshna, A. A review on south Indian edible leafy vegetables. *J Global Trends Pharm Sci.* 2013, *4*(4), 1248–1256.

Augustine, A.; Sreeraj, G. Medicinal properties of *Terminalia arjuna* (Roxb.) Wight & Arn.: A review. *J Tradit Complement Med.* 2017, *7*(1), 65–78.

Azgomi, R.N.D.; Zomorrodi, A.; Nazemyieh, H.; Fazlijou, S.M.B.; Bazargani, H.S.; Nejatbakhsh, F.; Jazani, A.M.; AsrBadr, Y.A. Effects of *Withania somnifera* on reproductive system: A systematic review of the available evidence. *BioMed Res Int.* 2018, *2018*(1), 1–17.

Bamola, N.; Verma, P.; Negi, C. A review on some traditional medicinal plants. *Int J Life Sci Scienti Res.* 2018, *4*(1), 1550–1556.

Bjelakovic, G.; Nikolova, D.; Gluud, C. Antioxidant supplements and mortality. *Curr Opin Clin Nutr Metabol Care.* 2014, *17*(1), 40–44.

Carocho, M.; Ferreira, I. C. F. R.; Morales, P.; Soković, M. Antioxidants and prooxidants: Effects on health and aging. *Oxid Med Cel Longev.* 2018, 2018: 1472708.

Cartea, M.E.; Francisco, M.; Soengas, P.; Velasco, P. Phenolic compounds in *Brassica* vegetables. *Molecules.* 2010, *16*(1), 251–280.

Cavallos-Casals, B.A.; Cisneros-Zevallos, L. Impact of germination on phenolic content and antioxidant activity of 13 edible seed species. *Food Chem.* 2010, *119*(4), 1485–1490.

Chakkilam, R.K. Review of *Lawsernia inermis*. *World J Pharm Pharm Sci.* 2017, *6*(4), 885–891.

Chikezie, P.C.; Ojiako, O.A. Herbal medicine: Yesterday, today and tomorrow. *Altern Integr Med.* 2015, *4*(3), 195.

Chouni, A.; Paul, S. A review on phytochemical and pharmacological potential of *Alpinia galanga*. *Pharmacog J.* 2018, *10*(1), 9–15.

Dhiman, A.; Goyal, J.; Sharma, K.; Nanda, A.; Dhiman, S. A review on medicinal prospectives of *Andrographis paniculata* Nees. *J Pharm Scient Innov.* 2012, *1*(1), 1–4.

Diwan, G.; Sinha, K.; Lal, N.; Rangare, N.R. Tradition and medicinal value of Indian gooseberry: A review. *J Pharmacog Phytochem.* 2018, *7*(1), 2326–2333.

Fatima, T.; Maqbool, K.; Hussain, S.Z. Potential health benefits of fenugreek. *J Med Plants Stud.* 2018, *6*(2), 166–169.

Ganjewala, D.; Srivastava, A.K. An update on chemical composition and bioactivities of *Acorus calamus*. *Asian J Plant Sci.* 2011, *10*(3), 182–189.

Gollen, B.; Mehla, J. *Evolvulus alsinoides*: An emerging antibacterial medicinal herb. *J Pharmacol Rep.* 2018, *3*(1), 139.

Gupta, B.M.; Ahmed, K.K.M.; Dhawan, S.M.; Gupta, R. *Aloe vera* (medicinal plant) research: A scientometric assessment of global publications output during 2007–16. *Pharmacog J.* 2018, *10*(1), 1–8.

Gupta, C.; Prakash, D. Phytonutrients as therapeutic agents. *J Compl Integrat Med.* 2014, *11* (3), 151–169.

Forman, H.J. Redox signaling: An evolution from free radicals to aging. *Free Radic Biol Med.* 2016, *97*, 398–407.

Haidan, Y.; Qianqian, M.; Li, Y.; Guangchun, P. The traditional medicine and modern medicine from natural products. *Molecules.* 2016, *21*(5), 559.

Hamden, K.; Jaoudi, B.; Salami, T.; Carreau, S.; Bejar, S.; Elfeki, A. Modulatory effect of fenugreek saponins on the activities of intestinal and hepatic disaccharidase and glycogen and liver function of diabetic rats. *Biotechnol Bioprocess Eng.* 2010, *15*, 745–753.

Hassan, W.; Kazmi, S.N.Z.; Noreen, H.; Riaz, A.; Zaman, B. Antimicrobial activity of *Cinnamomum tamala* leaves. *J Nutr Disorders Ther.* 2016, *6*, 190.

Jain, A.; Singh, B.; Solanki, R.; Saxena, S.; Kakani, R. Genetic variability and character association in fenugreek (*Trigonella foenum-graecum* L.). *Int J Seed Spices.* 2013, *2*, 22–28.

Jayakumar, T.; Hsieh, C.-Y.; Lee, J.-J.; Sheu, J.-R. Experimental and clinical pharmacology of *Andrographis paniculata* and its major bioactive phytoconstituent andrographolide. *Evid Based Complement Alternat Med.* 2013, *2013*: 846740.

Jin, Z.; Borjihan, G.; Zhao, R.; Sun, Z.; Hammond, G.B.; Uryu, T. Antihyperlipidemic compounds from the fruit of *Piper longum* L. *Phytother Res.* 2009, *23*(8), 1194–1196.

Joselin, J.; Jeeva, S. *Andrographis paniculata*: A review of its traditional uses, phytochemistry and pharmacology. *Med Aromat Plants.* 2014, *3*, 4–7.

Kajaria, D. Medicinal importance of *Albizia lebbeck*: A review. *Res Rev: J Herbal Sci.* 2015, *4*(3), 4–7.

Kanne, H.; Burte, N.P.; Prasanna, V.; Gujjula, R. Extraction and elemental analysis of *Coleus forskohlii* extract. *Pharmacog Res.* 2015, *7*(3), 237–241.

Kaushal, K. Review of pharmacological activities of Acacia nilotica (Linn.) willd W.S.R. to osteoporosis. *J Adv Res Ayur Yoga Unani Sidd Homeo.* 2017, *4*(1&2), 3–7.

Kavitha, C.; Rajamani, K.; Vadivel, E. *Coleus forskohlii*: A comprehensive review on morphology, phytochemistry and pharmacological aspects. *J Med Plants Res.* 2010, *4*, 278–285.

Khare, C.P. *Indian Medicinal Plants.* 1st ed. Berlin/Heidelburg: Springer Verlag, 2007, pp. 622–623.

Kota, V.; Ramana Aramati, B.; Reddy, M.; Ravi Kumar, N.V.; Sharad, M.; Singhal, S. Therapeutic potential of natural antioxidants. *Oxidat Med Cell Longev.* 2018, 2018: 9471051.

Koteswara Rao, Y.; Vimalamma, G.; Rao, C.V.; Tzeng, Y.M. Flavonoids and andrographolides from *Andrographis paniculata*. *Phytochemistry* 2004, *65*, 2317–2321.

Krup, V.; Prakash, L.H.; Harini, A. Pharmacological activities of turmeric (*Curcuma longa* Linn): A review. *J Homeop Ayurv Med.* 2013, *2*, 133.

Kumar, A.; Pandey, V.V.; Beg, S.; Rawat, J.M.; Singh, A. Biological, medicinal and toxicological significance of *Eucalyptus* leaf essential oil: A review. *J Sci Food Agric.* 2018, *98*(3), 833–848.

Kumar, N.; Khurana, S.M.P. Phytochemistry and medicinal potential of the *Terminalia bellerica* (Bahera). *Indian J Nat Prod Resour.* 2018, *19*(2), 97–107.

Kumar, S.; Kamboj, J.; Suman; Sharma, S. Overview for various aspects of the health benefits of *Piper longum* Linn. fruit. *J Acupunct Meridian Stud.* 2011, *4*(2), 134–140.

Lobo, V.; Patil, A.; Phatak, A.; Chandra, N. Free radicals, antioxidants and functional foods: Impact on human health. *Pharmacog Rev.* 2010, *4*(8), 118–126.

Mal, D.; Gharde, S.K.; Chatterjee, R. Chemical constituent of *Cinnamom umtamala*: An important tree spices. *Int J Curr Microbiol Appl Sci.* 2018, *7*(04), 648–651.

Matasyoh, J.C.; Maiyo, Z.C.; Ngure, R.M.; Chepkorir, R. Chemical composition and antimicrobial activity of the essential oil of *Coriandrum sativum*. *Food Chem.* 2009, *113*(2), 526–529.

Metha, D.; Belemkar, S. Pharmacological activity of *Spinacia oleracea* Linn: A complete overview. *Asian J Pharm Res Dev.* 2014, *2*(1), 83–93.

Milner, J.A.; Kaefer, C.M. The role of herbs and spices in cancer prevention. *J Nutri Biochem.* 2007, *19,* 347–361.

Moy, R.L.; Levenson, C. *Sandalwood album* oil as a botanical therapeutic in dermatology. *J Clin Aesthet Dermatol.* 2017, *10*(10), 34–39.

Nandkarni, K.M. Indian materia medica. Mumbai: Popular Prakashan, 2010, pp. 9–10.

Okhuarobo, A.; Falodun, J.E.; Erharuyi, O.; Imieje, V.; Falodun, A.; Langer, P. Harnessing the medicinal properties of *Andrographis paniculata* for diseases and beyond: A review of its phytochemistry and pharmacology. *Asian Pac J Trop Dis.* 2014, *4*(3), 213–222.

Olasupo, A.D.; Aborisade, A.B.; Olagoke, O.V. Phytochemical analysis and antibacterial activities of spinach leaf. *Am J Phytomed Clin Ther.* 2018, *6*(2), 8.

Pan, S.Y.; Chen, S.B.; Dong, H.G.; Yu, Z.L.; Dong, J.C.; Long, Z.X.; Fong, W.F.; Han, Y.F.; Ko, K.M. New perspectives on Chinese herbal medicine (Zhong-Yao) research and development. *Evid Based Complement Alternat Med.* 2011, *2011:* 403709.

Poprac, P.; Jomova, K.; Simunkova, M.; Kollar, V.; Rhodes, C. J.; Valko, M. Targeting free radicals in oxidative stress-related human diseases. *Trends Pharmacol Sci.* 2017, *38*(7), 592–607.

Pothuraju, R.; Sharma, R.K.; Chagalamarri, J.; Jangra, S.; Kumar Kavadi, P. A systematic review of *Gymnema sylvestre* in obesity and diabetes management. *J Sci Food Agric.* 2014, *94*, 834–840.

Promila; Madan, V.K. Therapeutic & phytochemical profiling of *Terminalia chebula* Retz. (Harad): A review. *J Med Plants Stud.* 2018, *6*(2), 25–31.

Raddatz-Mota, D.; Pérez-Flores, L.J.; Carrari, F.; Mendoza-Espinoza, J.A.; de León-Sánchez, F.D.; Pinzón-López, L.L.; Godoy-Hernández, G.; Rivera-Cabrera, F. Achiote (*Bixa orellana* L.): A natural source of pigment and vitamin E. *J Food Sci Technol.* 2017, *54*(6), 1729–1741.

Rahmani, A.H. *Cassia fistula* Linn. Potential candidate in the health management. *Pharmacognosy Res.* 2015, *7*, 217–224.

Rajput, S.B.; Tonge, M.B.; Karuppayil, S.M. An overview on traditional uses and pharmacological profile of *Acorus calamus* Linn. (Sweet flag) and other *Acorus* species. *Phytomedicine.* 2014, *21*, 268–276.

Ramachandran, A.; Das, A.K.; Joshi, S.R.; Yajnik, C.S.; Shah, S.; Kumar, K.M.P. Current status of diabetes in India and need for novel therapeutic agents. *J Assoc Phys India.* 2010, *58*(Suppl. 1), 7–12.

Roberts, K.T. The potential of fenugreek (*Trigonella foenum-graecum*) as a functional food and nutraceutical and its effects on glycemia and lipidemia. *J Med Food.* 2011, *14*(12), 1485–1489.

Sashini, D.; Uthpala P.; Jayawardena, A.; Chanika, D.; Jayasinghe. Potential use of *Euphorbia hirta* for dengue: A systematic review of scientific evidence. *J Trop Med.* 2018, 2018: 2048530.

Shadid Hossain, S.H.M.; Ali, R. Biological evaluation of ethanolic extract of *Aphanamixis polystachya* (Wall.) Parker leaf. *Int J Adv Multidiscip Res.* 2016, *3*(9), 13–21.

Shashikumar, J.N.; Champawat, P.S.; Mudgal, V.D.; Jain, S.K.; Deepak, S.; Mahesh, K. A review: Food, medicinal and nutraceutical properties of fenugreek (*Trigonella foenum graecum* L.). *Int J Chem Stud.* 2018, *6*(2), 1239–1245.

Singh, A.; Navneet. Ethnomedicinal, pharmacological properties and phytochemistry of *Sida spinosa* Linn.: A mini review. *J Phytopharmacol.* 2018, *7*(1), 88–91.

Singh, L.; Kumar, A.; Choudhary, A.; Singh, G. *Asparagus racemosus*: The plant with immense medicinal potential. *J Pharmacognosy Phytochem.* 2018a, *7*(3), 2199–2203.

Sinoriya, S.K.; Singh, R.; Singh, K. Review on medicinal and phytochemical properties of different extract of turmeric's rhizome. *J Pharmacognosy Phytochem.* 2018, *7*(3), 874–878.

Snehlata, H.S.; Payal, D.R. Fenugreek (*Trigonella foenum-graecum* L.): An overview. *Int J Curr Pharm Res.* 2012, *2*(4), 169–187.

Soni, H.; Singhai, A.K. Recent updates on the genus *Coleus*: A review. *Asian J Pharm Clin Res.* 2012, *5*, 12–17.

Vasuki, K.; Murugananthan, G.; Banupriya, C.; Ramya, R.; Mohana Priya, C.; Shenjudar, D. Investigation of immunomodulatory potential of whole plant of *Boerhavia erecta* Linn. *Pharmacog J.* 2018, *10*(2), 241–244.

Vats, V.; Yadav, S.P.; Biswas, N.R.; Grover, J.K. Anti-cataract activity of *Pterocarpus marsupium* bark and *Trigonella foenum-graecum* seeds extract in Alloxan-diabetic rats. *J Ethnopharmacol.* 2004, *93*, 289–294.

Yadav, N.K.; Saini, K.S.; Hossain, Z.; Omer, A.; Sharma, C.; Gayen, J.R.; Singh, P.; Arya, K.R.; Singh, R.K. *Saraca indica* bark extract shows in vitro antioxidant, antibreast cancer activity and does not exhibit toxicological effects. *Oxid Med Cell Longev.* 2015, *2015*: 205360.

Zaveri, M.; Khandar, A.; Patel, S. Chemistry and pharmacology of *Piper longum* L. *Int J Pharm Sci Rev Res.* 2010, *5*(1), 67–76.

CHAPTER 2

Herbal Food Product Development and Characteristics

AMRITA POONIA

Department of Dairy Science and Food Technology, Institute of Agricultural Sciences, Banaras Hindu University, Varanasi 221005, Uttar Pradesh, India

˙Corresponding author. E-mail: dramritapoonia@gmail.com, amritapoonia@yahoo.co.in

ABSTRACT

Herbs are well known for their diversity in foods. Various health benefits of herbal foods have been documented. Herbal foods have no side effects on the body and also replace synthetic antioxidants. Foods rich in antioxidants could lower the incidence of cancers, CVDs, hypertension, and diabetes. There are various applications of herbs in the form of extract or essential oils (EOs) in various food products like dairy products, bakery, meat and fish products, and edible packaging films. The major advantage of herbal foods is that they are easily available and affordable for the consumers. Some of the important herbs discussed in this chapter include garlic, maiden hair tree, ginger, dried root of panax ginseng, asafoetida, bael, brahmi, peppermint, aloe vera, basil, ashwagandha, arjuna, turmeric, sage, cinnamon, cumin, fenugreek, and peppermint. Herbal products are well known for their functional properties, that is, antimicrobial, antibacterial, antihypertensive, and angiotensin converting enzymes. Herbal products and their extracts have strong flavor and aroma which affect the sensory attributes of the herbal products. EOs become hardly water soluble and cause nonuniform distribution in food matrices due to their hydrophobicity.

2.1   INTRODUCTION

Herbs are not only used as food flavorings but are also used as a medicine as well as preservatives. India is the biggest producer of medicinal herbs and is referred to as the "Botanical Garden of the World" (Modak et al., 2007). Due to the coming out concept of fortification of foods, there is a remarkable increase in the health awareness about their importance and addition in food products. According to WHO, about 70%–80% of world population depends on the nonconventional medicines, mainly on herbal products in their primary health care (Chan, 2003). More than 80% of the developing countries population depends directly on the plants for their medicinal purposes. Herbs are well known for their use in treating diseases. They are mainly used in ayurvedic preparations, pharmaceuticals, confectionery, RTD mixes, seasonings, infant foods, and food products through value addition. Consumers are more health conscious about the use of natural antioxidant rich foods, rather than the synthetic antioxidants. They have been demanding a decrease in the use of chemicals in their food and beverages because of the associated possible health risks. Consumers from developed countries want to experience the trend towards natural foods, desiring fewer synthetic food additives as well as the products that show low impact on the environment. Synthetic antioxidants have limited application due to their carcinogenic effects (Zambonin et al., 2012). Many researchers have reported that the very limited consumption of antioxidant rich foods decreases the chances of some particular types of cancers, hypertension, cardiovascular diseases, and diabetes. These herbal products are best suited to the consumers of developing countries, where most of the people have limited resources and are unable to afford costly treatments (Marles and Farnsworth, 1994). Medicinal plants rich in natural antioxidants and phenolics are incorporated in various food products to improve the nutritional and therapeutic properties (Shori and Baba, 2011a, b). Antioxidants are used in controlling the excess formation of free radicals and to reduce the side effects of synthetic antioxidants to various body organs. Herbal products in various forms like essential oils (EOs) and extracts have been used as functional flavoring agents. Common herbs used as nutraceuticals are listed in Table 2.1.

2.2   HEALTH BENEFITS

Various health benefits of herbs including antimicrobial, antioxidants, anti-inflammatory, and anticarcinogenic properties have been reviewed by several researchers (Erkmen and Ozcan, 2001; Ozcan, 2009; Ozkan et al., 2007; Srinivasan, 2004; Tapsell et al., 2006).

TABLE 2.1 List of Herbs Used as Nutraceuticals in Food Product Development

Name of Herb	Constituents	Health Benefits
Garlic, *Allium sativum* (Liliaceae)	Alliin and allicin	Anti-inflammatory, antibacterial, antigout, nervine tonic
Maiden hair tree Leaves of Ginkgo biloba (Ginkgoaceae)	Ginkgolide and bilobalide	PAF antagonist, memory enhancer, antioxidant
Ginger, Rhizomes of *Zingiber officinale* (Zingiberaceae)	Zingiberene and gingerols	Stimulant, chronic bronchitis, hyperglycemia, and throat ache
Dried herb of *Echinacea purpurea* (Asteraceae), Echinacea	Alkylamide and echinacoside	Anti-inflammatory, immunomodulator, antiviral
Dried root of *Panax ginseng* (Araliaceae), Ginseng	Ginsenosides and Panaxosides	Stimulating immune and nervous system, and adaptogenic properties
Dried root of *Glycyrrhiza glabra* (leguminosae), Liquorice	Glycyrrhizin and liquirtin	Anti-inflammatory and antiallergic, expectorant
Rhizome of *Curcuma longa* (Zingiberacae), Turmeric	Curcumin	Anti-inflammatory, antiarthritic, anticancer, and antiseptic
Dried bulb of *Allium cepa* Linn. (Liliaceae), Onion	Allicin and alliin	Hypoglycemic activity, antibiotic, and antiatherosclerosis
Dried root of *Valeriana officinalis* Linn. (Valerianaceae), Valeriana	Valerenic acid and valerate	Tranquillizer, migraine and menstrual pain, intestinal cramps, bronchial spasm
Aloes, dried juice of leaves *Aloe barbadensis* Mill. (Liliaceae)	Aloins and aloesin	Dilates capillaries, anti-inflammatory, emollient, wound-healing properties
Goldenseal, dried root of *Hydrastis canadensis* (Ranunculaceae)	Hydrastine and berberine	Antimicrobial, astringent, antihemorrhagic, treatment of mucosal inflammation
Senna, dried leaves of *Cassia angustifolia* (Leguminosae)	Sennosides	Purgative
Asafoetida, Oleo gum resin of *Ferula assafoetida L.* (Umbelliferae)	Ferulic acid and umbellic acid	Stimulant, carminative, expectorant
Bael, unripe fruits of *Aegle marmelos* Corr. (Rutaceae)	Marmelosin	Digestive, appetizer, treatment of diarrhea, and dysentery
Brahmi, herbs of *Centella asiatica* (Umbelliferae)	Asiaticoside and madecassoside	Nervine tonic, spasmolytic, anti-anxiety

2.3   APPLICATION OF HERBS AND HERBAL NUTRACEUTICALS IN FOOD PRODUCTS

Herbs used in both forms, extracts and EO of herbs and spices, have been listed in Tables 2.2 and 2.3 in various food products such as oils, bakery products, meat products, dairy products, fish products, edible, and packaging film.

2.4   PROPERTIES OF HERBS

Phenolic compounds of herbs are a good choice for the synthetic antimicrobial agents that are used in food products. Various phenolic compounds present in herbs have been reported to inhibit the growth of pathogens, namely, *Salmonella enteritidis, Listeria monocytogenes, Staphylococcus aureus,* and fungi. Herbs are good sources of antioxidants and are considered safe. Antioxidants help in delaying the oxidation of molecules by inhibiting the initiation or propagation of oxidizing chain reactions by free radicals and may reduce oxidative damage to the human body (Namiki, 1990). Various functional properties of herbs are tabulated in Figure 2.1.

2.4.1   ANTIMICROBIAL ACTIVITY

Herbs are also known for their bactericidal effects and have a broad-spectrum activity against gram-positive and gram-negative bacteria. The EOs of various herbs have proved to be potentially useful sources of antimicrobial and antioxidant compounds (Abdelfadel et al., 2015). These herbal EOs have attained GRAS (generally recognized as safe) and the consumers can have these without restrictions and are usually accepted by consumers.

Many herbs and spice extracts contained increased levels of phenols and exhibited antibacterial activity against food-borne pathogens. Antibacterial substances can easily destroy the bacterial cell wall and cytoplasmic membrane resulting in a leakage of the cytoplasm (Shan et al., 2007).

EOs contain bioactive compounds that provide antimicrobial properties to EOs (Burt, 2004). Antimicrobial properties of essential oils are reviewed in Table 2.4. Antimicrobial activity of these oils can be due to individual effect of major components or due to a synergistic effect of its minor components. EOs is also used as food preservatives, and as antimicrobial, sedative, anti-inflammatory, analgesic, and spasmolytic remedies (Bakkali et al., 2005).

TABLE 2.2 Application of Herbs and Herbal Nutraceuticals in Dairy Products

Product	Herb	Remarks	References
Low fat yoghurt	Mint oil	Mint oil at 5–20 µl/g was effective against *S. enteritidis*	(Tassou et al., 1995)
Sandesh	Saffron (0.015%)	Shelf life increased up to 68 days of *Sandesh*	(Sen and Rajorhia, 1996)
Plain *dahi*	Banana juice	Banana-juice-based *dahi* (yoghurt) was acceptable up to 8 days of storage at refrigerated temperature	(Kamruzzaman et al., 2002)
Whole milk	Cinnamon, ginger, turmeric, pepper	Shelf life of commercial whole milk up to 5 days at 4 °C	(Khusniati and Yantyati, 2008)
Raw milk	Banana pseudo-stem	Banana pseudo-stem (0.1%, 0.2%, 0.3%, and 0.4%) showed controlled acidity development and remained acceptable up to 5 and 6 hours, respectively	(Ray, 2008)
Turkish Otlu-Herby Cheese	Allium sp., Ferula sp., Pranges sp., Thymus sp., Mentha sp., Chaerophylium macropodum) and Silene vulgaris	Improved cheese flavor and shelf life of the final product	(Tarakci and Temiz, 2009)
Whey based banana herbal beverage	Mentha avrensis extract	Added with 2.0% was stored at refrigeration temperature without adding any chemical preservatives, had good consumer acceptability of up to 15 days	(Ritika et al., 2010)
Khoa	Sage oil	Addition of the 0.1% level was found to be superior in physico-chemical and microbial characteristics	(Jadhav et al., 2011)
Paneer	Turmeric	0.6% of the milk before heating yielded a paneer with a shelf life of 12 days when stored at a temperature of 7 ± 1 °C	(Buch et al., 2012)
Ghee	*T. arjuna*	Antioxidative properties	(Rajnikant and Patil, 2005)

TABLE 2.2 *(Continued)*

Product	Herb	Remarks	References
Ghee (cow and buffalo)	T. arjuna bark (extract)	Retarding the autooxidation of both cow and buffalo ghee during storage	(Parmar et al.,)
Ghee and butter oil	Sage (*Salvia officinalis*) and Rosemary (*Rosmarinus officinalis*) extracts	The shelf life (accelerated test) of the Arjuna herbal ghee at 80 ± 1 °C was 8 days as compared to just 2 days for control ghee sample (devoid of herb)	(Ozcan, 2003)
Yoghurt	Cinnamon (6.0%) extract dissolved in 1.01 milk)	Shelf life increased up to 7 days, when stored under refrigeration	(Behrad et al., 2009)
Probiotic herbal yoghurt	ginger and garlic extract	Maximum body and texture and overall acceptability was found	(Simon et al., 2018)
Yoghurt	*Zingiber officinale* and *beta vulgaris*	Yoghurt shows high antioxidant properties which are beneficial for human health	(Srivastava et al., 2015)
Yoghurt	Essential oils (peppermint, basil, and zataria)	Peppermint and basil samples showed both good antiradical activity and sensory acceptability	(Azizkhani and Parsaeimehr, 2018)
Whey based watermelon beverage	Betel leaf distillate	Use of 2% level increased the shelf life of the product by about one month at 7 ± 1 °C	(Naik et al., 2009)
Labneh (concentrated yoghurt)	0.2 ppm each of thyme, marjoram, and sage essential oils	Extended shelf life (by 21 days over control) at 5 °C	(Otaibi and Demerdash, 2008)
Yoghurt spread	Mashed raw mint leaves were added at 2%, 4%, and 6%	Shelf life of the spread was 10 days when stored at 5 °C	(Kumar et al., 2013)
Ice cream	Ginger juice 4%, ginger shreds 4%, sugar syrup treated ginger shreds 6%, and ginger powder 1%	—	(Pinto et al., 2009)
Ice cream	*Ocimum sanctum* basil juice 6% and freeze dried basil powder 1% by weight	—	(Trivedi et al., 2014)
Ice cream	Herbal and crystal menthol extract	—	(Patil et al., 2017)

TABLE 2.2   *(Continued)*

Product	Herb	Remarks	References
Paneer	Turmeric 0.4% and 0.6% by weight	The *paneer* samples containing 0.6% turmeric by weight, remained acceptable up to 12 days	(Buch et al., 2014)
Sandesh	Paste of turmeric (*Curcuma longa*), coriander (*Coriandrum sativum*), curry leaf (*Murraya koenigii L*), spinach, (*Spinacia oleracea*), and aonla (*Emblica officinalis*) 10%	Herbal *sandesh* could be considered as value-added health food	(Bandyopadhyay et al., 2007)
Feta cheese	Clove essential oil	Antibacterial effect against *E. coli* and vancomycin-resistant *Enterococci* in feta cheese stored at 7 °C for 14 days	(Samy, 2013)
Burfi	Curry leaf (0.05–0.15 ppm) and clove bud (0.15–0.25 ppm) Essential oil	Antioxidative and antimicrobial potential of burfi	(Badola et al., 2018)
Functional *lassi*	*Aloe vera* (*Aloe barbadensis* Miller).	Better immune protective effects compared to control *lassi*	(Hussain et al., 2011)

TABLE 2.3 Application of Herbs in Meat, Fish, and Bakery Food Products

Product	Herb	Remarks	References
Meat and Poultry Products			
Raw chicken meat emulsion	Clove powder, ginger and garlic paste	0.2% clove powder could be utilized effectively as antioxidant in raw chicken meat emulsion	(Singh et al., 2014)
Lamb patties	0.05% rosemary extract (RE), and 0.50% ginger (GE)	Treated lamb patties significantly retarded oxidative process compared to control patties	(Baker et al., 2013)
Duck eggs	Clove extract	Treated eggs exhibited a higher docosahexaenoic acid content than that of control	(Harlina et al., 2018)
Fish Products			
Patties of Mackerel (cooked)	Ginger and coriander	Antioxidant activity increased in the order of coriander < ginger + coriander < ginger	(Normah et al., 2005)
Fish tissue	*Citrus grandis* (Pomelo) peels	Reduction in peroxide value, which indicated the inhibition of lipid oxidation in fish treated with pomelo peel	(Zarina and Tan, 2013)
Bakery Products			
Cakes	Clove essential oil	Results showed that the CEO was able to retard the oxidation rate and reduction of formed oxidation products in cakes compared with synthetic antioxidant	(Ibrahium et al., 2013)
Bread	Garlic, coriander, sumac, fennel, marjoram, thyme, and cardamom	Bread samples exhibited higher antioxidant activity than control	(Seleem and Mohamed, 2014)
Cake	Coriz	CEO at 0.05%, 0.10%, and 0.15% inhibited the rate of oxidation products formation in cake and their effects were almost equal to BHA at 0.02%	(Darughe et al., 2012)
Edible and Packaging Films			
Uwi starch (*Dioscorea alata* L)	Ginger essential oil	DPPH scavenging activity of the films increased significantly	(Herlina and Masril, 2013)
Soybean oil samples packed with cellulose-based pouches containing BHA	Cinnamon oil and clove oil	Can be potentially used as antioxidants for food	(Phoopuritham et al., 2012)

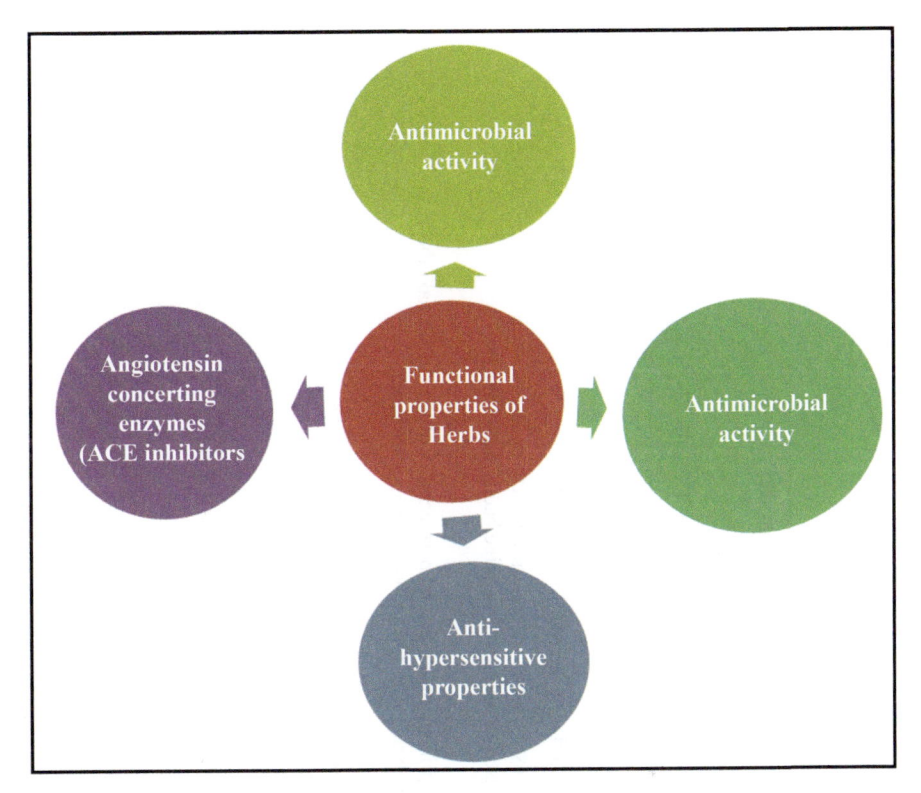

FIGURE 2.1 Functional properties of herbs used in food products.

2.4.2 ANTIOXIDANT ACTIVITY

Antioxidants play an important role in delaying the oxidation of molecules by inhibiting the commencement or propagation of oxidizing chain reactions by free radicals. This will lead to the reduction of oxidative damage to the living beings. Herbs contain high amounts of phenolic compounds which possess antioxidant properties. Najgebauer-Lejko et al. (2009) reported that the natural antioxidant properties of herbs have made their use in the formulation of functional foods particularly for the people suffering from cardio-vascular diseases. The mechanism of action involves oxidation of lipids that proceeds through three different stages, namely, initiation, propagation, and termination. Pokorny et al. (2001) reported that aldehydes, ketones, and acids are responsible for the rancid aroma and off-flavors in foods.

TABLE 2.4   Antimicrobial Properties of Herbs in Various Food Products

Herb	Micro Organisms	References
Plant essential oil	Effective for inhibiting the growth of *S. aureus*	(Burt and Reinders, 2002)
Ginger extract	Reducing microbial load	(Belewu et al., 2005)
Ethanolic extracts of Piper Betel leaves	Gram-positive bacteria was more sensitive	(Kriangkrai and Penkhae, 2009)
Phenolic compounds include monophenols (pcresol), diphenols (hydroquinone), and triphenols (gallic acid)	Interaction of the cytoplasmic membrane and activity is selectively increased against gram-positive bacteria and fungi	(Sharma et al., 2010)
Piper Betel	Alteration in the primary structure and permeability of the cell wall and membrane pore formation and degradation of bacterial components	(Chakraborty and Shah, 2011)
Extract grape and pomegranate seeds	Natural phenolic compounds had a negative effect on the starter culture and total viable counts	(Ersoz et al., 2011)
Ginger (*Zingiber Officinale*) and Pepper (*Piper Guinenses*)	Extended the shelf life of soymilk from 12 to 24 hours	(Odom et al., 2012)
Cloves (*Syzygium Aromaticum Myrtaceae*)	Increased the shelf life of soy milk by 2 and 8 days at refrigerated temperature	(Kabiru et al., 2012)

Medicinal herbs are reported to have hypotensive/antihypertensive potential. These herbs may help in regulation of blood pressure by stimulating the physiological systems in humans. The antioxidant properties of herbs are due to the presence of some vitamins, terpenoids, carotenoids, flavonoids, and phytoestrogens. Herbs and spices containing antioxidants are thyme, basil, saffron, cinnamon, clove, ginger, mint, oregano, dill, rosemary, and sage. Garlic, celery, tea, lavender, murungai, basil, kudzu, ginger, radish, ajwain, rauwolfia, and sesame are few examples of herbs having hypotensive properties. The antioxidant activity of herbs and spices is most often due to phenolic acids (gallic, protocatechuic, caffeic and rosmarinic acids, rosmanol, and rosmadial), flavonoids (quercetin, catechin, naringenin, kaempferol, epicatechin), phenolic diterpenes (carnosol, carnosic acid gallate, epigallocatechin gallate, and rutin), volatile oils (eugenol, carvacrol, thymol, menthol, safrole, 1,8-cineole, p-cymene, cinnamaldehyde, α-terpineol, myristicin, and piperine), and phenylpropanoids (thymol, eugenol, carvacrol, p-cymene) (Frankel, 2012). Angiotensin converting enzyme (ACE) inhibitors present in herbs may inhibit ACE (component of the blood pressure-regulating renin–angiotensin system) and thus lowering the blood pressure.

2.5 CONSTRAINTS IN DEVELOPMENT OF HERBAL FOOD PRODUCTS

2.5.1 STRONG FLAVOR AND AROMA

The strong flavor and aroma of the herbal products and their extracts are some of the factors which affect the sensory attributes of the herbal products. By deodorizing these extracts, the problem can be reduced to some extent. Shahidi (2015) reported that the whole spices have increased activity then their extracts and the isolated compounds. So, the minor compounds are critical to the activity due to their synergistic effect or potentiating influence on the activity of the minor components.

2.5.2 REGULATORY CONCERN

Some herb and spice extracts and oleoresins are GRAS. Many of them are used as indirect additives. For example, solvents permitted for the extraction process and solvent residues allowed are specified. FDA also regulates the standards about some herbal extracts, resins, and concentrates. Dietary

Supplement Health and Education Act of 1994 considered dietary ingredients for use to supplement the diet by increasing the total dietary intake.

Spices and herbs can also be used in milk and milk products such as shrikhand, cheese and cheese spread, ice cream, milk ice and frozen desserts. Spices may also be used in bread, pasta products, ready-to-eat vegetables, frozen vegetables, soybeans sauce, frozen fruits/fruit products, fruits and vegetable sauces, tomato ketchup and tomato sauce, mango chutney, vegetable juices and soups, pickles, culinary pastes, synthetic vinegar, fruits and vegetable chutney, and meat products (canned meat and canned cooked ham) (FSSAI, 2011).

2.5.3   NONUNIFORM DISTRIBUTION

The external application of these essential oils to the fresh food products enhances its shelf life. The technological challenges restricted the direct incorporation of essential oils in food. Due to their hydrophobicity, these oils become hardly water soluble and cause nonuniform distribution in food matrices.

2.6   FUTURE PERSPECTIVES

Herbal food products are associated with some pungent taste, a typical flavor, and color. Due to these particular characteristics, herbal foods are not fit for the daily use. Hence, there is a need to develop some technological changes to retain their sensory attributes. There is lack of research about the interactions among herbal and food components on the human well-being. Systematic studies are required to explore the processing conditions which cause less or no damage to the functional components during their incorporation into various foods. Scientific studies and documentation of these products are also required.

2.7   CONCLUSION

Since long time, herbs and spices are used for flavor as well as for their medicinal properties. The antimicrobial and antioxidant components present in herbs make them suitable to be used as effective preservatives. Inclusion of herbs in food products may result in the improvement of the health and medical state of human being. Herbal products that show potential health benefit should also fulfill regulatory requirements with respect to safety, quality testing, effectiveness, and marketing authorization procedures. It should be devoid of any side effects. There is a need of systematic scientific studies and documentation in this field.

KEYWORDS

- **herbs**
- **functional food**
- **dairy products**
- **phenolic compounds**
- **antioxidants**

REFERENCES

Abdelfadel, M.M; Alaf, H. H.; Sharoba, A.M; Assous M. T. M. Effect of extraction methods on antioxidant and antimicrobial activities of some spices and herbs extracts International. *J Adv Res.* 2015, 3, 165–179.

Abdel-Hameed, E.S.; Nagaty, M.A.; Salman, M.S.; Bazaid, S.A. Phytochemicals, nutritionals and antioxidant properties of two prickly pear cactus cultivars (*Opuntia ficus indica* Mill.) growing in Taif, KSA. *Food Chem.* 2014, 160, 31–38.

Ansari, M.M.; Kumar, D.S. Fortification of food and beverages with phytonutrients. *Food Pub Health.* 2012, 6, 241–253.

Azizkhani, M.; Parsaeimehr, M. Probiotics survival, antioxidant activity and sensory properties of yoghurt flavored with herbal essential oils. *Int Food Res J.* 2018, 25, 921–927.

Badola, R.; Panjagari, N.R.; Singh, R.R.B.; Singh, A.K.; Prasad, W.G. Effect of clove bud and curry leaf essential oils on the anti-oxidative and anti-microbial activity of burfi, a milk-based confection. *J Food Sci Technol.* 2018, 55, 4802–4810.

Baker, I.A.; Alkass, J.E.; Saleh, H.H. Reduction of oxidative rancidity and microbial activities of the Karadi lamb patties in freezing storage using natural antioxidant extracts of rosemary and ginger. *Int J Agri Food Res.* 2013, 2, 31–42.

Bakkali, F.; Averbeck, S.; Averbeck, D.; Zhiri, A.; Idaomar, M. Cytotoxicity and gene induction by some EOs in the yeast Saccharomyces cerevisiae. *Mutat Res.* 2005, 585,1–13.

Bandyopadhyay, M.; Chakraborty, R.; Raychaudhar, U. Incorporation of herbs into sandesh, an Indian sweet dairy product, as a source of natural antioxidants. *Int J Dairy Technol.* 2007, 60, 228–233.

Behrad, S.; Yusof, M.Y.; Goh, K.L.; Baba, A.S. Manipulation of probiotics fermentation of yogurt by cinnamon and licorice: Effects on yogurt formation and inhibition of *Helicobacter pylori* growth in vitro. *World Acad Sci Eng Technol.* 2009, 60, 590–594.

Belewu, M.A, Belewu, K.Y., Nkwunonwo, C.C. Effect of biological and chemical preservatives on the shelf life of West African soft cheese. *African J Biotechnol.* 2005, 4.

Buch, P.S.; Aparnathi, K.D. Evaluation of efficacy of turmeric as a preservative in paneer. *J Food Sci Technol.* 2014, 51(11), 3226–3234 ("https://pubmed.ncbi.nlm.nih.gov/?term= Buch+S&cauthor_id=26396315).

Burt, S. Essential oils: their antibacterial properties and potential applications in foods—a review. *Int J Food Microbiol.* 2004, 94, 223–253.

Burt, S.A, and Reinders, R.D. Antibacterial activity of selected plant essential oils against Escherichia coli O157: H7. *Appl Microbiol.* 2002, 36, 162–167

Chan, K. Some aspects of toxic contaminants in herbal medicines. *Chemosphere.* 2003, 52, 1361–1371.

Chakraborty, D.; Shah, B. Antimicrobial, anti-oxidative and anti-hemolytic activity of Piper betel leaf extracts. *Int J Pharmaceutical Technol.* 2011, 3, 192–199.

Chauhan, B.; Kumar, G.; Kalam, N.; Ansari, S.H. Current concepts and prospects of herbal nutraceutical: A review. *J Adv Pharma Technol Res.* 2013, 3, 423–435.

Darughe, F.; Barzegar, M.; Sahari, M.A. Antioxidant and antifungal activity of coriander (*Coriandrum sativum* L.) essential oil in cake. *Int Food Res J.* 2012, 19, 1253–1260.

Erkmen, O.; Ozcan, M. Antimicrobial activities of essential oils from Turkish spices. *Eur Food Res Technol.* 2001, 212, 658–660.

Ersöz, E.; Kınık, O.; Oktay Yerlikaya,O.; Açu, M. Effect of phenolic compounds on characteristics of strained yoghurts produced from sheep milk. *African J Agricul Res.* 2011, 6, 5351–5359.

Frankel, E.N. 2012. Antioxidants in food and biology—facts and fiction. *Woodhead Publishing in Food Science, Technology and Nutrition.* University of California, California, USA. pp. 25–33.

FSSAI. *Food Product Standard and Food Additives.* New Delhi: Ministry of Health and Family Welfare, 2011, pp. 210–211.

Gandhi, K.; Lal, D. Potential of herbal nutraceuticals in ghee—a review. 2014, http://www. fnbnews.com/FB-specials/Herbs-Bioactive-ingredients-in-traditional-dairy-products.

Harlina, P.W.; Meihu, M.; Shahzad, R.; Gouda, M.M.; Qiu, N. Effect of clove extract on lipid oxidation, antioxidant activity, volatile compounds and fatty acid composition of salted duck eggs. *J Food Sci Technol.* 2018, 55, 4719–4734.

Herlina, M.; Masril, K.I. Antibacterial and antioxidant of Uwi (Dioscorea Alata L) starch edible film incorporated with ginger essential oil. *Int J Biosci Biochem Bioinform.* 2013, 3, 354–356.

Hussain, S.A.; Sharma, P.; Singh, R.R.B. Functional dairy foods—an overview. In: *Souvenir of International Conference on Functional Dairy Foods,* Nov 16–19, Karnal, India, 2011, pp. 7–12.

Ibrahium, M.I.; Abd El-Ghany, M.E.; Ammar, M.S. Effect of clove essential oil as antioxidant and antimicrobial agent on cake shelf life. *World J Dairy Food Sci.* 2013, 8, 140–146.

Jadhav, M.V. Sakhale, B.K.; Pawar, V.D.; Solanki, S.G. Studies on effect of preservatives on keeping quality of Khoa. *Agarkar Food Sci Res J.* 2011, 2, 4–7.

Kabiru, Y.A.; Makun, H.A.; Saidu, A.N.; Muhammad, L.H.; Nuntah, L.C. Amoo, S.A. Soymilk preservation using extracts of cloves (syzygium aromaticum myrtaceae) and guinea-pepper (zylopia aethiopica annonaceae). *J Pharm Biol Sci.* 2012, 3, 44–50.

Kamruzzaman, M.; Islam, M.N.; Rahman, M.M. Shelf life of different types of dahi at room and refrigeration temperature. *Pakistan J Nutrition.* 2002. 5.

Khusniati, T.; Yantyati, W. Antibacterial effects of aromatic materials produced in Indonesia on the preservation of skimmed and whole milk in storage. *Int Food Res J.* 2008. 15, 109–118.

Kriangkrai, P.; Penkhae, W. Antimicrobial activity of thai herb extracts against coconut milk spoilage microorganisms. *Kasetsart J. (Nat. Sci.).* 2009. 43, 752–759.

Kumar, S.T.; Arvindakshan, P.; Sangeetha, A.; Pagote, C.N.; Rao, J.K. Development of mint flavoured yoghurt spread. *Asian J Dairy Food Res.* 2013, 32, 19–24.

Landge, U.B.; Pawar, B.K.; Choudhari, D.M. Preparation of shrikhand using ashwagandha powder as additive. *J Dairying Foods Home Sci.* 2011, 30, 79–84.

Marles, R.J.; Farnsworth, N. Antidiabetic plants and their active constituents. *Phytomedicine.* 1994, 2, 137–189.

Martins, A.; Barros, L.; Carvalho, A.M.; Santos-Buelga, C.; Fernandes, I.P.; Barreiro, F.; Ferreira, I.C.F.R. Phenolic extracts of *Rubus ulmifolius* Schott flowers: Characterization, microencapsulation and incorporation into yogurts as nutraceutical sources. *Food Funct.* 2014, 5, 1091–1100.

Meenakshi, S.; Manicka Gnanambigai, D.; Tamil Mozhi, S.; Arumugam, M.; Balasubramanian, T. Total flavanoid and in vitro antioxidant activity of two seaweeds of Rameshwaram coast. *Global J Pharmacol.* 2009, 3, 59–62.

Modak, M.; Dixit, P., Londhe, J., Ghaskadbi, S., Devasagayam, T.P. Indian herbs and herbal drugs used for the treatment of diabetes. *J Clin Biochem Nutr.* 2007, 40, 163–173.

Najgebauer-Lejko, D.; Grega, T.; Sady, M.; Domagała, J. The quality and storage stability of butter made from sour cream with addition of dried sage and rosemary. *Biotech Animal Husbandry.* 2009, 25, 753–761.

Naik, Y.K., A. Khare, P.L. Choudhary, B.K. Goel and Shrivastava, A. Studies on physico-chemical and sensory characteristics of whey based watermelon beverage. *Asian J Res Chem.* 2009. 2, 57–59

Namiki, M. Antioxidant/antimutagens in food. *Crit Rev Food Sci Nutr.* 1990, 29, 273–300.

Normah, I.; Diana, W.; Norisuliana, I.; Azura, A. Antioxidative activity of ginger and coriander in cooked patties of mackerel. *Sci Lett.* 2005, 2, 71–77.

Odom, T.C.; Udensi, E.A.; Dike, C.O.; Ogbuji, C.A.; Kanu, A.M.; Aji, R.U. Studies of ginger (*Zingiber officinale*) and black pepper (*Piper guinenses*) extracts at different concentrations on the microbial quality of soymilk and kunuzaki. *African J Biotechnol.* 2012, 11, 13494–13497.

Otaibi, M.A.; Demerdash, H.E. Improvement of the quality and shelf life of concentrated yoghurt (labneh) by the addition of some essential oils. *Afr J Microb Res.* 2008, 2, 156–161.

Ozcan, M. Antioxidant activity of rosemary, sage and sumac extracts and their combinations on stability of natural peanut oil. *J Med Food.* 2003, 6, 267–270.

Ozcan, M. Effect of some essential oils on rheological properties of wheat flour dough. *Int J Food Sci Nutr.* 2009, 60, 176–181.

Ozcan, M.; Akgül, A. Antioxidant activity of extracts and essential oils from Turkish spices on sunflower oil. *Acta Alimentaria.* 1995, 24, 81–90.

Ozkan, G.; Sagdıç, O.; Ozcan, M. Phenolic compounds, antioxidant and antibacterial activities of Turkish endemic *Origanum sipyleum L.* extract. *J Food Lipid.* 2007, 14, 157–169.

Pankaj, P.; Kaushik K.; Devaraja, H.C.; Singh, R.R.B. The effects of alcoholic extract of Arjuna (*Terminalia arjuna* Wight & Arn.) bark on stability of clarified butterfat. *J Med Plant Res.* 2011, 7, 2545–2550.

Parmar, P.; Kaushik, K.; Devaraja, H.C and Singh, R.R.B. The effects of alcoholic extract of Arjuna (Terminalia arjuna Wight & Arn.) bark on stability of clarified butterfat. *J. Medicinal Plants Res.* 2013, 7(35), 2545–2550.

Patil, Y.; Prabhakar, P.; Ramesh, P.; Suryawanshi, D.B. Comparative study on physico-chemical attributes of ice-cream prepared by using herbal menthol and crystal menthol. *J Pharma Phytochem.* 2017, (SP1), 133–136.

Phoopuritham, P., Thongngam, M., Yoksan, R., Suppakul, P. Antioxidant properties of selected plant extracts and application in packaging as antioxidant cellulose-based films for vegetable oil. *Packag Technol Sci.* 2012, 25, 125–136.

Pinto, S.V.; Patel, A.M.; Jana, A.H.; Solanky, M.J. Evaluation of different forms of ginger as flavouring in herbal ice cream. *Int J Food Sci Technol Nutr.* 2009, 3, 73–83.

Pokorny, J., Yanishlieva, N., Gordon, M. *Antioxidants in Food-Practical Applications.* Woodhead Publishing Limited, Cambridge, England, 2001, pp. 1–34.

Rajanikant; Patil, G.R. Development of process for herbal ghee. National Dairy Research Institute. 2005, 10, 8.

Ray, P.R. Department of Dairy Chemistry Faculty of Dairy Technology West Bengal University of Animal & Fishery Sciences Mohanpur, Nadia, 2008 (thesis submitted).

Ritika, B.; Yadav, B.S. Development and storage studies on whey-based banana herbal (*Mentha arvensis*) beverage. *Am J Food Technol.* 2010, 5, 121–129.

Samy, S. Antimicrobial activity of essential oils against vancomycin-resistant enterococci and E. coli 0157: H7 in Feta soft cheese and minced beef meat. *Braz J Microb.* 2013 42, 187–196.

Seleem, H.A.; Mohamed, Z.E.O.M. Influence of some medicinal and aromatic plants addition on pan bread quality. *World J Dairy Food Sci.* 2014, 9, 299–307.

Sen, D.C.; Rajorhia, G.S. Efficiency of cardamom as natural preservatives for sandesh. *Indian J Dairy Sci.* 1996. 49, 433–440.

Shan, B.; Cai, Y. Z.; Brooks, J. D.; Corke, H. The in-vitro antibacterial activity of dietary spice and medicinal herb extracts. *Int J Food Microbiol.* 2007, 117, 112–119.

Sharma, R.; Young, C.; Neu, J. Molecular modulation of intestinal epithelial barrier: contribution of microbiota. *J Biomed Biotechnol.* 2010, 1–15

Singh, P., Sahoo, J., Chatli, M.K., Biswas, A.K. Shelf life evaluation of raw chicken meat emulsion incorporated with clove powder, ginger and garlic paste as natural preservatives at refrigerated storage (4±1°C). *Int Food Res J.* 2014, 21: 1363–1373.

Shahidi, F. *Handbook of Antioxidants for Food Preservation.* Woodhead Publishing Series in Food Science, Technology and Nutrition, UK, 2015, pp. 251–285.

Shori, A.B.; Baba, A.S. Antioxidant activity and inhibition of key enzymes linked to type-2 diabetes and hypertension by *Azadirachta indica*-yogurt. *J Saudi Chem Soc.* 2011a, 17, 295–301.

Shori, A.B.; Baba, A.S. Comparative antioxidant activity, proteolysis and in vitro a-amylase and a-glucosidase inhibition of Allium sativum-yogurts made from cow and camel milk. *J Saudi Chem Soc.* 2011b, 18, 456–463.

Simon, H.P., Chandra, S.; Shukla, S.; Singh, S.S. Sensory evaluation of probiotic herbal yoghurt with ginger and garlic extract. *Pharma Innov J.* 2018, 7, 605–607.

Srinivasan, K. Spices as nutraceuticals with multi-beneficial health effects. *J Herbs Spices Med Plants.* 2004, 11–17.

Srivastava, P.; Prasad, S.G.M.; Ali, M.N.; Prasad, M. Analysis of antioxidant activity of herbal yoghurt prepared from different milk. *Pharma Innov J.* 2015, 4, 18–20.

Tapsell, L.; Hemphill, I.; Cobiac, L.; Patch, C.; Sullivan, D.; Fenech, M. Health benefits of herbs and spices: The past, the present, the future. *Med J Australia.* 2006, 185, 4–24.

Tarakci. Z.; Temiz, H. A review of the chemical, biochemical and antimicrobial aspects of Turkish Otlu (herby) cheese. *Int J Dairy Technol.* 2009, 62, 354–360.

Tassou, C.C.; Drosinos, E.H and Nychas, G.J.E. Effects of essential oil from mint (*Mentha piperita*) on *Salmonella enteritidis* and *Listeria monocytogenes* in model food systems at 40 and 10°C. *J Appl Bacteriol.* 1995, 78, 593–600.

Thabet, H.M.; Nogain, Q.A.; Abdoalaziz, O.; Qasha, A.S.; Omar, A. et al. Evaluation of the effects of some plant derived essential oils on shelf life extension of Labneh. *Merit Res J Food Sci Technol.* 2014, 2, 8–14.

Trivedi, V.; Prajapati, J.; Pinto, S.; Darji, V. Use of basil (tulsi) as flavouring ingredient in the manufacture of ice cream. *Am Int J Contemp Res.* 2014, 1, 28–43.

WHO. Diet, nutrition and the prevention of chronic diseases. *Report of the Joint WHO/FAO Expert Consultation.* 2002.

Writdhama, G.P. Effect of clove bud and curry leaf essential oils on the antioxidative and anti-microbial activity of burfi, a milk-based confection. *J Food Sci Technol.* 2018, 55, 4802–4810.

Zambonin, L.; Caliceti, C.; Vieceli Dalla Sega, F.; Fiorentini, D.; Hrelia, S.; Landi, L.; Prata, C. Dietary phenolic acids act as effective antioxidants in membrane models and in cultured cells, exhibiting proapoptotic effects in leukemia cells. *Oxidative Med Cell Longevity.* 2012, 20, 839–889.

Zarina, Z.; Tan, S.Y. Determination of flavonoids in *Citrus grandis* (Pomelo) peels and their inhibition activity on lipid peroxidation in fish tissue. *Int Food Res J.* 2013, 20, 313–317.

CHAPTER 3

Herbal Supplements and Health

HIMANGINI BANSAL* and SAKSHI BAJAJ

*Department of Pharmaceutical Chemistry,
Delhi Institute of Pharmaceutical Sciences and Research,
Sector 3, Pushp Vihar, Delhi 100017, India*

*Corresponding author. E-mail: himanginibansal@gmail.com

ABSTRACT

A dietary supplement is a product taken by mouth and intended to supplement the diet. These products may include vitamins, minerals, herbs or other botanicals, amino acids, and substances such as enzymes, organ tissues, glandular, and metabolites. Patented prescribed drugs have changed over the counter herbs or "roots," which have been both discovered too weak or risky. Medicinal plants especially herbs like drugs have several actions within the body. Sometimes, even if you take a herb or supplement for a certain reason, there can be other unintended reactions. Being natural does not necessarily mean being safe. Herbal and dietary products have chemical properties just as manufactured drugs have. Like anything that we ingest (eat) or apply, there can be side effects. Today, one in all the foremost issues with several herbal products in the market is that the numbers and therefore the purity of active ingredients are different from company to company. The legislature regulates herbal drugs just like food and nutritional supplements, but they do not follow the same standard as they do in prescription drugs.

3.1 INTRODUCTION

Over the past 100 years, scientific discoveries have benefitted the world's population by discovering the mechanisms of diseases and improved therapies that followed. Many believe herbs and botanicals to be regular and, in this way, more beneficial and gentler than customary medications, despite the fact that 35% of physician-recommended medications are of natural origin. Many use herbs and botanicals for the assortment of ailments, in addition to common health and well-being. Natural herbs and preparations are available as nutritional dietary supplements that are regulated in contrast to pharmaceutical medications and are sold as "natural remedies," "botanicals," "herbal products," and "herbal medicines." Dietary supplements are not required to experience the similar stringent testing as over the counter-drugs and medicine prescribed by doctors, and are not regulated as carefully by means of the Food and Drug Administration (FDA) (Bellows and Moore, 2011). There are no data to suggest that herbs are more beneficial than conventional drugs for treating illnesses. However, there may be sufficient proof to help the confined use of natural herbs and botanicals underneath the guidance of a scientific expert.

3.2   THREAT OF HERBAL SUPPLEMENTS

Despite the fact that numerous herbs are viewed as safe, some have risky reactions particularly in kids, pregnant ladies, or those with hidden ailments and illnesses. Herbals and natural drugs can possibly interact with physician-recommended medicines, over-the-counter medications, nutrients, and minerals. Herbal products are not as firmly regulated as medications and different prescriptions; however, they are regularly utilized for similar purposes. Herbs are not prescribed instead of therapeutic treatment or ordinary medication for chronic conditions or sicknesses, for example, hypertension, psychosis, coronary disease, and diabetes. Herbs likewise are not prescribed for the individuals who might be immuno-suppressed (e.g., cancer and HIV), kids younger than six years, pregnant or lactating women, those with liver or kidney damage, or those who are undergoing surgery.

3.3   EFFICACY OF HERBS

The mainly used herbs are as follows.

3.3.1   *ASTRAGALUS*

Astragalus membranaceus (Latin), *ogi* (Japanese), *hwanggi* (Korean), *huang qi* (Chinese), and membranous milk-vetch root (English), is one in all the essential "Qi tonifying" adaptogenic herbs present in the Chinese materia medica. It has been prescribed for hundreds of years for general weakness, chronic diseases, and to extend overall vitality. The class *Astragalus* has more than 2000 species around the world and is often called as milk-vetch root. The main constituent of *A. membranaceus* includes polysaccharides, saponins, flavonoids, amino acids, and trace elements. Currently, much of the pharmacological research on *Astragalus* is focused on its immune-stimulating polysaccharides and other active ingredients useful in treating immune deficiency conditions (Alternative Medicine Review, 2003; Ma et al., 2002)

It has been described that *A. membranaceus* might attenuate intestinal irritation; in any case, the basic mechanism for its medicinal activity in intestinal epithelial cells (IECs) remains unclear. In this investigation, Adesso and his team assessed *A. membranaceus* extract (5–100 μg/mL) in a model for inflammation and oxidative stress for IECs. They demonstrated that *Astragalus membranaceus* concentrate decreased the inflammatory action induced by means of lipopolysaccharides from *Escherichia coli* and interferon-γ, reducing tumor necrosis factor-α (TNF-α) release, inducible nitric oxidesynthase expression, nuclear factor-κB (NF-κB) activation, cycloxygenase-2 and nitrotyrosine formation, and reactive oxygen species (ROS) release in the nontumorigenic IEC line (IEC-6). The antioxidant potential of *A. membranaceus* extract was also evaluated in a model of hydrogen peroxide (H_2O_2)-induced oxidative stress in IEC-6, indicating that this extract reduced ROS release and increased the nuclear factor (erythroid-derived 2)-like 2 (Nrf2) activation and the expression of antioxidant cytoprotective factors in these cells. The outcomes added to explain that the mechanisms associated with *A. membranaceus* extract decreased inflammation and thus raised the use of this concentrate as an anti-inflammatory and antioxidant herb remedy for intestinal ailments (Adesso et al., 2018).

In most of the Chinese antidiabetic formulas, *A. membranaceus* has been used as a key component for the herbal antidiabetic compound. It has been logically examined for its antidiabetic actions. Ethnopharmacological studies have established its potential to alleviate diabetes mellitus. Recent investigation has tried to match its chemical constituents (specially polysaccharides) to type 1 and 2 diabetes mellitus (Agyemang et al., 2013).

3.3.2 Chaparral

Chaparral (*Larrea tridentate* (DC) *Coville, Larrea divaricata* Cav) and its chemical constituent nordihydroguaiaretic acid (NDGA) have been accounted to have antioxidant properties, mainly free-radical scavenging action. In spite of the fact that it has been proposed for the treatment of malignant growth, action has not been taken in clinical preliminaries. Chaparral has been related with instances of liver failure, cirrhosis, renal cysts, renal cell carcinoma, and hepatitis. Because of these reports, the US FDA expelled chaparral from its "generally recognized as safe" (GRAS) list in 1970. Chaparral and NDGA are generally considered dangerous and are not often recommended to be used (Ulbricht et al., 2003).

Chaparral is used as an intestinal cleanser, a blood purifier, a liver stimulant, and an anti-arthritic, diuretic, antirheumatic, expectorant, antiseptic, and tonic. It is mainly used in canker sores (when combined with sarsaparilla), venereal nodes, and rheumatism. It helps in decreasing the extent and growth of cancerous tumors, malignant melanomas, pimples, and skin cancers, persistent backache, cancer, arthritis, and warts and blotches. Chaparral builds up hair growth, improves vision, and although it is not a laxative, it still helps in increasing bowel elimination. It is also used in kidney infections, prostate gland disorders, leukemia, sinus problems, skin cancer, throat ailments, stomach disorders, and obesity. The twigs and leaves of *Larrea mexicana* are saturated with boiling water, and then it is applied as an antiseptic lotion. It is also used in the sores of human and pet animals. The flower buds are regularly preserved in vinegar and consumed as capers. In North America, the plant is taken into account as a healing aid for rheumatism (Syphard et al., 2018).

3.3.3   Chrysanthemum

In China, *Chrysanthemum indica* and *Chrysanthemum moriflorum* are being used for very long for the treatment of cardiovascular diseases, respiratory diseases, and inflammation. Although traditionally used in tea preparations, other preparations include tinctures, creams, and lotions. Extracts of *Chrysanthemum indicum* (*C. indicum*) exhibit anti-inflammatory properties in acute and persistent cutaneous inflammation (Lee et al., 2009).

The flower of *C. indicum* is a conventional Chinese drug with strong aroma and numerous past studies concentrated on its essential oil. It is a great supply of herbal quercitrin and myricetin, which is considerable for the improvement of potential pharmaceuticals. The chemical composition of its oil acquired from *C. indicum* L. was characterized by gas chromatography–mass spectrometry analysis. Seventy-two compounds accounting for 96.63% of the extracted essential oil were distinguished. The main compositions in the oil were camphor (10.12%), germacrene D (10.6%), 1,8-cineole (10.4%), bornylacetate (6.1%), α-thujone (6.05%), β-caryophyllene (5.1%), α-pinene (4.4%), borneol (3.6%), terpinen-4-ol (3.4%), *cis*-chrysanthenol (3.4%), and α-cadinol (3%). The essential oil of *C. indicum* has potent antibacterial activity against all oral cavity bacteria (minimum inhibitory concentrations [MICs], 0.1–1.56 mg/mL; minimum bactericidal concentrations [MBCs], 0.256–3.2 mg/mL) than their major compounds. Furthermore, the MICs/MBCs have been reduced to at least one-half to one-sixteenth due to the combinations of essential oil with antibiotics, for example, gentamicin or ampicillin for all oral bacteria. A sturdy bactericidal effect turned into exerted in drug combos. The *in vitro* records recommended that the essential oil of *C. indicum* with different antibiotics might be microbiologically gainful and synergistic (Jung, 2009; Wu et al., 2010).

C. indicum ethanolic extricate (CIE) essentially suppressed the multiplication and intrusion of MHCC97H cells, one of the hepatocellular carcinoma (HCC) cell lines with strong metastatic potential, in the dose-dependent type. CIE especially diminished matrix metallopeptidase-2 (MMP-2) and matrix metallopeptidase-9 (MMP-9) expressions, expanded all the while TIMP-1 and TIMP-2 expressions, further re-establishing their balance in the malignancy cells. The present examination demonstrates that CIE diminished MHCC97H cell's metastatic capacity to a limited extent at any rate through abatement of the MMP expression, synchronous increment of the TIMP expression, and further re-establishing their balance as therapeutic objective in HCC. It is recommended that *C. indicum* is a potential new medicinal herbal plant that is used for the treatment of HCC or malignancy and metastasis (Wang et al., 2010). Relative to versatile immune reactions, a few concentrates were additionally ready to adjust immunoglobulin production by mouse spleen cells after cyclophosphamide treatment, with more elevated amounts of precise IgG and IgM reaction to sheep red blood cells (Cheng et al., 2005).

3.3.4 *Dioscorea*

Dioscorea bulbifera (*D. bulbifera*) possesses profound therapeutic potential. It is located in the hotter parts of India, also known as Yam or air potato. Later pharmacological discoveries demonstrate that its tubers have significant activities such as laxative, deflatulent, reviving and tonic, aphrodisiac, anthelmintic, and it is utilized in scrofula, hematological issue, hemorrhoids, syphilis, looseness of the bowels, general debility, diarrhea, worm invasions, polyuric, and skin issue, which consent to the cases made within the ancient medicative texts (Subasini et al., 2013).

The oral management of aqueous and methanol concentrates caused noteworthy anti-inflammatory activity on paw edema induced by formalin, serotonin, and histamine. The present outcomes demonstrate that the bulbils of *D. bulbifera var sativa* have significant analgesic and anti-inflammatory properties. These activities may additionally be affected by the inhibition of inflammatory hormones such as prostaglandins, serotonin, and histamine. Consequently, the pain-relieving action of the bulbils of *D. bulbifera* might be partially linked to its anti-inflammatory properties (Mbiantcha et al., 2011).

The anticancer impacts of different parts extricated from *D. bulbifera* Linn. have dynamic anticancer components which are chiefly extricated by petroleum ether. Later, a few researchers investigated the mechanism of liver damage caused by *D. bulbifera* Linn. at the dimension of gene expression, and they discovered that when mice are dealt with *D. bulbifera* Linn. for half month, 82 genes are expressed differentially and serum alanine aminotransferase (ALT), aspartate aminotransferase (AST), and tissue total protein are expanded. However, when the mice are dealt for 30 days, 1657 genes are differentially expressed, and tissue absolute protein and the serum ALT are expanded with noteworthy decline in AST and ALP showing the adjustment in the liver mouse cell expression profile (Yu et al., 2004). The ethyl acetate dissolvable fraction of the 75% alcoholic concentrate of rhizome and their phyto-constituents, to be specific Kaempferol-3,5-dimethyl ether, Caryatin, (+)-Catechin, Myricetin, Quercetin-3-*O*-galactopyranoside, Myricetin-3-*O*-galactopyranoside, Myricetin-3-*O*-glucopyranoside, and Diosbulbin B, have demonstrated antitumor advancing action in opposition to the tumor promotion in JB6 cells actuated by 12-*O*-tetradecanoylphorbol-13-acetate (TPA) (Gao et al., 2002).

3.3.5 EPHEDRA

Ephedra (*Ephedra sinica*), also called Ma-Huang, is a natural stimulant medication made out of two active components (pseudoephedrine and ephedrine) that are present in numerous over-the-counter items. Ephedrine is a sympathomimetic agent that has boundless uses as an adrenergic stimulant. It originates from plants of the class *Ephedra* sp., being an alkaloid for the most part present in the specie *Ephedra sinica*. Ephedrine can be synthesized in the industry as a result of biotransformation or by extraction of the plant. Different subordinates of ephedrine can likewise be found in Ephedra, for example, pseudoephedrine that additionally has adrenergic properties. The most pure form of ephedrine is used for the remedial reason, basically as a bronchodilator and decongestant. Its pharmacological properties are associated with its adrenergic action, which thus happens because of its stimulating action on α-, β1-, and β2-adrenergic receptors via direct and indirect outcomes. These days, there is a worry about the aimless utilization of ephedrine, since it began to be utilized for weight reduction and improve athletic performance. Related with this propensity, an enormous number of side effects related with cardiovascular issues began to be accounted for (Limberger et al., 2013).

Ephedrine and pseudoephedrine are additionally utilized in dietary supplements that guarantee to increase weight reduction and upgrade athletic performance. The active components are basically associated with amphetamines; they play similar, although less potent, roles in stimulating the central nervous system. Ephedrine is usually the first ingredient present in illegally synthesized medicine, consisting of methamphetamines (Shekelle et al., 2003). In vitro studies demonstrate that there was also associated decreased specific antigen-induced T(H)2 cytokine secretion by polarized splenocytes. Distillates of *E. sinica* were able to help arthritis symptoms. It was also referred to the expression of C3 and C9; myeloperoxidase activity increased when Ephedra was administered to animals after traumatic spinal cord damage, thus suggesting it may be helpful for controlling complement activation and inflammation all through major trauma and injury (Dasgupta and Hammett-Stabler, 2010). Recent studies demonstrate that dietary supplements containing ephedrine can cause severe and possibly fatal adverse effects particularly when combined with caffeine or different stimulants.

3.3.6 GERMANDER

Generally, *Teucrium polium* L. *(Calpoureh)* has been utilized for various pathological conditions. In customary Iranian drug, the tea of *Teucrium polium* L. is utilized for treating numerous ailments, for example, type II diabetes. It is accepted that this plant has useful biological actions. The outcomes demonstrated that infusion of *T. polium* with various dosages had toxic effect on renal tubule cells; nonetheless, the toxicity changed into significant after end of medication administered for 28 days. It tends to be presumed that *T. polium* injections no longer display their outcomes immediately following drug injection; yet, following a time of 28 days, it has demonstrated its effectiveness. The outcomes likewise demonstrated that the destructive rate was dose based. The plant *T. polium* is utilized worldwide in conventional and natural prescription. But, it has been observed to cause hepatotoxicity in people. Despite the fact that the mechanism of *T. polium* hepatotoxicity is doubtful, teucrin A and a few neoclerodanediterpenoids, present within the aerial parts of the plant, have been mentioned as the likely hepatotoxic precursors of this herb. In certain reports, the liver damage has been related to the presence of autoantibodies in the blood (Baradaran et al., 2013). Generally, people take germander for treating fever, gall bladder conditions, mild diarrhea and stomachaches, germ-killer, as a digestive aid, for gout and for weight loss.

A few people use germander as a mouthwash to refresh the breath and eliminate germs in the mouth. In production, germander is utilized as a flavoring agent in alcoholic drinks. Germander (*Teucrium chamaedrys*) is a sweet-smelling plant inside the "mint family" belonging to genus Teucrium. Its blooms are utilized as a traditional drug to treat diabetes, dyspepsia, and gout. However, constant utilization of germander results in hepatotoxicity. In fact, the toxic results of germander are first visible inside 9 weeks of use and are manifested by way of jaundice and accelerated liver enzymes (ALT and AST). After cessation of the drug, restoration may take 6 weeks to half a year. The mechanism of toxicity is believed to be identified with the diterpenoid component present in plants that are converted into stronger toxins inside the liver. The main hepatotoxic diterpene present in germander is teucrin A, that is, bioactivated through the cytochrome P450 enzymes within the liver. Another hepatotoxic element of germander is teuchmaedryn A (Dasgupta and Sepulveda, 2013; Mattéi et al., 1995; Polymeros et al., 2002).

3.3.7   Ginkgo biloba

The ginkgo tree is the main enduring individual from *Ginkgoaceae* family, class of *Ginkgoatae*, rediscovered in Asian graced sanctuary plants by Kaempfer in 1670. The class of *Ginkgoatae* comprises of around 15 genera, and among these, Ginkgo, Baiera, and Ginkgoites are the most significant. The name ginkgo comes from the Chinese words sankyo or yin-kuo, which means a hill apricot or silver fruit, due to their apricot-shaped mature fruits and yellow color. Englbert Kaempfer, a German specialist, first utilized the expression "Ginkgo" in 1712; yet, it was Linnaeus who named it *Ginkgo biloba* in 1771. Each of the leaves and the nuts of this tree had been used for as far back as a few centuries in conventional Chinese drugs. Actually, the nuts are known to have a longer record of usage, being first referred to in herbals in the Yuan dynasty. For more than 5000 years, the seeds (nuts) have been known to treat respiratory diseases (like bronchial enuresis and cough), alcoholism abuse, and bladder irritation while the leaves have been mostly used to treat cardiac and pulmonary dysfunctions and skin contaminations. Notwithstanding, it was distinctly in the last 20–30 years that the utilization of the ginkgo leaf and its standardized concentrate formulation, EGb 761, began in Germany, and now is the most utilized type of enhancement for cognitive sicknesses in the United States. Different employments of this tree incorporate the fruit, prepared via fermentation and cooking, being a delicacy in weddings and galas. The roasted or stewed ginkgo seeds are likewise viewed as a gourmet delicacy in Korea, Japan, Malaysia, and China. The tree is likewise developed in numerous parts of Europe and the United States principally for its ornamental value. It develops well in many places because of properties such as bug, contamination, and sickness resistance (Mahadevan and Park, 2007).

The flavoglycosides present in ginkgo are its most active component and have shown surprising pharmacological capacities. These chemical compounds have free radical properties and act as an antioxidant agent. These flavonoids consist of kaempferol, isorhamnetine, and quercitin. The terpene substance of ginkgo, which incorporates the ginkgolides and the bilobalides, helps to reduce inflammation by inhibiting platelet activating factor (PAF) in the blood. This activity facilitates to reinforce circulation. PAF assumes a role in various disease, for example, heart attacks, strokes, atherosclerosis, and asthma (Nash and Shah, 2015).

Ginkgo biloba leave concentrate is among the most broadly sold natural dietary supplements in the United States. Its implied therapeutic impacts include reducing oxidative stress, scavenging free radical, diminishing platelets aggregation, decreasing neural harms, anti-inflammation, antitumor activities, and antiaging. Clinically, it has been endorsed to treat CNS problems, for example, Alzheimer's illness and cognitive deficits. It applies hypersensitivity and changes in bleeding time. However, its mutagenicity or anticancer activity has not been accounted for; its constituents, rutin, kaempferol, and quercetin, have been demonstrated to be genotoxic. There are no guidelines or rules managing the constituent parts of *Ginkgo biloba* leave extract nor are exposure limits obligatory. Safety evaluation of *Ginkgo biloba* leave concentrate is being led by the US National Toxicology Program (Chan et al., 2007).

3.3.8 Ginseng

Ginseng, the root of *Panax* species, is an outstanding traditional drug. It has been utilized as conventional natural medication in Korea, Japan, and China for many years, and today it is a famous and worldwide used herbal remedy.

The active elements of ginseng are ginsenosides which are additionally called ginseng saponins. Currently, there is growing evidence in the literature on the pharmacological and physiological activities of ginseng. Ginseng had been used primarily as a tonic to invigorate week bodies and help the restoration of homeostasis (Radad et al., 2004). Ginseng is the most well-known herb. Ginseng is frequently referred to as a definitive tonic, athletic performance, immune functions, the herb boosts general well-being, and libido. Ginseng is prominently utilized for its immunomodulatory, adaptogenic, cardiovascular, antineoplastic, endocrine, central nervous system, and ergogenic impacts; yet, these utilizations have not been affirmed by clinical preliminaries. Various ginseng species are utilized in natural items worldwide. A portion of these plants incorporates Sanchi ginseng, American ginseng, Chikusetsu ginseng, and Korean ginseng. Ginseng is otherwise called Siberian ginseng, eleuthero, devil's shrub, wild pepper, and touch me not. It has been utilized to improve the body's protection from stress and to increase energy. But, the mechanisms underlying ginseng's outcomes remain to be examined. Therapeutical importance of ginseng is because of its antineurological impact, anti-inflammatory effects, and hypoglycemia effect. Research has demonstrated that consuming a hot ginseng tea has an ani-inflammatory impact (Seervi et al., 2010).

Ginseng, an old and acclaimed therapeutic herb in the Orient, has been utilized as a significant tonic and for the treatment of different ailments including liver disorders. Ginseng saponins, ordinarily known as ginsenosides, are important constituents and have been accepted to be responsible for numerous ginseng health advantages. There are around 40 ginsenosides so far isolated from ginseng. So far, treatment choices for common liver diseases such as chronic hepatitis, cirrhosis, and fatty liver remain problematic. In such a manner, ginseng concentrates and individual ginsenosides have widely exhibited the advantageous role in the guideline of normal liver functions and the remedy of liver issue of hepatitis, acute/chronic hepatotoxicity, HCC, hepatic fibrosis/cirrhosis, etc., in different pathways and mechanisms (Huu-Tung et al., 2012).

Panax ginseng is frequently known as an adaptogen, which has shifted activities and impacts on the body that help nonspecific protection from biochemical and physical stressors, upgrade mental capacity, and improve energy and longevity. Surveys propose *Panax ginseng* has an immuno-modulating action by influencing the hypothalamic-pituitary-adrenal (HPA) pivot. In vitro analyses showed upgraded natural killer (NK) cell action and expanded immune cell phagocytosis after ginsenoside introduction. A double-blind, placebo-controlled 8-week study analyzed the resistant impacts of 100 mg Ginsana (G115), 100 mg liquid ginseng concentrate, or placebo treatment two times every day in 60 healthy volunteers. Blood tests gathered at baseline, week 4, and week 8 inspected polymorphonuclear (PMN) cell chemotaxis, total lymphocytes, phagocytosis, T-suppressor cells, NK-cell, and T-helper activity. The groups consuming ginseng experienced steady improvement in immune system action at week 4 and statistically critical differences at week 8, confirmed by betterment in phagocytosis, PMN cell chemotaxis, and total number of T-suppressor cells and T-helper. It was reasoned that ginseng concentrate animates the immune system and the standardized concentrate is more powerful than the liquid ginseng extract (Kumar et al., 2012).

3.3.9 Kava

Kava, a herbal sedative with antianxiety or calming effects, is prepared by extracting the rhizomes of *Piper methysticum*, a south pacific plant. There are at least 72 different cultivars of this species, which differ both in appearance and in chemical composition. The active chemicals of the plants, known as kavalactones, are concentrated in the rhizomes. Inhabitants of the south pacific islands prepare a kava-based drink by mixing fresh or dried rhizomes with cold water or coconut milk. Among more than 18 kavalactones characterized, 6 are considered the primary constituents of kava extracts: kawain, dihydrokawain, methysticine, dehydromethysticine, yangonin, and desmethoxyyangonin. Quite a considerable lot of these compounds, particularly those with a methylenedioxyphenyl derivatives (methysticine and dihydromethysticine), have been found to restrain various cytochrome P450s: CYP2C19, CYP1A2, CYP2C9, CYP3A4, CYP2D6, and CYPA4. It is therefore astonishing to discover that pharmacokinetic interactions among kava and Western medications are generally rare and are not very much reported in the literature. There is a case report that kava decreases the viability of levodopa (Dasgupta and Hammett-Stabler, 2010).

Kava concentrates are commonly very much tolerated; yet reports of hepatotoxicity required a worldwide reappraisal of its safety. Hepatotoxicity can occur as an acute, severe form or a chronic, mild form. Inflammation seems to be involved in both forms and may end result from the activation of liver macrophages (Kupffer cells), both at once or through kava metabolites (Rowe et al., 2011). In an ongoing randomized, double-blind, placebo-controlled examination done in Germany, the adequacy of kava was surveyed in 58 individuals with anxiety syndrome not brought about by mental issue.

Contrasted with the placebo group, the kava gathering showed a critical decrease in uneasiness side effects dependent on a few methods for evaluation. The specialists presumed that the kava concentrate was clinically successful in decrease of conditions of uneasiness, excitedness, and tension. Another ongoing randomized, placebo-controlled, double-blind study examination was directed for 25 weeks on 101 outpatients experiencing uneasiness and tension of nonpsychotic origin. The specialists presumed that the outcome supports the utilization of the kava extract "as a treatment alternative to tricyclic antidepressants and benzodiazepines in anxiety disorders, with proven long-term efficacy." The overall tolerability of the concentrate was incredible (Rouse 1998).

3.3.10   LIQUORICE

There is an expanding interest for natural drugs, health products, and prescription drugs. *Glycyrrhiza glabra* Linn is a plant utilized in conventional drug over the world for its ethnopharmacological reason. It is observed to contain significant phytoconstituents, for example, glycyrrhizinic acid, glycyrrhizin, isoflavones, and glabrin A and B. It is effectively used as anti-inflammatory, antibacterial, antifungal, antidiabetic, antiviral, anti-ulcer, antitussive, antioxidant, skin whitening, and antidiuretic agent. The prevailing article is a push to accumulate the available literature on *Glycyrrhiza glabra* concerning its conventional uses, bioactive elements, and pharmacologic activities. This might be valuable in finding potential therapeutical effects and growing new details (Damle, 2014). Glycyrrhizin had an antiviral activity against human immunodeficiency virus, herpes simplex virus, hepatitis B, and hepatitis C viruses; however, the mechanism is still not known. It might be credited to the hindrance of viral adherence to host cell, replication, different transduction mechanism, incitement of immune system, or complex mechanism, including at least one of the past (Fiore et al., 2008). In Japan, glycyrrhizin when given by intravenously improved liver capacity and serum hepatic transaminases (Coon and Ernst, 2004). Hepatoprotective impact of glycyrrhizin was inspected in various animals, and in vitro models proposed appearing in actuated liver toxicity with carbon tetrachloride estimating the possible mechanism to be connected with the restraint of the hepatic cytochrome P450 2E1 activation of tetrachloride (Jeong et al., 2002). Different investigations indicated a decrease in induced liver cells irritation on initiated by TNF-α by glycyrrhizin (Hong-Jhan et al., 2014), which needs higher affirmation with other distinctive hepatic-induced injuries or different transgenic or knock out creatures. Recent investigations appeared beneficial for the alcoholic root concentrate of *Glycyrrhiza glabra* on induction of autophagy-associated cell death in harmful prostatic cell lines and down directed the cell division in breast malignant cell instigated by endocrine aggravating chemicals; yet, at the same time, these discoveries need more research endeavors. In adjusting and improving immune reaction, test results demonstrated some logical inconsistency where glycyrrhizin had an up-managing impact over dendritic cells and immune responses through T helper 1 response (Bordbar et al., 2012), whereas in another test, 18β-glycyrrhetinic acid impaired that up-regulation (Kim et al., 2013), which still remains a question of exploration. The antitussive impact of water concentrate of *Glycyrrhiza glabra* on guinea pigs was more compelling than codeine (Saha et al., 2011). Glycyrrhizic acid recoupled the interleukin (IL)-4, IL-5, IL-13 levels in ovalbumin-actuated asthma animal model (Ma et al., 2013). Glycyrrhizin additionally improved the action of both epoxide hydrolase and thioredoxin reductase compounds, thereby re-establishing the NF-κB (Qamar et al., 2012).

3.4 SAFETY

Potential dangers in using natural drugs are commonly unrecognized. A standout amongst the most genuine dangers related to natural medications is the way that the customers erroneously accept that since herbs are natural, they are protected. Natural does not imply free from harm. Numerous natural medications are known to deliver poisonous responses, conceivable mutagenic impacts, or allergic reactions (Ernst, 1998; Senior, 1998). Regarding this, a case report from Belgium portrays an account of 30 ladies treated with a Chinese herbal slimming formulation. These ladies died from renal failure in 1991–1992. This happened with the use of *Aristolochia spp.* that is considered a nonharmful herb (Marwick, 1995). It is hard to figure out which botanicals contain lethal substances that have intense side effects in a huge fraction of users. It is more difficult, however, to recognize adverse effects that develop over a long period of time. Some of the effective herbal drugs with their side effects are summarized in Table 3.1.

3.5 HOW HERBS ARE REGULATED

Herbal supplements are managed by the FDA, but not as medications or as food materials. They fall under a class known as dietary supplements.

TABLE 3.1   Top Most Commonly Used Herbs with Their Side Effects

Common Name, Source	Uses	Side Effects
Echinacea (*Echinacea angustifolia*)	Reduced duration of cold Boosts immune system Heals wounds	GI symptoms include nausea, vomiting, abdominal pain, and diarrhea
Evening primrose oil *(Oenothera biennis)*	Reduce menopausal symptoms Reduce breast pain Treat eczema Treat rheumatoid arthritis	Stomach upset and headache
Feverfew thin *(Tanace tumparthenium)*	Reduce migraines, headache Treat arthritis	GI discomfort, mouth ulcer
Garlic *(Allium sativum)*	Reduce the risk of heart disease Lower high blood pressure Treat athlete's foot	Rarely increase in bleeding and nausea and vomiting
Ginger *(Zingiber officinale)*	Prevents/reduced motion sickness Used as digestive aid Treatment for rheumatoid arthritis	High dose causes GI discomfort, heartburn
Ginkgo biloba *(Ginkgo biloba)*	Improve age-related memory impairment and dementia Improve visual field in glaucoma and diabetic retinopathy Lowering blood pressure	Mild headache, dizziness, constipation, and allergic skin reactions
Ginseng *(Panax ginseng)*	Improve cognitive function Enhance athletic performance Lower blood glucose	Insomnia, headache, menstrual abnormalities
St. John's wort *(Hypericum perforatum)*	Treat depression Improve premenstrual syndrome Treat obsessive compulsive disorder Topically used for wounds (inflammation)	Insomnia, anxiety, irritability, diarrhea, fatigue, dry mouth, photosensitivity
Saw palmetto *(Serenoa repens)*	Treat benign prostatic hyperplasia Improve overall prostate health	GI discomfort such as nausea, vomiting, constipation, and diarrhea

These guidelines provide affirmation that:

- herbal supplements fulfill certain quality guidelines;
- the FDA can mediate to expel toxic formulations from the market.

In any case, the standards do not ensure that herbal supplements are secure for anybody to utilize. These items can present unforeseen dangers in light of the fact that various supplements contain active compounds that have strong impacts on the body. For instance, taking a combination of herbal supplements or utilizing supplements together with chemically synthesized drugs could be harmful, even dangerous (Tanaka et al., 2004).

In India, the customary herbal drugs, for example, Ayurveda, Siddha, and Unani (ASU), are viewed as safe on account of their long history of utilization. Natural medications are directed under the Drug and Cosmetics Act 1940 and Rules 1945 in India. Division of AYUSH is the regulatory authority that gives any manufacturing rights or promotes herbal medications after acquiring license, as relevant. In that capacity, no safety and viability studies are required for marketing endorsement, according to the Drugs and Cosmetics Act of 1940 (Kumar, 2017).

3.6 CONCLUSION

Herbal supplements comprise the principle part of traditional medication, which have been utilized since long back. They have made critical commitment to human well-being through their health promotive, therapeutic, and rehabilitative properties and in the counteractive action against ailments. Indeed, many herbal remedies used traditionally have become modern medicines through drug development. Ephedra, ginseng, and liquorice are some notable examples. Long convention of utilization of numerous natural cures and encounters passed on from ages has led to dependence of the general population on natural cures. At present, the utilization of therapeutic plants for medical advantages is expanding around the world.

The demand for herbal and nutritional supplements is high. By and by, around 82% of the total populace utilize herbal prescriptions and dietary supplement use is additionally normal. Evidence regarding the benefits and risks of supplements is becoming apparent. They are increasingly gaining acceptance and their sales have soared in recent years. Most individuals report using herbal supplement as a means of improving health or preventing illness. Herbal supplements are used by all kinds of population and ethnicities across all ages, though the type of them favored varies. In spite of their unregulated status, natural supplements tend to have poisonous quality leading to death, even in little dosages or momentary use. People ought to be educated so as to shield themselves from sketchy health items and services. Coming up next are tips planned to enable one to turn out to be increasingly mindful of the dangers related to natural and health supplements:

- Decide whether a herbal supplement is surely required.
- Stay educated with the aid of researching the product to decide: well-being, counter uses with different supplements or drugs, validity of claims, species, measurement, time span of use, most efficient form, plant part, adverse effects, and sensible price.
- Inform a specialist, drug specialist, and other human service experts of any herbs being considered or routinely utilized. Counsel them with any inquiries.
- Select brands which have been tried for reliability in dosage by searching for the National Formulary, United States Pharmaceopia, or Consumer Lab symbols.
- Read the label on the product and adhere to the guidelines intently.
- Use herbal medication only for minor situations and just on a short-term premise. In the event that a condition is severe or persistent, counsel a therapeutic expert.
- Discontinue its use if unfavorable side effects are seen.
- Do not consume natural products known to be toxic. Check the sources indexed in the fact sheet regularly, for extra knowledge of toxic herbs.
- The best remedy for ailment prevention is a healthy way of lifestyle that incorporates eating routinely whole-grains, foods that are low in fat, and vegetables and fruits. Physical excercises likewise play a significant role in lessening one's risk for illness.

KEYWORDS

- **herbal**
- **health**
- **supplement**

REFERENCES

Adesso, S.; Russo, R.; Quaroni, A.; Autore, G.; Marzocco, S. *Astragalus Membranaceus* Extract Attenuates Inflammation and Oxidative Stress in Intestinal Epithelial Cells via NF-KB Activation and Nrf2 Response. *Int J Mol Sci.* **2018**, 19(3): 800.

Agyemang, K.; Han, L.; Liu, E.; Zhang, Y.; Wang, T.; Gao, X. Recent Advances in *Astragalus Membranaceus* Anti-Diabetic Research: Pharmacological Effects of Its Phytochemical Constituents. *Evid Based Complement Alternat Med.* **2013**, 1–9.

Alternative Medicine Review Monograph.*Astragulus Membranaceus. Altern Med Rev.* **2003**, 8(1): 72–77.

Baradaran, A.; Madihi, Y.; Merrikhi, A.; Rafieian-Kopaei, M.; Nematbakhsh, M.; Asgari, A.; Khosravi, Z.; Haghighian, F.; Nasri, H. Nephrotoxicity of hydroalcoholic extract of *Teucrium polium* in Wistar rats. *Pak J Med Sci.* **2013**, 29(1) Suppl: 329–333.

Bellows, L.; Moore, R. Dietary Supplements: Herbals and Botanicals. **2011**, *Colorado State University Extension Publication* no. 9.370 (10/13).

Bordbar, N.; Karimi, H.M.; Amirghofran, Z. The effect of glycyrrhizin on maturation and T cell stimulating activity of dendritic cells. *Cell Immunol.* **2012**, 280: 44–49.

Chan, P.; Qingsu, X.; Peter, P.FU. "*Ginkgo Biloba* Leave Extract: Biological, Medicinal, and Toxicological Effects." *J Environ Sci Heal C.* **2007**, 25(3): 211–44.

Cheng, W.; Li, J.; You, T.; Hu, C. Anti-Inflammatory and Immunomodulatory Activities of the Extracts from the Inflorescence of *Chrysanthemum Indicum Linné. J Ethnopharmacol.* **2005**, 101(1–3): 334–37.

Coon, J.T.; Ernst, E. Complementary and Alternative Therapies in the Treatment of Chronic Hepatitis C: A Systematic Review. *J Hepatol.* **2004**, 40(3): 491–500.

Damle, M. "*Glycyrrhiza Glabra* (Liquorice)-a Potent Medicinal Herb." *Int J Herb Med.* **2014**, 2(2): 132–36.

Dasgupta, A.; Hammett-Stabler, C.A. Herbal Supplements: Efficacy, Toxicity, Interactions with Western Drugs, and Effects on Clinical Laboratory Tests. New York, NY: *A John Wiley & Sons Publication,* Inc., **2010**, p. 470.

Dasgupta, A.; Sepulveda, J. Accurate Results in the Clinical Laboratory : A Guide to Error Detection and Correction. New York, NY: *Elsevier,* **2013**, p. 382.

Ernst, E. Harmless Herbs? A Review of the Recent Literature. *Am J Med.* **1998**, 104(2): 170–78.

Fiore, C.; Eisenhut, M.; Krausse, R.; Ragazzi, E.; Pellati, D.; Armanini, D.; Bielenberg, J. Antiviral effects of Glycyrrhiza species. *Phytother Res.* **2008**, 22(2): 141–8.

Gao, H.; Kuroyanagi, M.; Wu, L.; Kawahara, N.; Yasuno, T.; Nakamura, Y. Antitumor-Promoting Constituents from *Dioscorea Bulbifera L.* in JB6 Mouse Epidermal Cells. *Biol Pharm Bull.* **2002**, 25(9): 1241–43.

Hong-Jhan, C.; Shih-Pei, K.; I-Jung, L.; Yun-Lian, L. Glycyrrhetinic Acid Suppressed NF-KB Activation in TNF-α-Induced Hepatocytes. *J Agr Food Chem.* **2014**, 62(3): 618–25.

Huu-Tung, N.; Uto, T.; Morinaga, O.; Kim, Y.H.; Shoyama, Y. Pharmacological Effects of Ginseng on Liver Functions and Diseases: A Minireview. *Evid Based Complement Alternat Med.* **2012**, 2012: 1–7.

Jeong, H.G.; You, H.J.; Park, S.J.; Moon, A.R.; Chung, Y.C; Kang, S.K.; Chun, H.K. Hepatoprotective Effects of 18beta-Glycyrrhetinic Acid on Carbon Tetrachloride-Induced Liver Injury: Inhibition of Cytochrome P450 2E1 Expression. *Pharmacol Res.* **2002**, 46(3): 221–27.

Jung, E.K. Chemical Composition and Antimicrobial Activity of the Essential Oil of *Chrysanthemum Indicum* Against Oral Bacteria. *J Bacteriol Virol* **2009**, 39(2): 61.

Kim, M.E.; Kim, H.K.; Kim, D.H.;Yoon, J.H.;Lee, J.S. 18β-Glycyrrhetinic Acid from Licorice Root Impairs Dendritic Cells Maturation and Th1 Immune Responses. *Immunopharmacol Immunotoxicol.* **2013**, 35(3): 329–35.

Kumar, D.; Arya, V.; Kaur, R.; Bhat, Z.A.; Gupta, V.K.; Kumar, V. A Review of Immunomodulators in the Indian Traditional Health Care System. *J Microbiol Immunol Infect.* **2012**, 45(3): 165–84.

Kumar, V. Herbal Medicines: Overview on regulations in India and South Africa. *World J Pharm Res.* **2017**, 6(8): 690–698.

Lee, D.Y.; Choi, G.; Yoon, T.; Cheon, M.S.; Choo, B.K.; Kim, H.K. Anti-Inflammatory Activity of *Chrysanthemum Indicum* Extract in Acute and Chronic Cutaneous Inflammation. *J Ethnopharmacol.* **2009**, 123(1): 149–54.

Limberger, R.P.; Jacques, A.L.B.; Schmitt, G.C.; Arbo, M.D. Pharmacological Effects of Ephedrine. *Nat Prod.* Springer Berlin Heidelberg. **2013**, 1217–37.

Ma, C.; Ma, Z.; Liao, X.L.; Liu, J.; Fu, Q.; Ma, S. Immunoregulatory Effects of Glycyrrhizic Acid Exerts Anti-Asthmatic Effects via Modulation of Th1/Th2 Cytokines and Enhancement of CD4(+)CD25(+)Foxp3+ Regulatory T Cells in Ovalbumin-Sensitized Mice. *J Ethnopharmacol.* **2013**,148(3): 755–62.

Ma, X.Q.; Shi, Q.; Duan, J.A.; Dong, T.T.X.; Tsim, K.W.K. Chemical Analysis of *Radix Astragali* (Huangqi) in China: A Comparison with Its Adulterants and Seasonal Variations. *J Agr Food Chem.* **2002**, 50(17): 4861–66.

Mahadevan, S.; Park, Y. Multifaceted Therapeutic Benefits of *Ginkgo Biloba L.*: Chemistry, Efficacy, Safety, and Uses. *J Food Sci.* **2007**, 73(1): 14–19.

Marwick, C. Growing Use of Medicinal Botanicals Forces Assessment by Drug Regulators. *JAMA: J Am Med Assoc.* **1995**, 273(8): 607.

Mattéi, A.; Pierre, R.; Didier, S.; Cyril, F.; Michel, R.; Henri, B. Liver Transplantation for Severe Acute Liver Failure after Herbal Medicine (*Teucrium Polium*) Administration. *J Hepatol.* **1995**, 22(5): 597.

Mbiantcha, M.; Kamanyi A.; Teponno, R.B.; Tapondjou, A.L.; Watcho, P.; Nguelefack T. B. Analgesic and Anti-Inflammatory Properties of Extracts from the Bulbils of *Dioscorea Bulbifera L.* Var Sativa (Dioscoreaceae) in Mice and Rats. *Evid Based Complement Alternat Med.* **2011**, 2011: 912935.

Nash, K.M.; Shah, Z.A. Current Perspectives on the Beneficial Role of *Ginkgo Biloba* in Neurological and Cerebrovascular Disorders. *Integr Med Insights.* **2015**, 10: 1–9.

Polymeros, D.; Demetrios K.; Vassilios T. Acute Cholestatic Hepatitis Caused by *Teucrium Polium* (Golden Germander) with Transient Appearance of Antimitochondrial Antibody. *J Clin Gastroenterol.* **2002**, 34(1): 100–101.

Qamar, W.; Khan, R.; Khan, A.Q.; Rehman, M.U.; Lateef, A.; Tahir, M.; Ali, F.; Sultana, S. Alleviation of Lung Injury by Glycyrrhizic Acid in Benzo(a)Pyrene Exposed Rats: Probable Role of Soluble Epoxide Hydrolase and Thioredoxin Reductase. *Toxicol.* **2012**, 291: 25–31.

Radad, K.; Gille, G.; Rausch, W.D. Use of Ginseng in Medicine: Perspectives on CNS Disorders. *Iran J Pharm Ther.* **2004**. 3: 30–40.

Rouse, J. Kava: A South Pacific Herb for Anxiety, Tension, and Insomnia. *Clini Nutr Insights.* **1998**, 6(10): 1–2.

Rowe, A.; Zhang, L.Y.; Ramzan, I. Toxicokinetics of Kava. *Adv Pharmacol Sci.* **2011**, 2011: 1–6.

Saha, S.; Ova, G.N.; Ghosh, D.; Flešková, D.; Capek, P.; Ray, B. Structural Features and in Vivo Antitussive Activity of the Water Extracted Polymer from *Glycyrrhiza Glabra. I J Biol Macromol.* **2011**, 48(4): 634–38.

Seervi, C.; Kirtawade, R.; Dhabale, P.; Salve, P. Ginseng-Multipurpose Herb. *J Biomed Sci Res.* **2010**, 2(1): 6–10.

Senior, K. Herbal Medicine under Scrutiny. *The Lancet.* **1998**, 352(9133): 1040.

Shekelle, P.G.; Hardy, M.L.; Morton, S.C.; Maqlione, M.; Mojica, W.A.; Suttorp, M.J.; Rhodes, S.L.; Junqviq, L.; Gagné, J. Efficacy and Safety of Ephedra and Ephedrine for Weight Loss and Athletic Performance A Meta-Analysis. *J Am Med Assoc..* **2003**, 289(12): 1537–45.

Subasini, U.; Thenmozhi, S.; Sathyamurthy, D.; Vetriselvan, S.; Rajamanickam, G.V.; Dubey, G.P. Pharmacognostic and Phytochemical Investigations of *Dioscorea Bulbifera L. I J Pharm Life Sci.* **2013**, 4(5): 2693–2700.

Syphard, A.D.; Brennan, T.J.; Keeley, J.E. Chaparral Landscape Conversion in Southern California. Valuing Chaparral. Springer Series on Environmental Management. *Springer, Cham.* 2018, 323–46

Tanaka, H.; Kaneda, F.; Suguro, R.; Baba, H. Current System for Regulation of Health Foods in Japan. *Jpn Med Assoc J.* **2004**, 47(9): 436–50.

Ulbricht, C.; Basch, E.; Vora, M.; Sollars, D. Chaparral Monograph. *J Herb Pharmacother.* **2003**, 3(1): 121–33.

Wang, Z.; Li, J.; Ji, Y.; An, P.; Zhang, S.; Li, Z. *Chrysanthemum Indicum* Ethanolic Extract Inhibits Invasion of Hepatocellular Carcinoma via Regulation of MMP/TIMP Balance as Therapeutic Target. *Oncol Rep.* **2010**, 23(2): 413–21.

Wu, L.U.; Gao, H.Z.; Wang, X.L.; Ye, J.H.; Lu, J.L.; Liang, Y.R. Analysis of chemical composition of *Chrysanthemum indicum* flowers by GC/MS and HPLC. *J Med Plant Res.* **2010**, 4(5): 421–26.

Yu, Z.L.; Liu, X.R.; McCulloch, M.; Gao, J. Anticancer Effects of Various Fractions Extracted from *Dioscorea Bulbifera* on Mice Bearing HepA. *Zhongguo Zhong Yao Za Zhi.* **2004**, 29(6): 563–67.

CHAPTER 4

Herbal Therapies

H. SHAHRUL[1*], M. L. TAN[2], A. H. AUNI[1], S. R. NUR[3], and S. M. N. NURUL[1]

[1]*Oncological and Radiological Sciences Cluster, Advanced Medical and Dental Institute, Universiti Sains Malaysia, Kepala Batas, Penang 13200, Malaysia*

[2]*Healthy Lifestyle Sciences Cluster, Advanced Medical and Dental Institute, Universiti Sains Malaysia, Kepala Batas, Penang 13200, Malaysia*

[3]*Faculty of Applied Sciences, Universiti Teknologi Mara (UiTM), Arau Campus, Arau Perlis 02600, Malaysia*

Corresponding author. E-mail: shahrulbariyah@usm.my

ABSTRACT

Herbs have been used traditionally for centuries worldwide. Some regions of Asia (East Asia, Southeast Asia, and Asia-Pacific), America, and Mediterranean are known for their diversity in plant taxonomies. There are many herbal plants that are locally used to treat ailments but have not been fully explored of their healing properties. This chapter provides an overview of usage of herbal therapies for various organ systems in human. Cultural practices in different regions have led to the discovery of numerous herb-based therapies. As such, the chapter highlights several local herbs and their mechanisms of action. It also describes about some plants that are used among the community but need to be further explored of their potential benefits to human health.

4.1 INTRODUCTION

Herbs provide nutrients for a healthy mind and body. Tropical and subtropical continents have primarily forest. Terrestrial herbs are important elements of tropical forests. However, there is limited research on their diversity patterns and effects of different intensities of forest use (Gómez-Díaz et al., 2017). Herbal medicines include herbs and their products that contain as active ingredients or combinations (WHO, 2018). Approximately half (125,000) of the flowering plant species live in the tropical forests. Tropical rain forests continue to be a reservoir of therapeutic plant species. A majority of people on this planet depend on their cultural materia medica (plant products and other materials) for their everyday healthcare needs. It is reported that one quarter of all medical prescriptions are plants or plant-derived synthetic analogs. Furthermore, almost 80% of the population, primarily those of developing countries, rely on plant-derived medicines for their healthcare (Gurib-Fakim, 2006). According to reports, herbs provide numerous benefits to health. Herbs are consumed for certain ailments by local communities. The chapter highlights the benefits of various plants to specific organs and different cultural practices.

4.2 HERBS AS CARDIOPROTECTIVE

Cardiovascular disease remains as major cause of death worldwide. Cardiovascular diseases (CVDs) comprise coronary heart disease, cerebrovascular disease, rheumatic heart disease, heart failure and related diseases. CVDs cause the death of an average17.9 million people every year, constituting 31% of all global deaths. Most of the CVD related deaths are related to heart attacks and strokes (WHO, 2018). Epidemiological studies reported risk factors such as environment and genetic origin were linked to atherosclerosis, namely increased blood pressure, elevated fasting blood glucose, hypertriglyceridemia, hypercholesterolemia, central obesity, smoking and sedentary lifestyle (Hajar, 2017; Smith, 2007). A major complication is the development of atherosclerotic plaque that results in myocardial infarction, a multiple consequence of foam cells formation, oxidation of low-density lipoprotein (LDL), production of free radicals, and disruption of endothelial function in artery.

The interest in phytonutrients from herbs and plants with cardioprotective effects has been gaining much attention. Plant bioactive compounds contain multitude of cardioprotective effects, indicating potential applications in managing CVDs. Compounds that have extensively studied include berberines from the genus *Berberis* (Wang et al., 2014), tanshinone IIA (Tan IIA), and salvianolic acids (Sal B) from the *Salvia miltiorrhiza* (Chang et al., 2016a) *and* astragalosides from *Astragalus membranaceus* (Li et al., 2017b).

Other plants that have also demonstrated cardioprotective effects include *Crocus sativus* (saffron) (Efentakis et al., 2017), *Platycodon grandiflorum* (Hao et al., 2017), *Morus alba* L. and *Schisandra chinensis* (Kim et al., 2017), *Delonix regia* (Wang et al., 2016), and many more.

Berberine is an alkaloid with pharmacological effects (Neag et al., 2018). Berberine is found in numerous herbs; however, the major natural source of berberine is derived from Berberis. Alkaloid content in its bark is more than 8% with berberine as main alkaloid (Arayne et al., 2007; Imanshahidi and Hosseinzadeh, 2008). Plants rich with berberine content are used traditionally worldwide for healing ailments. Interestingly, its cardiovascular protective role includes hypotensive effects, inotropic, anti-inflammatory effects, dilation of coronary artery, anticoagulation, lowering pulse, and lowering of elevated low-density lipoprotein cholesterol (Neag et al., 2018). As a cardioprotective compound, berberine up-regulates endothelial nitric oxide synthase (eNOS) mRNA expression. It also increases production of nitric oxide (NO) from arginine. This is facilitated by the activity of eNOS. It is reported to play an important role in vasodilation (Wang et al., 2009). NO mediates the endothelium-dependent relaxation in large conduit arteries. Abnormal production of NO would cause vascular and endothelial dysfunction (Furchgott, 1999; Tawfik et al., 2006). Apart from that, the accumulation of lipid-laden foam cells initiates the progression of atherosclerosis because of augmented inflammation and impaired cholesterol metabolism within vascular walls. Berberine's anti-atherogenic effects are manifested by suppression of ox-LDL-induced foam cell growth and cholesterol accumulation in macrophages by regulation of surtuin 1 (SIRT1) and peroxisome proliferator-activated receptor-γ (PPAR-γ) (Chi et al., 2014). Accumulation of fatty acid is reduced by SIRT1. It would repress PPAR-γ which is involved in atherogenesis (Picard et al., 2004). Overexpression of SIRT1 could hinder atherosclerosis by improving vascular function (Zhang et al., 2008b).

In animal studies, berberine protects the heart from ischemia-reperfusion injury by mediating its effect through AMPK activation, AKT phosphorylation, and GSK3βinhibition in the nonischemic areas of the heart (Chang et al., 2016b). AMP-activated kinase (AMPK) is a stress responsive kinase that plays an important function in cellular metabolism. It provides protection in ischemic conditions by regulating the metabolism of carbohydrate and lipid, organelles, and cell death (Zaha et al., 2016). The intrinsic activation of AMPK is critical to prevent excess mitochondrial reactive oxygen production and consequent JNK signaling during reperfusion, thereby protecting against the mitochondrial permeability transition pore (mPTP) opening, irreversible mitochondrial damage and myocardial injury (Zaha et al., 2016). Interestingly, berberine's protective effects on cardiac failure were demonstrated in human clinical trials. Patients with NYHA (New York Heart Association) III/IV cardiac failure who had left ventricular ejection fraction (LVEF) and premature ventricular contractions and/or ventricular tachycardia were given berberine (1.2 g/day). After treatment, there was an increase in LVEF and decrease in the frequency as well as complexity of premature ventricular contractions (Zeng and Zeng, 1999). Patients with chronic congestive heart failure (CHF) were either given berberine 1.2-2.0 g/day or conventional treatment. After treatment with berberine, there was a significant increase in LVEF, exercise capacity, dyspnea–fatigue index, and presented with the reduction in frequency and complexity of ventricular premature complexes in comparison to the control group. Furthermore, mortality was significantly decreased in the berberine-treated patients during long-term follow-up. There was an improvement in quality of life of patients with CHF who are on treatment with berberine (Zeng et al., 2003).

S. miltiorrhiza is from the *Labiatae* family. Addressed as "Danshen" in traditional Chinese medicine, it is used to treat CVDs (Wang et al., 2007), ischemic heart disease (Hung et al., 2015b), and ischemic stroke (Hung et al., 2015a). This medicinal herb exhibits protective effect on cardiac myocytes against oxidative stress (Fu et al., 2007), modulatory effect on endothelial cell permeability, and inhibitory effect on platelet aggregation (Liu et al., 2011). It also prevents damage to human umbilical vein endothelial cells from homocysteine-induced endothelial dysfunction (Chan et al., 2004). Among the many chemical components found in this plant, tanshinone IIA (Tan IIA) and salvianolic acid B (Sal B) are the major components with pharmacological activities.

As a cardioprotective compound, Tan IIA had anti-atherosclerotic effect as it caused the reduction in the expression of vascular cell adhesion molecule-1 (VCAM-1), intercellular adhesion molecule-1 (ICAM-1), and CX3CL1. It inhibited the NF-κB signaling pathway in vascular endothelial cells (Chang et al., 2014). Treatment with Tanshinone IIA had significant inhibitory effect compared to *S. miltiorrhiza* extract alone on adhesion of monocytes to vascular endothelial cells. Besides, *S. miltiorrhiza* root extract also has antioxidant activity. Its hydrophilic components, protocatechuic aldehyde and Sal B could inhibit the TNF-α-induced expression of ICAM-1, VCAM-1, and the NF-κB and activator protein-1 DNA binding activities in human umbilical vein endothelial cells (Zhou et al., 2005). Adhesion molecules play a crucial function in the progression of atherogenesis.

Endothelial cells produce the adhesion molecules after stimulation with various inflammatory cytokines.

In hypercholesterolemic rabbits, Sal B showed the inhibition of LDL oxidation and neointimal hyperplasia by halting reactive oxygen species (ROS) generation. Similarly, in human aortic endothelial cells, Sal B reduced oxidative stress, LDL oxidation, oxidized LDL-induced cytotoxicity, and ROS production (Yang et al., 2011), whereas the mixture of Sal B and ginsenoside Rg1 increased the viability of cardiac myocytes and reduced infarct size. This caused improvement in the functional parameters of the heart against I/R injury in rats (Deng et al., 2015). Another study also reported that compounds present in *S. miltiorrhiza* improved the I/R-induced vascular damage synergistically (Han et al., 2008). Reperfusion of ischemic tissue after MI would supply oxygen and nutrients for the tissues; however, it also causes I/R injury and generation of ROS, as well as inflammatory mediators (Carden and Granger, 2000). Findings suggest that Tan IIA may also have protective effect on myocardial I/R injury by several mechanisms such as ROS production inhibition, lowering of the oxidative stress level, and also the expression of high mobility group box B1(HMGB1) protein (Hu et al., 2015). It induced vasodilation in coronary arterioles via up-regulation of NO and activation of conductance Ca(2+) activated K(+) channels (BK_{Ca})channels (Wu et al., 2009). High conductance BK_{Ca} channels in vascular smooth muscles will control the vascular tone. It will regulate membrane potential level and Ca(2+) influx via voltage gated Ca^{2+} channels (Yang et al., 2008).

So far, clinical trials involving medicinal herb were done using the combination of herbs. Meta-analysis of trials was done to compare the effect of danshen dripping pills and isosorbide dinitrate among patients with angina pectoris. Findings revealed that danshen was more effective than isosorbide dinitrate (Jia and Leung, 2017; Jia et al., 2012). In another clinical trial, *S. miltiorrhiza* and *Puerariae lobata* medicinal herbs were combined as a Traditional Chinese Medicine (TCM) preparation and given to postmenopausal women with early hypercholesterolemia for 12 months. It had significant anti-atherosclerotic effects (Kwok et al., 2014). There was an improvement in the carotid intima thickness, the LDL and total cholesterol levels were lowered, and the treatment was well tolerated (Kwok et al., 2014). In another randomized, double-blind, 12-week, active-controlled, parallel-group study, a fixed-dose Fufang Danshen extract capsule containing the purified *S. miltiorrhiza* extract, mixed with the *Rhodiola rosea* extract, chysanthemum extract, and *Pueraria* extract as a TCM preparation was administered to patients with uncontrolled essential hypertension (Yang et al., 2012). The outcome of this trial showed that blood pressure lowered to a greater extent than placebo while maintaining an acceptable safety profile (Yang et al., 2012).

Radix Astragali Mongolici is a root of *A. membranaceus* (Fisch.) Bunge. It belongs to the family of Fabaceae. This plant part is known as a qi-tonifying drug in China (Li et al., 2017a). The constituents of Radix *Astragali Mongolici* include saponins, polysaccharides, and flavonoids. Astragalosides (AGs) are a vital group of triterpenoid saponins and constitute approximately 0.13% of *Radix Astragali Mongolici*. They are major component to be responsible for its pharmacological and therapeutic efficacy (Yu et al., 2007). Pharmacological studies indicate they exhibit antioxidant (Li et al., 2017a), anti-inflammatory (He et al., 2014) and reduces endothelial cell injury and atherosclerosis (Zhu et al., 2013). Among more than 40 constituents of Astragalus saponins, astragaloside IV (ASI) is the main compoundand shows pharmacological activities. It is used as a quality-control marker component of Huangqi in the Chinese Pharmacopeia (Ren et al., 2013). Another component such as Astragalus polysaccharide (APS) is also widely investigated for various properties.

ASI has protective effects in multiple human disorders including CVDs, hepatitis, kidney disease, and skin diseases (Ren et al., 2013). The cardioprotective effect of ASI was investigated extensively. It plays an important role in the treatment of CVDs by regulating the myocardial energy metabolism, antioxidation, and anti-apoptosis effect (Tu et al., 2013, Yin et al., 2014b). It could prevent LPS-induced injury in cardiomyocytes by increasing the activities of antioxidant enzymes, inhibiting lipid peroxidation, and down-regulating inflammatory mediators (Wang et al., 2015). ASI was also found to reduce doxorubicin-induced cardiomyocyte loss and markedly ameliorated doxorubicin-caused cardiomyocyte dysfunction via restoring the beating cell ratio and beating rate in cardiomyocytes (Jia et al., 2014). Furthermore, it caused reduction in ROS production and lactate dehydrogenase (LDH), creatine kinase-MB isoenzyme (CK-MB) and cytochrome c (CytC) release, and restored the reduced ATP level, succinate dehydrogenase (SDH) and ATP synthase activities induced by doxorubicin, suggesting that ASI significantly attenuated DOX-induced mitochondrial damage and dysfunction. All these observations indicate that ASI appears to be an efficient cardioprotective agent for cardiotoxicity due to doxorubicin (Jia et al., 2014).

In addition to ASI, *Astragalus polysaccharide* (APS) shows similar cardioprotective properties. Previous studies showed that APS could improve autophagic flux and heart function by regulating AMPK/mTOR pathway in doxorubicin-induced cardiotoxicity. Doxorubicin as a broad-spectrum anticancer chemotherapeutic drug causes the generation of ROS and DNA mutagenesis. It was also linked with cardiotoxicity (Smith et al., 2010; Takemura and Fujiwara, 2007). In a separate study, APS was reported to protect cardiomyocytes from doxorubicin-induced oxidative stress. It reduced ROS generation and apoptosis through the PI3K/Akt signaling pathway. It also impairs p38MAPK activation, hence attenuates cell inflammation (Cao et al., 2014). Replacement of differentiated cardiomyocytes with fibrotic tissue would lead to transition from cardiac hypertrophy to functionally decompensated heart failure (Basuray et al., 2014), Meanwhile, activation of the PI3K/Akt signaling pathway is needed for cardiomyocyte cell survival (Walsh, 2006).

In clinical trials, Astralagus as Traditional Chinese Medicine (TCM) preparation helps alleviate calcium overload-induced myocardial damage and improve both systolic and diastolic functions of heart in chronic heart failure (CHF) patients (Yang et al., 2010). After therapy, the level of TNF-α in patients given the TCM preparation was lower compared with the conventional treatment group. Findings showed that the improvement in cardiac function was negatively correlated with TNF-α level (Yang et al., 2010). Another study investigated effects of different doses of *A. membranaceus* on diastolic function in postmenopausal hypertensive women with metabolic syndrome. The TCM therapy improved diastolic function in these patients probably through its antioxidant properties, anti-inflammatory, and endothelial protection (Li et al., 2018). In previous experimental study in patients with CHF, ASI injection alleviates symptoms of chest distress such as dypsnea and reinforced the capability to exercise (Luo et al., 1995). The drug injection could improve the left ventricular modeling and ejection function among these patients (Luo et al., 1995). *A. membranaceus* as TCM preparation could also strengthen the left ventricular function in acute myocardial infarction patients (Chen et al., 1995). In addition, previous clinical trials revealed that *A. membranaceus* injection along with the conventional therapy appeared to be more efficacious than conventional therapy only for treating viral myocarditis (Piao and Liang, 2014).

A wide range of phytochemicals and herbal plants exhibit cardioprotective effects. The key pathways modulated by phytochemicals in cardiomyocytes usually involve the oxidative stress, inflammatory and cell death signaling pathways. However, the potentials of herbal plants and active compounds need to be explored for clinical utility. Well-designed clinical trials involving these TCM preparation, pharmacokinetic and pharmacodynamics profiles and dosage remain as issues to be addressed.

4.3 HERBS FOR GASTROINTESTINAL SYSTEM

According to World Health Organization, 70%–80% of the human population, mostly in the developing countries, relies on herbal therapies as primary healthcare in 2008. The knowledge related to traditional uses of herbs is totally in the custody of elder community members and local herbalists (Aziz et al., 2017). Common gastrointestinal disorders are stomach/abdominal pain, diarrhea, dysentery, gastroenteritis, constipation, and vomiting. The Asia-Pacific region has diversity in ethnicity, culture, and economic development. It comprises some of the least and most developed nations. In this region, gastrointestinal diseases are reported to be common. It is reported that the prevalence, presentation, and management vary considerably within region. Emerging evidence show the important role of gut microbiota in gastrointestinal health. Geographic variations in the composition of the gut microbiota may contribute to variations in both the prevalence and response to therapy of specific diseases (Ghoshal et al., 2018). Outbreaks of cholera in Haiti, Pakistan, and Zimbabwe suggest that global action plans against cholera are failing (Ryan, 2011). Besides that, diarrhea/dysentery/cholera/thyphoid outbreaks were reported in several other countries (Kamble, 2014; Manjusree et al., 2017; Sow et al., 2016).

The tropical countries have rich diversity of plant kingdom, whereby the Brazilian flora is one of the most diverse. It accounts for roughly 20% of plant biodiversity in the world. Fabaceae is among the largest families with ethnopharmacological importance. It is used by local communities living along the Savanna. Their local knowledge has been of great value for supporting phytochemistry and pharmacology studies for drug discovery (Macêdo et al., 2018). Medical species of the Fabaceae family include *Stryphnodendron ortundifolium* and *Bowdichia virgiloides*. *Stryphnodendron* rotundifolium had therapeutic properties. Its most common uses are for general inflammation, respiratory and gastrointestinal disorders, gynecological inflammation, healing and for the treatment of injuries, using the bark and stem bark (Macêdo et al., 2018).

In Nepal, similar families are used to treat gastrointestinal disorders. They include Fabaceae, Asteraceae, Lamiaceae, Rosaceae, and Ranunculaceae. These plants are used in Africa, New Zealand and Ecuador based ethnobotanical studies (Bennett and Husby, 2008, Saslis-Lagoudakis et al., 2011). It is also found in North and South America. It is reported that the indigenous group living in the Canadian boreal forest used Asteracea, Rosaceae and Ranunculaceae for gastrointestinal diseases. It was taken together other families from Liliaceae, Ericaceae, Betulaceae, Caprifoliaceae, and Salicaceae (Uprety et al., 2012). An ethnobotanical review of the Mapuche medicinal flora of South America (Argentina and Chile) showed that Asteraceae, Rosaceae, Solanaceae, Apiaceae and Fabaceae were frequently used in gastrointestinal problems (Molares and Ladio, 2009).

Artemisia dracunculus L (Tarragon) is from the Asteracaceae and it has been used to treat gastrointestinal problems in several parts of India. The leaves of *Artemisia scoparia* are used for gastric disorder, treating intestinal parasites and indigestion. *Matricaria recutita* L. (syn. *Matricaria chamomilla L., Chamomilla recutita (L.)* consumed for various gastrointestinal conditions. *Senecio rufinervis* and *Tanacetum gracile* are used for stomach ailment and anti-helmintic, respectively. Locally, the powder of *Tanacetum longifolium* root is used for stomach ailment (Joshi et al., 2016).

In Nepal, despite rich diversity in herbal plants, there is still mortality occurring in suburban and rural areas due to diarrhea, dysentery, and cholera. Some of the reasons were treatment of gastrointestinal disorders were not necessarily available everywhere and limited information on the required herbal formulations against different disorders due. Furthermore, there was lack of information sharing within different communities (Rokaya et al., 2012). It was reported that medicinal plants and associated herbal knowledge was not uniformly distributed across the country (Rokaya et al., 2012). The second, probably most convincing explanation could be related to poverty, low sanitation standards and lack of effective communication (low schooling, lack of health workers, no guidelines available) in many rural areas (Rokaya et al., 2014).

A study was done based on interview with 200 participants who consumed a total of 104 fresh medicinal plant species within the last five years (Siew et al., 2014). Combination ingredients were often added to the preparation to complement the therapeutic effect of other plants. Many users believed that fresh plants were "cooler" in nature, in comparison to their dried counterparts. They had used Chinese red dates in brewed herbal preparations to counter the cooling effect of the plants. Majority of the users (171 users, 85.5%) stated that they did not experience any undesirable effect after ingesting fresh medicinal plants. Others (28 users, 14.0%) reported to have mild self-resolving symptoms with a total of 16 species of plants. There was no report of any significant adverse effect which required medical attention, except in one incident of hyponatremia. However, it was not clinically confirmed whether the adverse effect was due to *Clinacanthus nutans*. Most of the participants reported that their conditions improved or symptomatic relief was achieved (118, 59.0%), whereas 98 participants (49%) achieved a sense of well-being. Five users informed that their medical condition and state of well-being worsened after taking the plants, prompting them to stop the usage, whereas some participants failed to experience any effect from using the plants.

Weeds grow rampantly and commonly used as feed for livestock. However, the use of weeds for treatment of diseases is rare. In Thailand, a study was conducted to gather information on the types of herbal weeds used for treating gastrointestinal disorders and knowledge and Fidelity Level (FL) (Neamsuvan and Ruangrit, 2017). They found a total of 49 species in 46 genera and 28 families of weed used as therapies. The locals consumed weeds as herbs or the whole plant. It was taken either in the form of decoction or drinks. According to this study, highest FLs (100%) were in 12 plant species which included *Amaranthus spinosus*, *Amaranthus viridis*, *Alternanthera sessilis*, *Sauropus androgynus* and *Plantago major*. Generally, study indicated the presence of pharmacological activity in 31 species of weeds. However, 20 species of weeds had toxicity effect. Therefore, awareness of the use of herbs is necessary to ensure safety of users.

A recent study investigated lysophosphatidic acid (LPA) content in 21 herbs that are traditionally used in Japan (Afroz et al., 2018). Herbs that are taken for various digestive disorders were selected based on the descriptions in the oldest Chinese traditional herbal medicine book, the Shennong Ben Cao Jing. Effects of LPA and herbal lipids on NSAID-induced gastric ulcer were investigated. Findings suggest that peony root lipid contained highly concentrated LPA. It had an ameliorative effect on NSAID-induced gastric ulcer and enhanced prostaglandin E2 (PGE2) production in gastric cells. Study also showed that in addition of PGE2 enhancement, LPA protects against NSAID-induced acute cell toxicity and stimulates the proliferation of gastric cells. LPA2 in gastric mucosal cells could be involved in these LPA actions.

Furthermore, LPA could induce numerous cellular responses including proliferation, protection of cells from apoptosis, and migration of cells (Moolenaar et al., 2004). These cellular responses are mediated through six LPA-specific G-protein-coupled receptors, LPA1–6 (Yung et al., 2014). LPA is a phospholipid present in plants. It is a metabolic intermediate of de novo lipid synthesis and glycerophospholipid storage. In animals, it also has diverse cellular effects, that is, from brain development to wound healing through the activation of G protein-coupled LPA receptors (Lee et al., 2016a). In addition, it exhibits important actions in the mammalian gastrointestinal

(GI) tract (Tokumura, 2011; Yun and Kumar, 2015). These include inhibition of diarrhea, integrity of gastrointestinal tract epithelium and wound healing (Tanaka et al., 2005; 2009, Thompson et al., 2018).

Another condition which is treated with herbs is peptic ulcer, it is a major GI disorder that occurs due to an imbalance in mucosal offensive (gastric acid secretion) and defensive (gastric mucosal integrity) factors (Umukoro and Ashorobi, 2006). Among the factors causing peptic ulcer are *Helicobacter pylori* infection, smoking, drinking alcohol, and chronic ingestion of drugs (Zhang et al., 2010).

Herbal plants such as corydalis tuber and ginseng also contain LPAs (Lee et al., 2016a). A Chinese medicine called Antyu-san (AS) consists of corydalis tuber, licorice, fennel, cinnamon bark, amomum seed, minor of galangal and oyster shell. A mice model study indicated without corydalis tuber, AS did not show much anti-gastric ulcer activity. It had lower LPA level. This indicated that LPAs are the main active ingredients for anti-gastric ulcer activity. MALDI-TOF-MS analysis showed presence of LPA C18:2>LPA C16:0>LPA C18:1>LPA18:3>LPA18:0 in AS (Adachi et al., 2011).

Ginseng is a well-known traditional herbal medicine due to its numerous benefits. Although ginseng exhibits a variety of physiological and pharmacological activities, ginseng saponin (or ginsenoside) has no specific receptor. Therefore, ginseng saponin alone may not explain all of ginseng's actions (Nah et al., 2007). Recent studies found presence of unique form of LPAs, designated gintonin. It was co-isolated with ginseng proteins such as ginseng major latex-like protein 151 and ginseng major storage protein. The complex of LPAs and ginseng proteins could be involved in the physiological and pharmacological actions of LPAs as the free form of LPAs are labile to hydrolysis by lipid phosphate phosphatase (Salous et al., 2013). The protein component might protect LPAs from hydrolysis. They may also play roles in storage and transportion of LPAs to receptors at target organs. The most abundant LPA species in gintonin are LPA C18:2>LPA C16:0>LPA C18:1. The study also indicated that ginseng contained high LPAs content, where it was 10-fold than the amount present in corydalis tuber and other foodstuff.

Curcuma sp grows in tropical climates, where it requires substantial humidity for its growth. It is mostly grown in India, Southeast Asia, and tropical regions of Australia. The entire *Curcuma* genus comprises more than a hundred different species, which are distributed across different continents. Plants belonging to *Curcuma* genus are known for their antiallergic, anticancer, antidiabetic, anti-inflammatory, antivenom, cardioprotective, digestive stimulant, hepatoprotective, hypolipidemic, and neuroprotective properties. Curcumin is a naturally occurring polyphenolic phytochemical derived from *Curcuma longa Linn.* The chemopreventive mechanisms of curcumin have been extensively studied (Miura et al., 2015).

Gastric cancer evolves through multistep mechanism, in which *H. pylori* is linked with the disease. This organism has an association with histological gastritis, gastric atrophy, gastric cancer, and mucosa-associated lymphoma in the stomach (Sugiyama and Asaka, 2004). The current PPI-based triple regimens for treatment of *H. pylori* face uprising resistance problem demanding for the search of novel candidates. A study was done to determine the anti-*H. pylori* effect of 50 commonly used Unani (traditional) medicine plants from Pakistan that are widely utilized for the cure of gastrointestinal diseases (Zaidi et al., 2009). The study showed that the 70% aqueous-ethanol extracts of *Curcuma amada Roxb, Mallotus phillipinesis (Lam) Muell., Myrisctica fragrans Houtt.*, and *Psoralea corylifolia L.* had strong anti-*H. pylori* effects. They found the most potent bactericidal activity was exhibited by *M. phillipinesis (Lam) Muell.*

4.4 HERBS FOR ENDOCRINE SYSTEM

Many pharmaceuticals commonly used today are structurally derived from the natural compounds that are found in traditional medicinal plants (El-Tantawy and Temraz, 2017). For example, the antihyperglycemic drug metformin can be traced to use of *Galega officinalis* traditionally to treat diabetes (Bailey, 2017). There are numerous preclinical studies in animal models of the anti-diabetic properties of herbs like sage, basil, bay leaf, dill, coriander, cumin and fennel (Bower et al., 2016). *Rosmarinus* officinalis (rosemary) has diverse varieties and is native to coastal Mediterranean areas. It grows in many regions in Asia, Europe, and America. Its extract is reported to be rich with rosmarinic acid (Neveu et al., 2010).

Apart from rosemary, fenugreek (*Trigonella foenum-graecum*) is a legume and used as spice (Wani and Kumar, 2018). The seeds and green leaves of fenugreek are also used for medicinal application as cultural practice. It is used to increase the flavoring, coloring, and modifying the texture of food materials. Seeds of fenugreek spice have health benefits such as hypocholesterolemic, lactation aid, antibacterial, gastric stimulant, for anorexia, antidiabetic, galactogogue, hepatoprotective effect and anticancer. The beneficial physiological effects including the antidiabetic and hypocholesterolemic actions are mainly due to intrinsic dietary fiber constituent (Srinivasan, 2006). It also contains fiber, gum, other chemical composition and volatile contents. Fenugreek seeds contain 45.4% dietary fiber (32% insoluble and 13.3% soluble), and the gum contains galactose and mannose. The latter compounds may play a role in the lowering of glucose. The hypoglycemic effect of fenugreek was found in both humans and animal based studies in type 1 and type 2 diabetes mellitus (Roberts, 2011).

Diabetes is frequently linked with dyslipidemia, one of the risk factors in the development of CVDs (Ghorbani, 2013). A literature search of clinical trial on diabetic and without-diabetic subjects was done to investigate the efficacy of selected herbs (Ghorbani, 2013). Data indicated that consumption of *Nigella sativa* (black seed) oil along with hypolipidemic and hypoglycemic drugs of patients with metabolic syndrome significantly decreased FBG and LDL but increased the HDL level (Najmi et al., 2008). Lowering of serum lipids was observed in an uncontrolled trial of patients with Type 2 diabetes treated with black seed (Kaatabi et al., 2012). The seeds are used as a natural remedy for various ailments. Yet, further well-designed studies using placebo group are needed to ascertain the hypolipidemic effects of black seed in diabetic patients.

Another plant with anti-diabetes properties is the *Silybum marianum* (milk thistle). It is known for its hepatoprotective action. Lipid lowering effects of milk thistle seed extract, silymarin (a flavonolignan), was seen in diabetic patients (Huseini et al., 2006). In a two-month clinical trial, silymarin (200 mg thrice daily) caused decrease in fasting blood glucose (15%), total cholesterol (12%), triglycerides (26%), LDL (12%), serum glutamic oxalacetic transaminase (22%) and serum glutamic pyruvic transaminase (37%) in Type 2 diabetic patients who received conventional therapy (Huseini et al., 2006). The reduction in glucose, cholesterol and LDL was consistent with the results of another trial in which patients received the same dose of silymarin for four months (Ramezani et al., 2008).

Mints (*Mentha* species) are widely used as food, medicine, spice, and flavoring agents (Figure 4.1). According to the plant list database, this genus contains around 56 accepted taxa including 42 species distributed world-wide especially in temperate and subtemperate regions. *Mentha* species are also rich in phenolic compounds, especially in phenolic acids and flavo-noids (Bahadori et al., 2018). This genus exhibits anti-diabetes effects. Other examples such as *Mentha arvensis* and *Citrullus colocynthus* are common medicinal plants consumed by the Arabs. Both *M. arvensis* and *C. colocynthus* are used in controlling diabetes and cholesterol lowering herbal medicines in Arab countries. A study was done to evaluate the components present in wild growing Mentha (Al-Sabahi et al., 2016). Findings showed that menthol is the major component present in essential oil of *M. arvensis*. Whereas the seed, flesh, and peel of *C. colocynthus* have high contents of Na, K, Ca, Li, and N. These findings suggest that mints could contribute in regulating blood glucose level.

FIGURE 4.1   Mentha species.

Extracts of *Ocimum basilicum* L. (Lamiaceae) also commonly known as "Holy basil" have different physiological effects, including blood glucose lowering and hepato-protective properties (El-Beshbishy and Bahashwan, 2012). The plant is widely used traditionally in different cultures and also for culinary uses (Figure 4.2). Reported phytochemical components present in *O. basilicum* extract include linalool, methylchavikol, methyl cinnamate, linolen, rosmarinic acid, citral, eugenol, and geraniol (Lalko and Api, 2006; Opalchenova and Obreshkova, 2003; Renzulli et al., 2004). Similarly, extract of aerial parts has antidiabetic effects, possibly mediated by limiting glucose absorption through inhibition of carbohydrate metabolizing enzymes and enhancement of hepatic glucose mobilization. Oral intake may not predispose to the risk of hepatotoxicity in the short term. Further studies are warranted to evaluate effects of oral intake of the extract in diabetic patients.

FIGURE 4.2 *Ocimum basilicum.*

4.5 HERBS FOR SKIN THERAPIES

Skin disease is one of the most common ailments and a major concern worldwide. It is reported that between 30% and 70% of individuals are at high risk of having skin diseases (Hay et al., 2014). Approximately 1-7 in every 10 people that visit to a primary care physician are due to skin problems (Gupta et al., 2010). The detrimental effects of skin disease on health may cause physical incapacity and even could lead to death (Hay et al., 2014). It remains the 18th leading cause of health burden to worldwide (Hay et al., 2014; Karimkhani et al., 2017). There are many types of skin diseases and the most common include eczema, psoriasis, acne vulgaris, pruritus, scabies, fungal and bacterial skin diseases, abscess and nonmelanoma skin cancer (Hay et al., 2014).

Side effect caused by existing drugs used in treating skin disease has led to explore potential of natural remedies. There is need to discover alternative compounds from the natural sources. The use of some medicinal plants as a remedy in treating disease was recommended by the World Health Organization (Delfan et al., 2014). Natural products have been used for medical and therapeutic purposes over centuries. The use of traditional medicine by ethnic groups varies owing to the cultural differences. The demand for herbal medicines is increasing rapidly as it has lack of side effects (Gupta et al., 2010). More than 80% of world population are using traditional medicine for treatment. Almost one third of all drugs are used for treating skin disease and wounds (Delfan et al., 2014). Herbs are key player in treating skin disorders (De Wet et al., 2013), Tabassum and Hamdani, 2014. It is used traditionally for treating dermatological problem and widely exploit as added value in cosmetics product to treat acne, dandruff, dry and rough skin and other conditions (Narayanaswamy and Ismail, 2015). Herbalists prescribe various preparations of plants in treating itch, eczema, scabies and skin problems (Gupta et al., 2010; Sivaperumal et al., 2009).

One of the herbs used in dermatology is the *Acalypha indica* (Figure 4.3). It is a weed plant used by elder generations in many countries, particularly in Asia and Africa. It is a traditional medicine for treating parasites, scabies and other skin disorders (Ramalashmi et al., 2018). Even though the plant is known for its therapeutic purposes, some locals consume this plant as vegetable or as fried snack. The ethno-medicinal purposes of *A. indica* plant toward skin can be either from the leaves or the whole plant. It can be consumed directly or in a combination with other ingredients as a treatment. The preparation of the plants whether fresh or dry is an important aspect for determining its therapeutic efficacy (Zahidin et al., 2017). The leaves are the most abundant part and easy to be separated. It can be either eaten raw or in the form of decoction. In addition, the leaves are also used to treat gum and teeth disease, insect bites, pimples and wound healing (Zahidin et al., 2017). Application of *A. indica* topically can reduce soreness of the insect bites as this herb has the potential to reduce pain from inflammation and act as a potential analgesic drug (Sudhakar et al., 2016). The saponin constituent found in the

FIGURE 4.3   *Acalypha indica.*

plant is believed to contribute the anti-inflammatory properties of *A. indica.* Furthermore, the leaves contain anti-bacterial, anti-fungal and antioxidant properties which are useful in protecting the skin from external hazards and accelerate healing properties (Selvamani and Balamurugan, 2015). Alkaloid content in *A. indica* has anti-bacterial activity (Pradeep et al., 2014). This could be due to the presence of tannin, flavonoids, polyphenol, saponin and protein in the herbs. The antioxidant and antibacterial properties of the plant are exhibited through the inhibition of bacterial growth (Batubara et al., 2016). In some practices, the whole plant is consumed in treating mouth ulcers. The mixture of its leaves with oil or other herbs such as black cumin and *Cardiospermum halicacabum* can be applied to treat skin ailments (Zahidin et al., 2017). The fresh *A. indica* possesses natural phytochemicals including fatty acid, volatile compound and essential oil that are beneficial for numerous therapeutic activities. During the drying process, the fresh *A. indica* leaves produce a strong smell. This is due to the volatile compounds present in this plant. They may play a role in the healing process. Loss of 80% weight after the drying process shows the high moisture and volatile compound composition in this plant. However, the remaining phytochemical in the dried plant is still beneficial and provides therapeutic properties for dermatological disorders (Zahidin et al., 2017). The aqueous extract of *A. indica* caused cell death in dermal cancer cells. The extract induces apoptosis and cell death by interacting with the cell membrane proteins and inducing cellular leakage and finally leading to cell death (Banala et al., 2017).

Aloe vera (*Aloe barbadensis* Miller) is from the *Liliaceae* family (Figure 4.4). Aloe vera gel is derived from the mesophyll. It has been used as a therapy for centuries (Tanaka et al., 2015). It still remains as important part in treatment especially in contemporary cultures. The main feature is its high level of water content. The remaining consist of solid material which has more than 75 different potentially active compounds including water and fat-soluble vitamins, minerals, enzymes, simple/complex polysaccharides, phenolic compounds, and organic acids (Radha and Laxmipriya, 2015). Some of the traditional uses are for burn injury, eczema, cosmetics, inflammation, and fever. Many studies have reported the gel extracts could accelerate wound healing. This is supported by *in vitro* study, where the extracts stimulated the proliferation of several cell types. It is manifested by increase in rate of contraction of wound area and collagen synthesis (Khan et al., 2013). This could be due to mannose-6-phosphate which is present in the gel. This polysaccharide promoted the proliferation of fibroblast, production of hyaluronic acid and hydroxyproline in fibroblasts that is involved in extracellular matrix remodeling during wound healing (Liu et al., 2010). Besides that, the antibacterial and antifungal properties of Aloe vera provide a protective effect against dandruff on head (Hashemi et al., 2015). It contains anthraquinones active compound that is able to inhibit bacterial protein synthesis. Therefore, the bacteria growth is inhibited in the presence Aloe vera extract (Prueksrisakul et al., 2015). Polysaccharide also stimulates phagocytic leucocytes to destroy the bacteria. Reduction of eczema conditions of skin dryness and scaling were reported after applying the Aloe vera gel. The gel provided moisturizing effect that facilitates in wound healing (Khiljee et al., 2011).

FIGURE 4.4 *Aloe vera.*

Another type of traditional herbs for skin therapy is the *Centella asiatica* or commonly known as "pegaga" among the Malaysians (Figure 4.5). It comes from the Apiaceae family where its common name is Gotu kola or Indian pennywort (Bylka et al., 2014). It is used for thousands of years in Indian Ayurveda, Malaysian, China and other parts of Asia (Belwal et al., 2019). It is consumed as herb, spice, vegetable or juice. Besides that, is used as nutraceutical and cosmetic products. The plant grows several continents such as Asia, Oceania, Africa and America. The *C. asiatica* is applied for the treatment of dermatitis and skin lesions such as excoriations, burns, hypertrophic scars, or eczema (Park et al., 2017). Apart from that, it has been accepted by Indian pharmacopeia for its healing property to treat skin conditions such as leprosy, lupus, varicose ulcers, eczema, and psoriasis (Mahmood et al., 2016). The Thailand National List of Essential Medicines includes *C. asiatica* for its antipyretic and wound healing effects. It is also referred as the "champion herbal product" to be developed that contribute to the economy (Puttarak et al., 2017). Other medicinal properties reported include neuroprotective activities, anti-inflammatory, antipsoriatic, anti-ulcer, anticonvulsant, antidiabetic, antifungal, antioxidant, antiviral, antibac-terial, sedative, immune stimulant, cardio-protective, and hepato-protective properties (Belwal et al., 2019). The *C. asiatica* extract is used as the main ingredient in some cosmetic products (Bylka et al., 2013). It has anti-aging properties. It increases the proliferation of keratinocyte and fibrolast skin cell by up-regulating the expression of Aquaporin 3 as well as improves the production of Type I and III collagen (Yulianti et al., 2016). The extracts also exhibits wound healing properties based on numerous reports. Bioactive compound such as asiaticoside, asiatic acid and medacassic acid facilitate the wound healing effects by increasing peptidic hydroxyproline content, tensile strength, collagen synthesis, angiogenesis, epitheliazation and skin cellular proliferation (Belwal et al., 2019).

4.6 HERBS FOR LIVER DISEASES

Liver is a vital organ that is a center for metabolism of carbohydrates, proteins, lipids and the excretion of metabolites. It is also the site for metabolism and excretion of drugs and other xenobiotic from the body. Thus, providing protection against foreign substances by detoxifying and eliminating them. Liver diseases is a major health problem in countries with high endemicity. Cirrhosis is a late-stage liver disease which occurs when scar tissue replaces healthy tissue. Liver cirrhosis caused over one million deaths in 2010,

FIGURE 4.5 *Centella asiatica.*

which is equivalent to approximately 2% of all deaths worldwide (Mokdad et al., 2014). Hepatocellular carcinoma (HCC) is the most common liver cancer that begins at the hepatocytes. Hepatitis is a viral infection defined by inflammation of the liver, whereas alcoholic liver disease is a damage to liver due to alcohol abuse. Nonalcoholic fatty liver disease (NAFLD) is the accumulation of triglycerides within hepatocytes that exceeds 5% of liver weight, which most commonly results from metabolic syndrome on hepatic metabolism (Kneeman et al., 2012). *Silybum marianum* is known as milk thistle (MT). It is one of the earliest and extensively studied for treating liver diseases. It grows as a stout thistle in areas with rock soils, with large purple-flowering heads. The leaf is characterized by its milky veins, from which the name of plant is derived (Abenavoli et al., 2010). MT is also known as blessed thistle, bull thistle, fructus cardui mariae, fructus silybi mariae, holy thistle, Lady's milk, Lady's thistle, marian thistle, St. Mary thistle, mild marian thistle, milk thistle, pternix, Silberdistil, silibinin, silybe, silybon, silybum, silymarin, thistle, and thistle of the Blessed Virgin. It has been used medicinally since the 4th century BC. Ancient practitioners used its extract to treating hepatitis, cirrhosis, and jaundice. Besides that, it is used for protecting the liver of chemical and environmental toxins from snake bites, insect stings, mushroom poisoning and alcohol. Its active component is a lipophilic extract which is derived from its seeds. It consists of three flavonolignan isomer of silymarin (Rambaldi et al., 2005). Silymarin is extracted from dried MT seeds because it exists in higher concentrations compared to other parts of the plant (Abenavoli et al., 2010). The active constituents of silymarin consist of silibinin, isosilybinin, silydianin, and silychristin. Silibinin is the major and most active component in silymarin, at about 60%–70% (Saller et al., 2001). Silymarin is used worldwide for many years as an alternative medicine for treatment of hepatic diseases. It prevents lipid peroxidation by scavenging free radical scavenging and increasing the glutathione (GSH) level. It regulates membrane permeability and increases membrane stability in the presence of xenobiotic damage. Besides that, it also regulates nuclear expression through steroid-like effects. Silymarin inhibits the transformation of stellate hepatocytes into myofibroblasts that mediate the deposition of collagen fibers that lead to liver damage (Polyak et al., 2010; Pradhan and Girish, 2006). Many studies suggest that silymarin has the potential of treating chronic liver diseases, especially NAFLD.

Another plant consumed traditionally for liver disease is the Gale of the wind, or scientifically known as *Phyllanthus niruri.* It belongs to the Euphorbiaceae family with a widespread distribution across tropical and subtropical continents of Asia, America and China. Another common name is stone breaker or seed-under-the-leaf. The Spanish name "chanca piedra" means "stone breaker or shatter stone." The taste is bitter and it exerts laxative effect. In South America, the plant has been used to eliminate gallbladder and kidney stones, treatment of gall bladder infections and cardiovascular problems (Kamruzzaman and Hoq, 2016). Meanwhile, in Malay traditional medicine, *P. niruri* or "dukong anak" is traditionally used for kidney disorders and cough (Burkill, 1966). It is used in the mixture of various ayurvedic formulations for gallstones and jaundice. In South India, it is called Bhumyamalaki and used for treating constipation, gonorrhea, and syphilis. Wheras in northern India, it is locally known as "pitirishi." It is used as remedy for asthma, bronchitis and even tuberculosis. Sometimes the young shoots used for infusion in treating chronic dysentery. In China, *P. niruri* or "zhu zi cao" is locally consumed for liver injury secondary to various hepatotoxic agents (Lee et al., 2016b). Studies show that the leaves and fruit extracts contain antioxidant effects. This was demonstrated by the inhibition of membrane lipid peroxidation and ROS following treatment with aqueous and methanolic extracts of *P. niruri*. It also showed hepatoprotective effect because it markedly decreased CCl4-induced hepatotoxicity in rats, based on the elevated serum enzymes, glutamate oxaloacetate transaminase (GOT) and glutamate pyruvate transaminase (GPT) (Harish and Shivanandappa, 2006). Besides that, the hepatoprotective nature was contributed by the presence of various bioactive compounds like lignans, alkaloids, terpenoids and tannins especially phyllanthin and hypophyllanthin, quercetin, astragalin, gallic acid, ellagic acid and corilagin. It also contains glycosides, flavonoids, flavonols, polyphenols, phenylpropanoids, which are well-known antioxidants (Markom et al., 2007; Rajeshkumar et al.; 2002, Syamasundar et al., 1985). Many studies have investigated the possible mechanism of hepatoprotetive effects of *P. niruri*. Recently, Amin et al. (2013) proposed that *P. niruri* regulates the transforming growth factor-β (TGFβ), collagenα1 (Collα1), matrix metalloproteinase-2 (MMP2), and tissue inhibitor of matrix metalloproteinase-1 (TIMP1) genes expression by down-regulating their mRNA expression level in rats with liver cirrhosis.

4.7 HERBS FOR REPRODUCTIVE SYSTEM

Malaysian plants are diverse and have been used widely valued for their therapeutic properties for many generations (Nadia et al., 2012). Among the common local herbs for reproductive health are *Labisia pumila* (Kacip

Fatimah), *Eurycoma longifolia Jack* (Tongkat Ali) and *Quercus infectoria* (Manjakani). These plants are show similarity in phytochemical properties that are protective against diseases and also reproductive health.

Kacip Fatimah or its scientific name *Labisia pumila*, also known by the locals as Selusuh Fatimah, Rumput Siti Fatimah, Akar Fatimah, Pokok Pinggang, and Belangkas Hutan (Ali et al., 2010). *Labisia pumila* var. *alata* is traditionally consumed by the Malay women to maintain healthy female reproductive function and as postpartum medicine in the form of water extract (Abdullah et al., 2013). This is due to the presence of phytoestrogens that acts as primary female sex hormone (Abdullah et al., 2013). Several reports have demonstrated the estrogenic activity of *L. pumila*. The water extract of *L. pumila* has the ability to displace estradiol and bind to antibodies raised against estradiol (Wahab et al., 2011). Apart from that, *L. pumila* is suggested as a potential alternative agent for hormone replacement therapy in postmenopausal women (Abdul Kadir et al., 2012). Findings suggested it resembled the effect of estrogen in the ovariectomised rats (Wahab et al., 2011). It also causes increase in the level of estrogen and testosterone levels. In contrast, it suppresses follicle stimulating hormone (FSH) and luteinizing hormone (LH). These abilities may be contributed due to presence of triterpene, saponins, and ardisiacrispin A (Avula et al., 2010). However, these activities have only been studied *in vitro* using cell and animal model studies. These findings require further investigation in order to be applicable to humans since *L. pumila* as an alternative treatment for estrogen deficient or postmenopausal-related diseases.

Eurycoma longifolia Jack popularly known as 'Tongkat Ali' is an indigenous plant species to South-East Asian countries especially Peninsular Malaysia and Indonesia (Bhat and Karim, 2010). Its root is the most valuable component of the plant as it contains important bioactive substances such as quassinoids, eurycomaoside, eurycolactone, eurycomalactone and eurycomanone (Ayob et al., 2013). These compounds were reported to contribute to its remedial properties for the treatment of aches, persistent fever, malaria, sexual insufficiency, dysentery, glandular swelling, and as health supplements (Rehman et al., 2016). Its root extract is administered for regaining energy, vitality, improving blood flow after child birth (Rehman et al., 2016). The rich extract of quassinoids could enhance the production of testosterone, improvised the spermatogenesis and enhanced fertility (Low et al., 2013). However, roles of *E. longifolia* on the female reproductive functions and disorders are the least studied compared in the males (Rahman et al., 2017). A previous study on *E. longifolia* extract stated that it ameliorates chronic effects of androgens and in sustaining female fertility in testosterone-induced amenorrhoea and polycystic ovary-like conditions in rats (Abdulghani et al., 2012). *E. longifolia* is a safe traditional herbal medicine, which has gained high demand and increased production of the Herbal Medicinal Products (HMP) (Abubakar et al., 2017). Although evidences show the medicinal values of *E. longifolia* active constituents, trials are required on its health benefits and safety aspects as a new drug candidate in the future.

Quercus infectoria is another popular medicinal plant which is used traditionally in postpartum. In Malaysia, the galls is known as "manjakani" nut. It has been studied over the years and most well accepted health supplement used during postpartum care (Shrestha et al., 2014). It has high content of active constituents which comprise tannins (50% and 70%), gallic acid, syringic acid, ellagic acid, sitosterol, amentoflavone, hexamethyl ether, isocryptomerin, methyl betulate, methyloleanate and hexagalloyl glucose (Iminjan et al., 2014). This "magic" nut is utilized as an astringent, anti-inflammatory agent, antiseptic and antidiarrheal agent (Baharuddin et al., 2015; Digrak et al., 1999) and also as uterotonic agents in the modulation of uterine contractility (Noureddini et al., 2018). However, further investigations are needed for production of standardized *Q. infectoria* gall extract for safe consumption and elucidating its molecular mechanisms.

4.8 HERBS FOR NEUROPROTECTION

Neurological diseases are characterized by progressive dysfunction and death of neurons and the main degenerative disorders manifest predominantly as movement disorders, those of cognition or a mixture of both (Vajda, 2002). Movement disorders include akinetic and rigid forms, predominantly extrapyramidal deficits, hyperkinetic dysregulation of movement, ataxic with features of cerebellar ataxia, and motor neuron disorders (Vajda, 2002). Neuronal damage may be contributed by excitotoxicity, cerebral ischaemia and target deprivation. Excitotoxicity refers to an excessive activation of neuronal amino acid receptors, leading to apoptotic DNA fragmentation and cellular fragmentation (Mark et al., 2001; Vajda, 2002). On the other hand, cerebral ischemia is a condition where cessation of blood supply to the brain tissue leads to necrosis and apoptosis (Lee et al., 2001). Neurological disease is reported as the third most common cause of death. It is the leading cause of adult neurological disability (Carter et al., 2007). Among the important pathological mechanisms include inflammatory reaction, blood-brain barrier (BBB) disruption, oxidative stress, and neuronal apoptosis. They are widely considered as the four major therapeutic targets for acute ischemic stroke (Lalkovičováand Danielisová,

2016). Competition between neurons for innervation of their targets may cause target deprivation-induced neuronal death and is mainly apoptotic and occurs by programmed cell death (Deshmukh and Johnson, 1997; Martin et al., 1998). Environmental factors in neurodegenerative disease comprise physical, toxic and infection related factors. Other possible factors include dietary excitotoxic and bacterial or viral infections (Vajda, 2002). Neuroprotection could possibly result in salvage, recovery or regeneration of the nervous system, its cells, structure and function. It is also used to refer to relative mechanisms that prevents the central nervous system (CNS) from neuronal injuries caused by chronic (e.g., Alzheimer's and Parkinson's diseases) or acute (e.g., stroke) neurodegenerative diseases (Elufioye et al., 2017). Numerous herbal plants contain neuroprotective and possess memory enhancing effects. Among well-studied herbs include *Ginkgo biloba*, *Panax ginseng*, and *Salvia officinalis* will be highlighted in this chapter.

Ginkgo biloba is a TCM that is well known for a multitude of disorders. *Ginkgo biloba* extract (EGb) is obtained from the leaves of the maidenhair tree. Its standardized extract, EGb761, contains different kinds of flavone glycosides and terpenoides. They are flavone glycosides 22%–27%, terpene lactones that include ginkgolides A, B and C (2.8%–3.4%) and bilobalide (2.6%–3.2%) and ginkgolic acids (Baron-Ruppert and Luepke, 2001; Zuo et al., 2017). EGb761 has been widely used throughout Europe, North America and Asia (Yin et al., 2014a). Its applications range from dementia and macular degeneration, to tinnitus and winter depression, and in modern Chinese Pharmacopeia, EGb761 is listed for treating the heart and lung diseases (Yin et al., 2014a). The effects of EGb extract have been extensively explored in research and clinical trials on age-associated diseases including brain dysfunction (Konczol et al., 2016; Zuo et al., 2017), central nervous system (CNS) and early dementia (Gauthier and Schlaefke, 2014; Ihl, 2013) as well as cerebrovascular and neuroprotective (Maclennan et al., 2002; Nash and Shah, 2015). Several evidences indicated that the EGb has neuroprotective effects by protecting neurons from ROS, $Ca2+$-overload, nitric oxide, glutamate, or beta-amyloid induced toxicity (Bastianetto et al., 2000; Oyama et al., 1993; Yao et al., 2001; Yin et al., 2013; Zhou and Zhu, 2000). It also helps improve neurological function and cognitive impairment induced by D-galactose, ischemia, and chronic restraint stress (Liu et al., 2013; Rocher et al., 2011; Takuma et al., 2012; Wang et al., 2013; Yin et al., 2014a; Zhang et al., 2012).

In addition to fundamental research, there are also a lot of clinical trials conducted on EGb. These trials aim to improve the cognition and memory among those with dementia and Alzheimer's disease (Nash and Shah,

2015). Studies show that EGb 761 could modulate excitotoxic glutamatergic neurotransmission (Williams et al., 2004), reduces amyloid-β aggregation and toxicity (Wu et al., 2006) and functions as a radical scavenger and reduces stress sensitivity (Kampkotter et al., 2007). These findings suggest its use in the various dementia pathologies. In one most recent meta-analysis, where five randomized, placebo-controlled clinical trials of EGb761 patients with mild to moderate dementia were analyzed, EGb761was clearly superior to placebo in alleviating both tinnitus and dizziness (Spiegel et al., 2018). This result was similar to earlier trials in patients with tinnitus or vertigo (Hamann, 2007; Von Boetticher, 2011). Tinnitus and dizziness are frequent in old age and often seen as concomitant symptoms in patients with dementia. A number of the patients enrolled for their diagnoses of dementia had such neurosensory symptoms (Spiegel et al., 2018). In another recent double-blind randomized trial where patients with subchronic or chronic tinnitus were enrolled, it was found that both 120 mg of EGb761 and 600 mg of pentoxifylline if administered twice a day over a 12-week period had a similarly effect in reducing the loudness and annoyance of tinnitus and the overall suffering of the patients. Interestingly, the number of adverse events was much lower in the EGb761 group (Prochazkova et al., 2018).

In another recent multicenter, prospective and randomized clinical trial enrolling patients with an onset of acute stroke where patients were either given 450 mg of EGb with 100 mg aspirin daily or the control group (100 mg aspirin daily) for 6 months, EGb in combination with aspirin treatment alleviated cognitive and neurological deficits without increasing the incidence of vascular events (Li et al., 2017c). EGb also seemed to slow down the cognitive deterioration in patients with vascular cognitive impairment (VCI), although the effect was shown in only one of the four neuropsychological tests administered (Clinical Global Impression score) (Demarin et al., 2017). VCI is a broad spectrum of syndromes ranging from mild cognitive impairment to dementia, which is influenced by risk factors for cerebrovascular disease (Battistin and Cagnin, 2010). VCI and dementia are two major cognitive dysfunctions caused by chronic cerebral hypoperfusion (Yang et al., 2017). In another randomized, double-blind, placebo-controlled, exploratory phase IV trial to determine the effects of EGb on spatio-temporal gait parameters of mild cognitive impairment (MCI) patients, patients received EGb 120 mg twice daily for six months showed improvement in dual-task-related gait stability and performance. In patients with MCI, gait instability, particularly in dual-task situations are associated with impairment of executive function and an increased fall risk, hence the study indicated that EGb could be an agent to improve gait stability (Gschwind et al., 2017).

As for the safety of the extract, a multicenter randomized, placebo-controlled, double-blinded trial (GiBiEx) revealed that doses of 120 mg of EGb two times daily for a period of six months did not produce adverse clinical effects or lead to any increase in liver injury markers, suggesting that EGb was safe at prescribed doses (Bonassi et al., 2018). EGb 761 had minimal toxicity. It was well-tolerated elderly patients taking multiple medications, for doses up to 240 mg/day (Nash and Shah, 2015). As a standardized extract, EGb produces consistent results and reliable dosing for the treatment of a wide range of conditions, particularly in cognitive and cerebrovascular diseases (Nash and Shah, 2015). However, more extensive trials are certainly needed to translate the promising preclinical findings into clinical application specifically on neutroprotective effects, which is currently lacking.

Ginseng is a popular herb used in China for thousands of years for wide array of health conditions and ailments. It is from the Araliaceae family and the genus of *Panax* (Helms, 2004). It is taken as an energy booster, nourishes the lungs, calms the heart, and tranquilizes the mind (Kim, 2012b; Sun et al., 2016). In Chinese, Panax means "cure all," and it describes the traditional belief that ginseng has properties that heal all bodily diseases (Kim et al., 2018). To date, 12 species and 2 infraspecific taxa have been classified under the genus *Panax*, and the three major commercial ginseng are the Korean ginseng (*Panax ginseng* Meyer), the Chinese ginseng [*Panax notoginseng* (Burk.) F. H. Chen], and the American ginseng (*Panax quinquefolius* L.). They have been used throughout the world (Kim, 2012a; Shin et al., 2015). Ginseng is reported to contain various functional constituents, including ginseng saponins or ginsenosides, polyacetylenes, phenolic compounds, sesquiterpenes, alkaloids, polysaccharides, and oligopeptide (Kim et al., 2018). Its major active ingredient is ginsenosides. Ginsenosides possess many biological activities (Zheng et al., 2018). Depending on the position and quantity of sugar moiety, ginsenosides are divided into three types: (a) Panaxadiol group (e.g., Rb1, Rb2, Rb3, Rc, Rd, Rg3, and Rh2), (b) Panaxatriol group (e.g., Re, Rg1, Rg2, and Rh1), and (c) Oleanolic acid group (e.g., Ro)—with each one of them playing different pharmacological roles (Zheng et al., 2018). The neuroprotective effect of ginsenosides has been gaining much interest particularly for treating neurological and neurodegenerative diseases.

Some studies suggest that ginseng and its ginsenosides have good preventive and therapeutic effect on neurological diseases (Ahmed et al., 2016; Gonzalez-Burgos et al., 2015). This valuable herb has demonstrated promising role as a protector for various neurological conditions including stroke, cognitive impairment diseases such as Alzheimer's disease and motor neuron diseases such as amyotrophic lateral sclerosis (ALS) (Jiang et al., 2000; Lee et al., 2009; Ye et al., 2013). For example, ginsenoside Rd exerts neuroprotective effect by promoting neurogenesis, increasing vascular endothelial growth factor (VEGF) and BDNF expression, and activating the PI3K/Akt and ERK1/2 pathways. It also attenuates ischemia/reperfusion injury in the rat brain (Liu et al., 2015). In addition, this compound is also capable to improve behavioral score in rats and viability of cultured neurons by inhibiting the hyperactive phosphorylation of *N*-methyl-D-aspartate receptor 2B (NMDAR 2B) subunit by decreasing its expression levels in cell membrane (Xie et al., 2016). NMDARs are ligand-gated ion channels with a high permeability to calcium and which can be activated by glutamine (Cull-Candy et al., 2001; Lau and Zukin, 2007; Prybylowski and Wenthold, 2004). After acute cerebral ischemia, excess glutamine is accumulated in synapses and activates NMDAR leading to the overinflux of calcium and excitotoxicity. The NR2B subunit is the predominant region for phosphorylation, which is essential for calcium influx and neuronal injury (Xie et al., 2016). In addition, ginsenoside Rd is thought to have neuroprotective effect via its antioxidant properties (Shang et al., 2013).

Ginsenoside Re also has showed protective role toward cerebral ischemia reperfusion injury. It exerts neuroprotection properties via reduction of the malondialdehyde (MDA) content, decreases mitochondrial swelling and retains H+-ATPase activity (Chen et al., 2008; Zheng et al., 2018). In addition to ginsenoside Re, the neuroprotective properties ginsenoside Rg1was also demonstrated in several experimental studies. It decreased the concentration of IL-1β, TNF-α, and high-mobility group box-1 (HMGB1), but also down-regulated expressions of cleaved caspase-3, cleaved caspase-9, and receptor for advanced glycation end product (RAGE) in model rats, indicating its role in protecting against apoptosis and inflammation (Yang et al., 2015). Similarly, in other study, ginsenoside Rg1 could attenuate H_2O_2-induced apoptosis and inhibit intracellular Ca^{2+} overload in astrocytes, loss of mitochondrial membrane potential, and ROS production in astrocytes (Sun et al., 2014). Rg1 also markedly reduced the neurological deficit scores, brain edema, and infarct volume in the model mice in vivo study, suggesting that Rg1 possesses significant neuroprotective effects, which might be related to the prevention of astrocytes from apoptosis (Sun et al., 2014). Interestingly, ginsenoside Rg1 is also found to have both inflammatory and antioxidant effects, which can inhibit free radicals and protein kinase activation and decrease GSH concentration, thereby providing a good microenvironment for the *in vivo* survival and differentiation of bone marrow mesenchymal stem cells (BMSC). BMSCs show self-renewal capacity that is a good option for stem cell transplantation, as the transplanted cells could differentiate into neurons and glial cells, which also improved cerebral ischemia. Hence, it would reduce morbidity and mortality in ischemic encephalopathy (Bao et al., 2015). Treatment with Rg1 also results in the decrease in cell apoptosis and improved ischemic conditions. Another important ginsenoside, Rg2 is also known to protect against memory impairment in rat models with vascular dementia. The cerebral ischemia and reperfusion injury-induced expression of pro-apoptotic factors BAX and P53 were inhibited by ginsenoside Rg2, suggesting that this compound protects the neuron possibly through prevention of the development of apoptosis (Zhang et al., 2008a). Similarly, ginseng total saponins and ginsenoside Rg3 show neuroprotective effect by inhibiting increase of intracellular concentrations of Ca^{2+} and spontaneous Ca^{2+} oscillations in cultured rat hippocampal neurons. Seizures and epilepsy are dependent on elevated intracellular calcium concentration and perturbed Ca^{2+} levels is forerunner of neuronal death. Hence, ginseng could play a neuroprotective role in perturbed homeostasis of intracellular Ca^{2+} and neuronal cell death (Kim and Rhim, 2004).

However, not many recent trials were conducted specifically to look at neuroprotective roles of ginseng or ginsenosides. The latest one was a randomized, double-blind, placebo-controlled trial that enrolled patients from multiple centers across China who had received a clinical diagnosis of primary acute ischaemic stroke (Liu et al., 2012). Selected patients were randomly given intravenous infusion of either Ginsenoside Rd or placebo. The results from the trial demonstrated that ginsenoside-Rd improved the disability in patients with acute ischemic stroke as measured by the modified Rankin scale (mRs) at 90 days. Besides, neurologic functioning also improved after 15 days based on the National Institutes of Health Stroke Scale (NIHSS). Ginsenoside-Rd appears to improve the primary outcome of acute ischemic stroke and had an acceptable adverse-event profile (Liu et al., 2012). However, other recent clinical trial usually involved combination of ginsenosides with other components. For example, in a double-blind, placebo-controlled, crossover design randomized controlled trial conducted to assess the effects of a combination of omega 3 essential fatty acids, green tea catechins, and ginsenosides on cognition and brain functioning in healthy older adults, one-month supplementation with a combination of all three substances was associated with changes in cognitive functioning as well as modification of brain activation and brain functional connectivity in cognitively healthy older adults (Carmichael et al., 2018). Unfortunately, in this trial, the exact effects of ginsenosides were unclear.

The genus *Salvia* (sage) belongs to the Lamiaceae family, comprising over 900 species distributed throughout the world. Some of the common species include *S. officinalis* (common sage), *S. miltiorrhiza* (Chinese sage or danshen), *S. lavandulaefolia* (Spanish sage*), S. fruticose* (Greek sage), *S. sclarea* (clary sage) and *S. hispanica* (chia) (Lopresti, 2017). *Salvia* genus constituents comprise an array of phenolic acids, flavonoids, terpenoids and essential oils. Phenolic compounds include caffeic acid and its derivatives, rosmarinic acid, salvianolic acids, sagecoumarin, lithospermic acids, sagernic acid, and yunnaneic acids, whereas flavonoids include luteolin, apigenin, hispidulin, kaempferol, and quercetin (Lu and Foo, 2002). Terpenoids include diterpenes and triterpenes such as carnosic acid, ursolic acid, carnosol, and tanshinones. Examples of essential oils include α- and β-thujone, camphor, 1,8-cineole, α-humulene, β-caryophyllene, and viridiflorol (Lopresti, 2017). Interestingly, these chemical constituents are thought to influence the multiple physiological process and pharmacodynamics pathways related to the brain. The pharmacodynamics effects include cholinergic activities, anti-inflammatory, antioxidant, neurotrophic, and protection against amyloid-β peptide activities.

Several cell and animal model studies indicate that *Salvia* species and their constituents are promising AChE inhibitors with cholinergic activities. Phenolic diterpenes, 7a-methoxyrosmanol, isorosmanol, rosmarinic acid, carnosic acid, quercetin, and tanshinones from *S. miltiorrhiza* all possess AChE inhibition activities (Marcelo et al., 2013; Merad et al., 2014; Sallam et al., 2016; Xu et al., 2016; Zhou et al., 2011). AChE catalyzes the breakdown of acetylcholine (ACh). Inhibition of AChE is hypothesized to improve the ACh signaling deficiency in multiple neurodegenerative disorders and Alzheimer's disease.

As anti-inflammatory agent, *S. officinalis* contains essential oils such as 1, 8-cineole and camphor. They significantly inhibited LPS-stimulated nitric oxide production in mouse macrophages (Abu-Darwish et al., 2013). Besides that, phenolic diterpenes (carnosol and carnosic acid) reduced the nitric oxide and prostaglandin E2 (PGE2) production and inhibited expression of iNOS, cytokines/interleukins (IL-1α, IL-6) and various chemokines genes of the inflammatory process (Schwager et al., 2016). In acute inflammatory, *S. officinalis* tincture reduced the percentages of total leukocyte, monocyte, and the activation of circulating phagocytes (Oniga et al., 2007). Tanshinones and salvianolic acids exhibit anti-inflammatory mechanisms by inhibiting cyclooxygenase-2, hypoxia-inducible factor-1α, and NF-κβ activity (Bonaccini et al., 2015). Tanshinones significantly inhibit the mRNA and protein expression of TNF-α, IL-1β, and IL-8 in LPS-stimulated macrophages (Ma et al., 2016). The other compounds such as caffeic acid, rosmarinic acid, and ursolic acid also show similar anti-inflammatory properties (Kashyap et al., 2016; Nabavi et al., 2015). As for antioxidant properties, most *Salvia* species demonstrate strong antioxidant capacity with *S. forsskaolii* and *S. verticillate*, and *S. officinalis* having the highest antioxidant potential (Sulniute et al., 2016). In addition, *S. miltiorrhiza* reduced the production of ROS and superoxide. It also inhibited the oxidative modification of LDLs and mitochondrial oxidative stress. Apart from that, it increased the activities of catalase, manganese superoxide dismutase, GSH peroxidase, and coupled eNOS. Some of the other functions include maintaining the integrity of the BBB, and by promoting self-renewal and proliferation of neural stem/progenitor cells in stroke (Chang et al., 2016a).

The neuroprotective effects of compounds could be related with the augmentation of brain-derived neurotrophic factor (BDNF) levels. Rosmarinic acid neuroprotection activities is seemingly involved neuronal loss suppression, increase of synaptophysin expression and increase of BDNF levels (Fonteles et al., 2016). BDNF is a neutrophin supports the survival of neurons, growth and differentiation of new neurons and synapses. It further enhances learning and memory capabilities (Bowling et al., 2016). Rosmarinic acid administration (10 mg/kg daily) reversed depressive-like behaviors in rats exposed to stress and restored hippocampal BDNF (Jin et al., 2013). Furthermore, *in vitro* study showed that the BDNF level increased following treatment with rosmarinic acid in cultured astrocytes (Jin et al., 2013). The flavonoid luteolin, showed significant effects in modulating mRNA expression and protein secretion of nerve growth factor (NGF), glial-derived neurotrophic factor (GDNF), and BDNF in cultured astrocytes (Xu et al., 2013). On the other hand, quercetin is also found to increase BDNF levels and decrease cell apoptosis in the focal cerebral ischemia rat brain (Yao et al., 2012). Carnosic acid, carnosol (Kosaka and Yokoi, 2003), and tanshinones (Zhao et al., 2015) all have shown to be neuroprotectant by enhancing the production of NGF in neuronal cells.

S. miltiorrhiza as a TCM preparation that protects mice against amyloid-β-induced neurotoxicity. It inhibits the increases in tumor necrosis factor-α (TNF-α), interleukin-6 (IL-6) levels. Furthermore, there was an increase in the expression of choline acetyltransferase (ChAT), activated protein kinase C1 (RACK1) receptor and the brain-derived neurotrophic factor (BDNF) (Teng et al., 2014). Tanshinones also protect against amyloid-β-induced toxicity by ameliorating the mRNA expression of inducible nitric oxide synthase (iNOS), matrix metalloproteinase 2 and nuclear transcription factor-κ, suggesting its potential neuroprotective activity for Alzheimer's disease (Jiang et al., 2014). Other *Salvia*constituents such as salvianolic acid, carnosic acid, and quercetin were also shown to have protective effects against amyloid-β-induced toxicity (Lee et al., 2013; Patil et al., 2003; Rasoolijazi et al., 2013).

The use of *Salvia* as standalone TCM in neurology disorders in larger clinical trials is still lacking although there are several trials provided some evidence of neuroprotective effects in certain conditions. For example, a systematic review of scientific literature indicated that intake of herbal preparations derived from *S. officinalis* and *S. lavandulaefolia* exhibited better cognitive performance, where the subject healthy and patients with dementia or cognitive impairment showed improvement in memory. Both herbs were found to be safe with no serious adverse effects compared with placebo (Miroddi et al., 2014). In another systematic review and meta-analyses, to evaluate the clinical effectiveness and safety of Danhong injection [combination of Danshen and Honghua (*Flos Carthami tinctorii*) extraction] and S. *miltiorrhiza* injection (SMI) in the treatment of cerebral infarction, where 12 randomized clinical trials were studied, Danhong injection combined with conventional therapy is more effective in improving the clinical total effective rate and neurologic impairment (Wang et al., 2017). A recent double-blind, randomized, placebo-controlled pilot study was the first to evaluate effects of a combination of sage, rosemary and melissa (*Salvia officinalis* L., *Rosmarinus officinalis* L. and *Melissa officinalis* L.; SRM) for memory and brain function. Findings showed SRM was more effective than a placebo among the supported verbal episodic memory in healthy subjects (Perry et al., 2018). More controlled studies on *Salvia* as standalone TCM in other neurodegenerative diseases or disorders are certainly warranted.

4.9 RECOMMENDATIONS FOR FUTURE RESEARCH

Standardization of the phytochemicals present in herbs such as the leaves, root, and stem may increase their efficacy. Further research is needed to select plant parts, as well as the method of extraction and suitable solvents would enable better segregaration of active compounds or constituents derived from herbs. Other aspects of stages of plant growth and harvesting of herbs and their appropriate storage would add value as these efforts could minimize loss of vital phytochemicals and nutrients. Furthermore, the distribution of herb-based products to communities across the regions needs to be facilitated.

4.10 CONCLUSION

Owing to their health benefits, herbs have been used by various communities worldwide. Each herb has its own unique characteristics. Geographical differences in soil composition and climate contribute to the growth of different genus and species in each continent. Herbal plant-based studies using in vivo and in vitro models, as well as community history of medicinal usages, provide evidence that herbs could play a significant role in the healing process. The chapter highlights diverse plant species and their use in treating diseases relating to the heart, gastrointestinal system, endocrine system, skin, liver, reproductive system, and nervous system. We have presented selected plant genus which are used traditionally among the local ethnic group in regions such as South East Asia, South America, West Asia, China, India, Japan, Middle East, and Europe. Plants contain numerous compounds that play important role in the healing process. Active compounds present in these plants and possible underlying mechanisms are described based on previous studies. Therefore, it is important to identify these compounds in order to obatin an in-depth understanding on chemical composition, suitable dosage according to the specific age group, preparation method, and standarization. Such information will enable the proper use of herbs. Furthermore, it would enable the standardization of herb-based products and provide sufficient information to consumers on the safe usage of herbs and their contents. Herbs with promising findings based on preclinical studies could be further investigated in clinical trials.

KEYWORDS

- **cardioprotective**
- **neuroprotective**
- **skin therapies**
- **hepatoprotective**
- **gastrointestinal therapies**
- **reproductive therapies**

REFERENCES

Abdul Kadir, A., Nik Hussain, N. H., Wan Bebakar, W. M., Mohd, D. M., Mohammad, W., Zahiruddin, W. M., Hassan, I. I., Shukor, N., Kamaruddin, N. A., & Wan Mohamud, W. N. (2012). The effect of *Labisia pumila* var. *alata* on postmenopausal women: A pilot study. *J Evid Based Compliment Alternat Med.*, 2012: 216525.

Abdulghani, M., Hussin, A. H., Sulaiman, S. A., & Chan, K. L. (2012). The ameliorative effects of *Eurycoma longifolia* jack on testosterone-induced reproductive disorders in female rats. *Reprod Biol.*, 12, 247255.

Abdullah, N., Chermahini, S. H., Suan, C. L., & Sarmidi, M. R. (2013). *Labisia pumila*: A review on its traditional, phytochemical and biological uses. *World App Sci J.*, 27, 1297–1306.

Abenavoli, L., Capasso, R., Milic, N., & Capasso, F. (2010). Milk thistle in liver diseases: Past, present, future. *Phytother Res.*, 24, 1423–1432.

Abu-Darwish, M. S., Cabral, C., Ferreira, I. V., Goncalves, M. J., Cavaleiro, C., Cruz, M. T., Al-Bdour, T. H., & Salgueiro, L. (2013). Essential oil of common sage (*Salvia officinalis* L.) from Jordan: Assessment of safety in mammalian cells and its antifungal and anti-inflammatory potential. *Biomed Res Int.*, 2013, 538940.

Abubakar, B. M., Salleh, F. M., & Wagiran, A. (2017). Chemical composition of *Eurycoma longifolia* (*tongkat ali*) and the quality control of its herbal medicinal products. *J Appl Sci.*, 17, 324–338.

Adachi, M., Horiuchi, G., Ikematsu, N., Tanaka, T., Terao, J., Satouchi, K., & Tokumura, A. (2011). Intragastrically administered lysophosphatidic acids protect against gastric ulcer in rats under water-immersion restraint stress. *Dig Dis Sci.*, 56, 2252–2261.

Afroz, S., Yagi, A., Fujikawa, K., Rahman, M. M., Morito, K., Fukuta, T., Watanabe, S., Kiyokage, E., Toida, K., & Shimizu, T. (2018). Lysophosphatidic acid in medicinal herbs enhances prostaglandin E$_2$ and protects against indomethacin-induced gastric cell damage in vivo and in vitro. *Prostaglandins Other Lipid Mediat.*, 135, 36–44.

Ahmed, T., Raza, S. H., Maryam, A., Setzer, W. N., Braidy, N., Nabavi, S. F., De Oliveira, M. R., & Nabavi, S. M. (2016). Ginsenoside Rb1 as a neuroprotective agent: A review. *Brain Res Bull.*, 125, 30–43.

Al-Sabahi, J. N., Hanif, M. A., Al-Maskari, A. Y., Al Busaidi, M. S. M., Al-Maskari, M. Y., & Al-Haddabi, M. H. (2016). Chemical report on wild growing *Mentha arvensis* and *Citrullus colocynthus* from Oman. *J Essent Oil Bear Pl.*, 19, 719–726.

Arayne, M. S., Sultana, N., & Bahadur, S. S. (2007). The berberis story: Berberis vulgaris in therapeutics. *Pak J Pharm Sci.*, 20, 8–92.

Avula, B., Wang, Y., Ali, Z., Smillie, T., & Khan, I. (2010). Quantitative determination of *Triperpene saponins* and alkenated-phenolics from *Labisia pumila* by LC-UV/ELSD method and confirmation by LC-ESI-TOF. *Planta Med.*, 76, P25.

Ayob, Z., Wagiran, A., & Abd Samad, A. (2013). Potential of tissue cultured medicinal plants in Malaysia. *J Teknol.*, 62, 111–117.

Aziz, M. A., Khan, A. H., Adnan, M., & Izatullah, I. (2017). Traditional uses of medicinal plants reported by the indigenous communities and local herbal practitioners of Bajaur agency, Federally administrated tribal areas, Pakistan. *J Ethnopharmacol.*, 198, 268–281.

Bahadori, M. B.; Zengin, G.; Bahadori, S.; Dinparast, L. & Movahhedin, N. (2018). Phenolic composition and functional properties of wild mint (*Mentha longifolia* var. *calliantha* (Stapf) Briq.). *Int J Food Prop.*, 21, 183-193.

Baharuddin, N. S., Abdullah, H., & Wahab, W. N. A. W. A. (2015). Anti-candida activity of *Quercus infectoria* gall extracts against Candida species. *J Pharm Bioallied Sci.*, 7, 15–20.

Bailey, C. J. (2017). Metformin: Historical overview. *Diabetologia.*, 60, 1566–1576.

Banala, R. R., Vemuri, S. K., Reddy, A. V. G., & Subbaiah, G. P. V. (2017). Aqueous extract of *Acalypha indica* leaves for the treatment of Psoriasis: In-vitro studies. *Int J Bioassays.*, 6, 5360–5364.

Bao, C., Wang, Y., Min, H., Zhang, M., Du, X., Han, R., & Liu, X. (2015). Combination of ginsenoside Rg1 and bone marrow mesenchymal stem cell transplantation in the treatment of cerebral ischemia reperfusion injury in rats. *Cell Physiol Biochem.*, 37, 901–910.

Baron-Ruppert, G., & Luepke, N. P. (2001). Evidence for toxic effects of alkylphenols from *Ginkgo biloba* in the hen's egg test (HET). *Phytomedicine.*, 8, 133–138.

Bastianetto, S., Zheng, W. H., & Quirion, R. (2000). The *Ginkgo biloba* extract (EGb 761) protects and rescues hippocampal cells against nitric oxide-induced toxicity: Involvement of its flavonoid constituents and protein kinase C. *J Neurochem.*, 74, 2268–2277.

Basuray, A., French, B., Ky, B., Vorovich, E., Olt, C., Sweitzer, N. K., Cappola, T. P., & Fang, J. C. (2014). Heart failure with recovered ejection fraction: Clinical description, biomarkers, and outcomes. *Circulation.*, 129, 2380–2387.

Battistin, L. & Cagnin, A. (2010). Vascular cognitive disorder: A biological and clinical overview. *Neurochem Res.*, 35, 1933–1938.

Batubara, I., Wahyuni, W. T., & Firdaus, I. (2016). Utilization of anting-anting (*Acalypha indica*) leaves as antibacterial. *IOP Conference Series: Earth and Environmental Science*, 31, 012038.

Belwal, T., Andola, H. C., Atanassova, M. S., Joshi, B., Suyal, R., Thakur, S., Bisht, A., Jantwal, A., Bhatt, I. D., & Rawal, R. S. (2019). Gotu Kola (Centella asiatica). In: Nabavi, S. M. & Silva, A. S. eds., Nonvitamin and Nonmineral Nutritional Supplements. New York, NY: Academic Press.

Bennett, B. C. & Husby, C. E. (2008). Patterns of medicinal plant use: An examination of the Ecuadorian Shuar medicinal flora using contingency table and binomial analyses. *J Ethnopharmacol.*, 116, 422–430.

Bhat, R. & Karim, A. (2010). *Tongkat Ali (Eurycoma longifolia* Jack): A review on its ethnobotany and pharmacological importance. *Fitoterapia.*, 81, 669–679.

Bonaccini, L., Karioti, A., Bergonzi, M. C., & Bilia, A. R. (2015). Effects of *Salvia miltiorrhiza* on CNS neuronal injury and degeneration: A plausible complementary role of tanshinones and depsides. *Planta Med.*, 81, 1003–1016.

Bonassi, S., Prinzi, G., Lamonaca, P., Russo, P., Paximadas, I., Rasoni, G., Rossi, R., Ruggi, M., Malandrino, S., Sanchez-Flores, M., Valdiglesias, V., Benassi, B., Pacchierotti, F., Villani, P., Panatta, M., & Cordelli, E. (2018). Clinical and genomic safety of treatment with *Ginkgo biloba* L. leaf extract (IDN 5933/Ginkgoselect(R)Plus) in elderly: A randomised placebo-controlled clinical trial [GiBiEx]. *BMC Complement Altern Med.*, 18, 22.

Bower, A., Marquez, S., & De Mejia, E. G. (2016). The health benefits of selected culinary herbs and spices found in the traditional Mediterranean diet. *Crit Rev Food Sci Nutr.*, 56, 2728–2746.

Bowling, H., Bhattacharya, A., Klann, E., & Chao, M. V. (2016). Deconstructing brain-derived neurotrophic factor actions in adult brain circuits to bridge an existing informational gap in neuro-cell biology. *Neural Regen Res.*, 11, 363–367.

Burkill, I. H. (1966). A dictionary of the economic products of the Malay Peninsula. *Nature.*, 137, 255.

Bylka, W., Znajdek-Awiżeń, P., Studzińska-Sroka, E., & Brzezińska, M. (2013). *Centella asiatica* in cosmetology. *Postępy Dermatol Alergol.*, 30, 46–49.

Bylka, W., Znajdek-Awiżeń, P., Studzińska-Sroka, E., Dańczak-Pazdrowska, A., & Brzezińska, M. (2014). *Centella asiatica* in dermatology: An overview. *Phytother Res.*, 28, 1117–1124.

Cao, Y., Ruan, Y., Shen, T., Huang, X., Li, M., Yu, W., Zhu, Y., Man, Y., Wang, S., & Li, J. (2014). Astragalus polysaccharide suppresses doxorubicin-induced cardiotoxicity by regulating the PI3k/Akt and p38MAPK pathways. *Oxid Med Cell Longev.*, 2014, 674219.

Carden, D. L., & Granger, D. N. (2000). Pathophysiology of ischaemia-reperfusion injury. *J Pathol.*, 190, 255–266.

Carmichael, O. T., Pillai, S., Shankapal, P., Mclellan, A., Kay, D. G., Gold, B. T., & Keller, J. N. (2018). A combination of essential fatty acids, panax ginseng extract, and green tea catechins modifies brain FMRI signals in healthy older adults. *J Nutr Health Aging.*, 22, 837–846.

Carter, A. M., Catto, A. J., Mansfield, M. W., Bamford, J. M., & Grant, P. J. (2007). Predictive variables for mortality after acute ischemic stroke. *Stroke.*, 38, 1873–1880.

Chan, K., Chui, S. H., Wong, D. Y., Ha, W. Y., Chan, C. L., & Wong, R. N. (2004). Protective effects of Danshensu from the aqueous extract of *Salvia miltiorrhiza* (Danshen) against homocysteine-induced endothelial dysfunction. *Life Sci.*, 75, 3157–3171.

Chang, C.C., Chang, Y.C., Hu, W.L., & Hung, Y.C. (2016a). Oxidative stress and *Salvia miltiorrhiza* in aging-associated cardiovascular diseases. *Oxid Med Cell Longev.*, 2016, 4797102.

Chang, C. C., Chu, C. F., Wang, C. N., Wu, H. T., Bi, K. W., Pang, J. H., & Huang, S. T. (2014). The anti-atherosclerotic effect of tanshinone IIA is associated with the inhibition of TNF-α-induced VCAM-1, ICAM-1 and CX3CL1 expression. *Phytomedicine.*, 21, 207–216.

Chang, W., Li, K., Guan, F., Yao, F., Yu, Y., Zhang, M., Hatch, G. M., & Chen, L. (2016b). Berberine pretreatment confers cardioprotection against ischemia-reperfusion injury in a rat model of type 2 diabetes. *J Cardiovasc Pharmacol Ther.*, 21, 486–494.

Chen, L. M., Zhou, X. M., Cao, Y. L., & Hu, W. X. (2008). Neuroprotection of ginsenoside Re in cerebral ischemia-reperfusion injury in rats. *J Asian Nat Prod Res.*, 10, 439–445.

Chen, L. X., Liao, J. Z., & Guo, W. Q. (1995). Effects of *Astragalus membranaceus* on left ventricular function and oxygen free radical in acute myocardial infarction patients and mechanism of its cardiotonic action. *Zhongguo Zhong Xi Yi Jie He Za Zhi.*, 15, 141–143.

Chi, L., Peng, L., Pan, N., Hu, X., & Zhang, Y. (2014). The anti-atherogenic effects of berberine on foam cell formation are mediated through the upregulation of sirtuin 1. *Int J Mol Med.*, 34, 1087–1093.

Cull-Candy, S., Brickley, S., & Farrant, M. (2001). NMDA receptor subunits: Diversity, development and disease. *Curr Opin Neurobiol.*, 11, 327–335.

De Wet, H., Nciki, S., & Van Vuuren, S. F. (2013). Medicinal plants used for the treatment of various skin disorders by a rural community in northern Maputaland, South Africa. *J Ethnobiol Ethnomed.*, 9, 51–51.

Delfan, B., Bahmani, M., Eftekhari, Z., Jelodari, M., Saki, K., & Mohammadi, T. (2014). Effective herbs on the wound and skin disorders: a ethnobotanical study in Lorestan province, west of Iran. *Asian Pac J Trop Dis.*, 4, S938–S942.

Demarin, V., Basic Kes, V., Trkanjec, Z., Budisic, M., Bosnjak Pasic, M., Crnac, P., & Budincevic, H. (2017). Efficacy and safety of *Ginkgo biloba* standardized extract in the treatment of vascular cognitive impairment: a randomized, double-blind, placebo-controlled clinical trial. *Neuropsychiatr Dis Treat.*, 13, 483–490.

Deng, Y., Yang, M., Xu, F., Zhang, Q., Zhao, Q., Yu, H., Li, D., Zhang, G., Lu, A., Cho, K., Teng, F., Wu, P., Wang, L., Wu, W., Liu, X., Guo, D. A., & Jiang, B. (2015). Combined Salvianolic acid B and ginsenoside Rg1 exerts cardioprotection against ischemia/reperfusion injury in rats. *PLoS One.*, 10.

Deshmukh, M., & Johnson, E. M., Jr. (1997). Programmed cell death in neurons: Focus on the pathway of nerve growth factor deprivation-induced death of sympathetic neurons. *Mol Pharmacol.*, 51, 897–906.

Digrak, M., Alma, M. H., İlçim, A., & Sen, S. (1999). Antibacterial and antifungal effects of various commercial plant extracts. *Pharm Biol.*, 37, 216–220.

Efentakis, P., Rizakou, A., Christodoulou, E., Chatzianastasiou, A., Lopez, M. G., Leon, R., Balafas, E., Kadoglou, N. P. E., Tseti, I., Skaltsa, H., Kostomitsopoulos, N., Iliodromitis, E. K., Valsami, G., & Andreadou, I. (2017). Saffron (*Crocus sativus*) intake provides nutritional preconditioning against myocardial ischemia-reperfusion injury in wild type and ApoE((-/-)) mice: Involvement of Nrf2 activation. *Nutr Metab Cardiovasc Dis.*, 27, 919–929.

El-Beshbishy, H., & Bahashwan, S. (2012). Hypoglycemic effect of basil (*Ocimum basilicum*) aqueous extract is mediated through inhibition of α-glucosidase and α-amylase activities: An in vitro study. *Toxicol Ind Health.*, 28, 42–50.

El-Tantawy, W. H., & Temraz, A. (2017). Management of diabetes using herbal extracts: A Review. *Arch Physiol Biochem.*, 124, 383–389.

Elufioye, T. O.; Berida, T. I. & Habtemariam, S. (2017). Plants-derived neuroprotective agents: Cutting the cycle of cell death through multiple mechanisms. *J Evid Based Complement Alternat Med.*, 2017: 3574012.

Fonteles, A. A., De Souza, C. M., De Sousa Neves, J. C., Menezes, A. P., Santos Do Carmo, M. R., Fernandes, F. D., De Araujo, P. R., & De Andrade, G. M. (2016). Rosmarinic acid prevents against memory deficits in ischemic mice. *Behav Brain Res.*, 297, 91-103.

Fu, J., Huang, H., Liu, J., Pi, R., Chen, J., & Liu, P. (2007). Tanshinone IIA protects cardiac myocytes against oxidative stress-triggered damage and apoptosis. *Eur. J. Pharmacol.*, 568, 213–221.

Furchgott, R. F. (1999). Endothelium-derived relaxing factor: Discovery, early studies, and identification as nitric oxide. *Biosci Rep.*, 19, 235–251.

Gauthier, S., & Schlaefke, S. (2014). Efficacy and tolerability of *Ginkgo biloba* extract EGb 761(R) in dementia: A systematic review and meta-analysis of randomized placebo-controlled trials. *Clin Interv Aging.*, 9, 2065–2077.

Ghorbani, A. (2013). Phytotherapy for diabetic dyslipidemia: Evidence from clinical trials. *Clin Lipidol.*, 8, 311–319.

Ghoshal, U. C., Gwee, K.A., Holtmann, G., Li, Y., Park, S. J., Simadibrata, M., Sugano, K., Wu, K., Quigley, E. M. M., & Cohen, H. (2018). The role of the microbiome and the use of probiotics in gastrointestinal disorders in adults in the Asia-Pacific region-background and recommendations of a regional consensus meeting. *J Gastroen Hepatol.*, 33, 57–69.

Gómez-Díaz, J. A., Krömer, T., Kreft, H., Gerold, G., Carvajal-Hernández, C. I., & Heitkamp, F. (2017). Diversity and composition of herbaceous angiosperms along gradients of elevation and forest-use intensity. *PLoS One.*, 12.

Gonzalez-Burgos, E., Fernandez-Moriano, C., & Gomez-Serranillos, M. P. (2015). Potential neuroprotective activity of Ginseng in Parkinson's disease: A review. *J. Neuroimmune Pharmacol.*, 10, 14–29.

Gschwind, Y. J., Bridenbaugh, S. A., Reinhard, S., Granacher, U., Monsch, A. U., & Kressig, R. W. (2017). *Ginkgo biloba* special extract LI 1370 improves dual-task walking in patients with MCI: A randomised, double-blind, placebo-controlled exploratory study. *Aging Clin Exp Res.*, 29, 609–619.

Gupta, A., Nagariya, A. K., Mishra, A. K., Bansal, P., Kumar, K., Gupta, V., & Singh, A. K. (2010). Ethno-potential of medicinal herbs in skin diseases: An overview. *J Pharm Res.*, 3, 435–441.

Gurib-Fakim, A. (2006). Medicinal plants: Traditions of yesterday and drugs of tomorrow. *Mol Aspects Med.*, 27, 1–93.

Hajar, R. (2017). Risk factors for coronary artery disease: Historical perspectives. *Heart Views.*, 18, 109–114.

Hamann, K. F. (2007). Special ginkgo extract in cases of vertigo: A systematic review of randomised, double-blind, placebo controlled clinical examinations. *HNO.*, 55, 258–263.

Han, J. Y., Fan, J. Y., Horie, Y., Miura, S., Cui, D. H., Ishii, H., Hibi, T., Tsuneki, H., & Kimura, I. (2008). Ameliorating effects of compounds derived from *Salvia miltiorrhiza* root extract on microcirculatory disturbance and target organ injury by ischemia and reperfusion. *Pharmacol Ther.*, 117, 280–295.

Hao, W., Liu, S., Qin, Y., Sun, C., Chen, L., Wu, C., & Bao, Y. (2017). Cardioprotective effect of *Platycodon grandiflorum* in patients with early breast cancer receiving anthracycline-based chemotherapy: Study protocol for a randomized controlled trial. *Trials.*, 18, 386.

Harish, R., & Shivanandappa, T. (2006). Antioxidant activity and hepatoprotective potential of *Phyllanthus niruri*. *Food Chem.*, 95, 180–185.

Hashemi, S. A., Madani, S. A., & Abediankenari, S. (2015). The review on properties of *Aloe vera* in healing of cutaneous wounds. *Biomed Res Int.*, 2015, 6.

Hay, R. J., Johns, N. E., Williams, H. C., Bolliger, I. W., Dellavalle, R. P., Margolis, D. J., Marks, R., Naldi, L., Weinstock, M. A., Wulf, S. K., Michaud, C., Murray, C. J. L., & Naghavi, M. (2014). The global burden of skin disease in 2010: An analysis of the prevalence and impact of skin conditions. *J Invest Dermatol.*, 134, 1527–1534.

He, Y. X., Du, M., Shi, H. L., Huang, F., Liu, H. S., Wu, H., Zhang, B. B., Dou, W., Wu, X. J., & Wang, Z. T. (2014). Astragalosides from *Radix astragali* benefits experimental autoimmune encephalomyelitis in C57BL /6 mice at multiple levels. *BMC Complement Alternat Med.*, 14, 313.

Helms, S. (2004). Cancer prevention and therapeutics: Panax ginseng. *Alternat Med Rev.*, 9, 259–274.

Hu, H., Zhai, C., Qian, G., Gu, A., Liu, J., Ying, F., Xu, W., Jin, D., Wang, H., Hu, H., Zhang, Y., & Tang, G. (2015). Protective effects of tanshinone IIA on myocardial ischemia reperfusion injury by reducing oxidative stress, HMGB1 expression, and inflammatory reaction. *Pharm Biol.*, 53, 1752–1758.

Hung, I. L., Hung, Y. C., Wang, L. Y., Hsu, S. F., Chen, H. J., Tseng, Y. J., Kuo, C. E., Hu, W. L., & Li, T. C. (2015a). Chinese herbal products for ischemic stroke. *Am J Chin Med.*, 43, 1365–1379.

Hung, Y. C., Tseng, Y. J., Hu, W. L., Chen, H. J., Li, T. C., Tsai, P. Y., Chen, H. P., Huang, M. H., & Su, F. Y. (2015b). Demographic and prescribing patterns of Chinese herbal products for individualized therapy for ischemic heart disease in Taiwan: Population-based study. *PLoS One.*, 10.

Huseini, H. F., Larijani, B., Heshmat, R., Fakhrzadeh, H., Radjabipour, B., Toliat, T., & Raza, M. (2006). The efficacy of *Silybum marianum* (L.) Gaertn. (silymarin) in the treatment of type II diabetes: A randomized, double-blind, placebo-controlled, clinical trial. *Phytother Res.*, 20, 1036–1039.

Ihl, R. (2013). Effects of *Ginkgo biloba* extract EGb 761 (R) in dementia with neuropsychiatric features: review of recently completed randomised, controlled trials. *Int J Psychiatry Clin Pract.*, 1(17 Suppl.), 8–14.

Imanshahidi, M., & Hosseinzadeh, H. (2008). Pharmacological and therapeutic effects of Berberis vulgaris and its active constituent, berberine. *Phytother Res.*, 22, 999–1012.

Iminjan, M., Amat, N., Li, X.H., Upur, H., Ahmat, D., & He, B. (2014). Investigation into the toxicity of traditional uyghur medicine quercus infectoria galls water extract. *PLoS One.*, 9, e90756.

Jia, Y., Huang, F., Zhang, S., & Leung, S. W. (2012). Is danshen (*Salvia miltiorrhiza*) dripping pill more effective than isosorbide dinitrate in treating angina pectoris? A systematic review of randomized controlled trials. *Int J Cardiol.*, 157, 330–340.

Jia, Y., & Leung, S. W. (2017). How efficacious is danshen (*Salvia miltiorrhiza*) dripping pill in treating angina pectoris? Evidence assessment for meta-analysis of randomized controlled trials. *J Alternat Complement Med.*, 23, 676–684.

Jia, Y., Zuo, D., Li, Z., Liu, H., Dai, Z., Cai, J., Pang, L., & Wu, Y. (2014). Astragaloside IV inhibits doxorubicin-induced cardiomyocyte apoptosis mediated by mitochondrial apoptotic pathway via activating the PI3K/Akt pathway. *Chem Pharm Bull (Tokyo).*, 62, 45–53.

Jiang, F., Desilva, S., & Turnbull, J. (2000). Beneficial effect of ginseng root in SOD-1 (G93A) transgenic mice. *J Neurol Sci.*, 180, 52–54.

Jiang, P., Li, C., Xiang, Z., & Jiao, B. (2014). Tanshinone IIA reduces the risk of Alzheimer's disease by inhibiting iNOS, MMP-2 and NF-kBp65 transcription and translation in the temporal lobes of rat models of Alzheimer's disease. *Mol Med Rep.*, 10, 689–694.

Jin, X., Liu, P., Yang, F., Zhang, Y. H., & Miao, D. (2013). Rosmarinic acid ameliorates depressive-like behaviors in a rat model of CUS and Up-regulates BDNF levels in the hippocampus and hippocampal-derived astrocytes. *Neurochem Res.*, 38, 1828–1837.

Joshi, R., Satyal, P., & Setzer, W. (2016). Himalayan aromatic medicinal plants: A review of their ethnopharmacology, volatile phytochemistry, and biological activities. *Medicines.*, 3, 6.

Kaatabi, H., Bamosa, A. O., Lebda, F. M., Al Elq, A. H., & Al-Sultan, A. I. (2012). Favorable impact of Nigella sativa seeds on lipid profile in type 2 diabetic patients. *J Family Community Med.*, 19, 155–161.

Kamble, S. M. (2014). Water pollution and public health issues in Kolhapur city in Maharashtra. *Int J Sci Res Publications.*, 4, 1–6.

Kampkotter, A., Pielarski, T., Rohrig, R., Timpel, C., Chovolou, Y., Watjen, W., & Kahl, R. (2007). The *Ginkgo biloba* extract EGb761 reduces stress sensitivity, ROS accumulation and expression of catalase and glutathione S-transferase 4 in Caenorhabditis elegans. *Pharmacol Res.*, 55, 139–147.

Kamruzzaman, H. M., & Hoq, M. O. (2016). A review on ethnomedicinal, phytochemical and pharmacological properties of *Phyllanthus niruri*. *J Med Plants.*, 4, 173–180.

Karimkhani, C., Dellavalle, R. P., Coffeng, L. E., Flohr, C., Hay, R. J., Langan, S. M., Nsoesie, E. O., Ferrari, A. J., Erskine, H. E., Silverberg, J. I., Vos, T., & Naghavi, M. (2017). Global skin disease morbidity and mortality: An update from the global burden of disease study 2013. *JAMA Dermatol.*, 153, 406–412.

Kashyap, D., Tuli, H. S., & Sharma, A. K. (2016). Ursolic acid (UA): A metabolite with promising therapeutic potential. *Life Sci.*, 146, 201–213.

Khan, A. W., Kotta, S., Ansari, S. H., Sharma, R. K., Kumar, A., & Ali, J. (2013). Formulation development, optimization and evaluation of aloe vera gel for wound healing. *Pharmacogn Mag.*, 9, S6–S10.

Khiljee, S., Rehman, N. U., Khiljee, T., Ahmad, R. S., Khan, M. Y., & Ahmad, Q. U. (2011). Use of traditional herbal medicines in the treatment of eczema. *J Pakistan Assoc Dermatol.*, 21, 112–117.

Kim, D. H. (2012a). Chemical diversity of *Panax ginseng*, *Panax quinquifolium*, and *Panax notoginseng. J Ginseng Res.*, 36, 1–15.

Kim, D. S., Irfan, M., Sung, Y. Y., Kim, S. H., Park, S. H., Choi, Y. H., Rhee, M. H., & Kim, H. K. (2017). *Schisandra chinensis* and *Morus alba* synergistically inhibit in vivo thrombus formation and platelet aggregation by impairing the glycoprotein VI pathway. *Evid Based Complement Alternat Med.*, 2017: 7839658.

Kim, J. H. (2012b). Cardiovascular diseases and *Panax ginseng*: A review on molecular mechanisms and medical applications. *J Ginseng Res.*, 36, 16–26.

Kim, K. H., Lee, D., Lee, H. L., Kim, C. E., Jung, K., & Kang, K. S. (2018). Beneficial effects of *Panax ginseng* for the treatment and prevention of neurodegenerative diseases: Past findings and future directions. *J Ginseng Res.*, 42, 239–247.

Kim, S., & Rhim, H. (2004). Ginsenosides inhibit NMDA receptor-mediated epileptic discharges in cultured hippocampal neurons. *Arch Pharm Res.*, 27, 524–530.

Kneeman, J. M., Misdraji, J., & Corey, K. E. (2012). Secondary causes of nonalcoholic fatty liver disease. *Therap Adv Gastroenterol.*, 5, 199–207.

Konczol, A., Rendes, K., Dekany, M., Muller, J., Riethmuller, E., & Balogh, G. T. (2016). Blood-brain barrier specific permeability assay reveals N-methylated tyramine derivatives in standardised leaf extracts and herbal products of *Ginkgo biloba. J Pharm Biomed Anal.*, 131, 167–174.

Kosaka, K., & Yokoi, T. (2003). Carnosic acid, a component of rosemary (*Rosmarinus officinalis* L.), promotes synthesis of nerve growth factor in T98G human glioblastoma cells. *Biol Pharm Bull.*, 26, 1620–1622.

Kwok, T., Leung, P. C., Lam, C., Ho, S., Wong, C. K., Cheng, K. F., & Chook, P. (2014). A randomized placebo controlled trial of an innovative herbal formula in the prevention of atherosclerosis in postmenopausal women with borderline hypercholesterolemia. *Complement Ther Med.*, 22, 473–480.

Lalko, J., & Api, A. M. (2006). Investigation of the dermal sensitization potential of various essential oils in the local lymph node assay. *Food Chem Toxicol.*, 44, 739–746.

Lalkovičová, M., & Danielisová, V. (2016). Neuroprotection and antioxidants. *Neural Regen Res.*, 11, 865–874.

Lau, C. G., & Zukin, R. S. (2007). NMDA receptor trafficking in synaptic plasticity and neuropsychiatric disorders. *Nat Rev Neurosci.*, 8, 413–426.

Lee, B.H., Choi, S.H., Kim, H.J., Jung, S.W., Kim, H.K., & Nah, S.Y. (2016a). Plant lysophosphatidic acids: A rich source for bioactive lysophosphatidic acids and their pharmacological applications. *Biol Pharm Bull.*, 39, 156–162.

Lee, M. S., Yang, E. J., Kim, J. I., & Ernst, E. (2009). Ginseng for cognitive function in Alzheimer's disease: A systematic review. *J Alzheimer's Dis.*, 18, 339–344.

Lee, N. Y. S., Khoo, W. K. S., Adnan, M. A., Mahalingam, T. P., Fernandez, A. R., & Jeevaratnam, K. (2016b). The pharmacological potential of *Phyllanthus niruri. J Pharm Pharmacol.*, 68, 953–969.

Lee, S. H., Kim, M., Yoon, B. W., Kim, Y. J., Ma, S. J., Roh, J. K., Lee, J. S., & Seo, J. S. (2001). Targeted hsp70.1 disruption increases infarction volume after focal cerebral ischemia in mice. *Stroke.*, 32, 2905–2912.

Lee, Y. W., Kim, D. H., Jeon, S. J., Park, S. J., Kim, J. M., Jung, J. M., Lee, H. E., Bae, S. G., Oh, H. K., Son, K. H., & Ryu, J. H. (2013). Neuroprotective effects of salvianolic acid B on an Abeta25-35 peptide-induced mouse model of Alzheimer's disease. *Eur J Pharmacol.*, 704, 70–77.

Li, H. F., Xu, F., Yang, P., Liu, G. X., Shang, M. Y., Wang, X., Yin, J., & Cai, S. Q. (2017a). Systematic screening and characterization of prototype constituents and metabolites of total astragalosides using HPLC-ESI-IT-TOF-MS(n) after oral administration to rats. *J Pharm Biomed Anal.*, 142, 102–112.

Li, L., Hou, X., Xu, R., Liu, C., & Tu, M. (2017b). Research review on the pharmacological effects of astragaloside IV. *Fundam Clin Pharmacol.*, 31, 17–36.

Li, N. Y., Yu, H., Li, X. L., Wang, Q. Y., Zhang, X. W., Ma, R. X., Zhao, Y., Xu, H., Liang, W., Bai, F., & Yu, J. (2018). *Astragalus membranaceus* improving asymptomatic left ventricular diastolic dysfunction in postmenopausal hypertensive women with metabolic syndrome: A prospective, open-labeled, randomized controlled trial. *Chin Med J (Engl).*, 131, 516–526.

Li, S., Zhang, X., Fang, Q., Zhou, J., Zhang, M., Wang, H., Chen, Y., Xu, B., Wu, Y., Qian, L., & Xu, Y. (2017c). *Ginkgo biloba* extract improved cognitive and neurological functions of acute ischaemic stroke: A randomised controlled trial. *Stroke Vasc Neurol.*, 2, 189–197.

Liu, A. H., Bao, Y. M., Wang, X. Y., & Zhang, Z. X. (2013). Cardio-protection by *Ginkgo biloba* extract 50 in rats with acute myocardial infarction is related to Na^+-Ca^{2+} exchanger. *Am J Chin Med.*, 41, 789–800.

Liu, J. Q., Lee, T. F., Miedzyblocki, M., Chan, G. C., Bigam, D. L., & Cheung, P. Y. (2011). Effects of tanshinone IIA, a major component of *Salvia miltiorrhiza*, on platelet aggregation in healthy newborn piglets. *J Ethnopharmacol.*, 137, 44–49.

Liu, L. Y., Chen, X. D., Wu, B. Y., & Jiang, Q. (2010). Influence of Aloe polysaccharide on proliferation and hyaluronic acid and hydroxyproline secretion of human fibroblasts in vitro. *J Chin Integr Med.*, 8, 256–262.

Liu, X., Wang, L., Wen, A., Yang, J., Yan, Y., Song, Y., Liu, X., Ren, H., Wu, Y., Li, Z., Chen, W., Xu, Y., Li, L., Xia, J., & Zhao, G. (2012). Ginsenoside-Rd improves outcome of acute ischaemic stroke: A randomized, double-blind, placebo-controlled, multicenter trial. *Eur J Neurol.*, 19, 855–863.

Liu, X. Y., Zhou, X. Y., Hou, J. C., Zhu, H., Wang, Z., Liu, J. X., & Zheng, Y. Q. (2015). Ginsenoside Rd promotes neurogenesis in rat brain after transient focal cerebral ischemia via activation of PI3K/Akt pathway. *Acta Pharmacol Sin.*, 36, 421–428.

Lopresti, A. L. (2017). Salvia (Sage): A review of its potential cognitive-enhancing and protective effects. *Drugs R D.*, 17, 53–64.

Low, B.S., Das, P. K., & Chan, K.L. (2013). Standardized quassinoid-rich *Eurycoma longifolia* extract improved spermatogenesis and fertility in male rats via the hypothalamic–pituitary–gonadal axis. *J Ethnopharmacol.*, 145, 706–714.

Lu, Y., & Foo, L. Y. (2002). Polyphenolics of Salvia—a review. *Phytochemistry*, 59, 117–140.

Luo, H. M., Dai, R. H., & Li, Y. (1995). Nuclear cardiology study on effective ingredients of *Astragalus membranaceus* in treating heart failure. *Zhongguo Zhong Xi Yi Jie He Za Zhi.*, 15, 707–709.

Ma, S., Zhang, D., Lou, H., Sun, L., & Ji, J. (2016). Evaluation of the anti-inflammatory activities of tanshinones isolated from *Salvia miltiorrhiza* var. alba roots in THP-1 macrophages. *J Ethnopharmacol.*, 188, 193–199.

Macêdo, M. J. F., Ribeiro, D. A., Santos, M. D. O., Macêdo, D. G. D., Macedo, J. G. F., Almeida, B. V. D., Saraiva, M. E., Lacerda, M. N. S. D., & Souza, M. M. D. A. (2018). Fabaceae medicinal flora with therapeutic potential in Savanna areas in the Chapada do Araripe, Northeastern Brazil. *Rev Bras Farmacogn.*, 28, 738–750.

Maclennan, K. M., Darlington, C. L., & Smith, P. F. (2002). The CNS effects of *Ginkgo biloba* extracts and ginkgolide B. *Prog Neurobiol.*, 67, 235–257.

Mahmood, A., Tiwari, A. K., Sahin, K., Kucuk, O., & Ali, S. (2016). Triterpenoid saponin-rich fraction of *Centella asiatica* decreases IL-1βand NF-κB, and augments tissue regeneration and excision wound repair. *Turk J Biol.*, 40, 399–409.

Manjusree, T., Joseph, S., Petitta, M., & Thomas, J. (2017). Integrated approach for identifying the factors controlling groundwater quality of a tropical coastal zone in Kerala, India. *Environ Earth Sci.*, 76, 486.

Marcelo, F., Dias, C., Martins, A., Madeira, P. J., Jorge, T., Florencio, M. H., Canada, F. J., Cabrita, E. J., Jimenez-Barbero, J., & Rauter, A. P. (2013). Molecular recognition of rosmarinic acid from *Salvia sclareoides* extracts by acetylcholinesterase: A new binding site detected by NMR spectroscopy. *Chemistry*, 19, 6641–6649.

Mark, L. P., Prost, R. W., Ulmer, J. L., Smith, M. M., Daniels, D. L., Strottmann, J. M., Brown, W. D., & Hacein-Bey, L. (2001). Pictorial review of glutamate excitotoxicity: Fundamental concepts for neuroimaging. *Am J Neuroradiol.*, 22, 1813.

Markom, M., Hasan, M., Daud, W. R. W., Singh, H., & Jahim, J. M. (2007). Extraction of hydrolysable tannins from *Phyllanthus niruri* Linn.: Effects of solvents and extraction methods. *Sep Purif Technol.*, 52, 487–496.

Martin, L. J., Al-Abdulla, N. A., Brambrink, A. M., Kirsch, J. R., Sieber, F. E., & Portera-Cailliau, C. (1998). Neurodegeneration in excitotoxicity, global cerebral ischemia, and target deprivation: A perspective on the contributions of apoptosis and necrosis. *Brain Res Bull.*, 46, 281–309.

Merad, M., Soufi, W., Ghalem, S., Boukli, F., Baig, M. H., Ahmad, K., & Kamal, M. A. (2014). Molecular interaction of acetylcholinesterase with carnosic acid derivatives: A neuroinformatics study. *CNS Neurol Disord Drug Targets.*, 13, 440–446.

Miroddi, M., Navarra, M., Quattropani, M. C., Calapai, F., Gangemi, S., & Calapai, G. (2014). Systematic review of clinical trials assessing pharmacological properties of Salvia species on memory, cognitive impairment and Alzheimer's disease. *CNS Neurosci Ther.*, 20, 485–495.

Miura, K., Satoh, M., Kinouchi, M., Yamamoto, K., Hasegawa, Y., Kakugawa, Y., Kawai, M., Uchimi, K., Aizawa, H., Ohnuma, S., Kajiwara, T., Sakurai, H., & Fujiya, T. (2015). The use of natural products in colorectal cancer drug discovery. *Expert Opin Drug Discov.*, 10, 411–426.

Mokdad, A. A., Lopez, A. D., Shahraz, S., Lozano, R., Mokdad, A. H., Stanaway, J., Murray, C. J., & Naghavi, M. (2014). Liver cirrhosis mortality in 187 countries between 1980 and 2010: A systematic analysis. *BMC Med.*, 12, 145.

Molares, S., & Ladio, A. (2009). Ethnobotanical review of the Mapuche medicinal flora: Use patterns on a regional scale. *J Ethnopharmacol.*, 122, 251–260.

Moolenaar, W. H., Van Meeteren, L. A., & Giepmans, B. N. (2004). The ins and outs of lysophosphatidic acid signaling. *Bioessays*, 26, 870–881.

Rasadah Mat Ali, R.M., Samah, Z.A., Mustapha, N.M., & Hussein, N. *L. pumila* (Benth. & Hook) (2010). In: ASEAN Herbal and Medicinal Plants (page 172) (PDF). Jakarta, Indonesia..

Nabavi, S. F., Tenore, G. C., Daglia, M., Tundis, R., Loizzo, M. R., & Nabavi, S. M. (2015). The cellular protective effects of rosmarinic acid: From bench to bedside. *Curr Neurovasc Res*, 12, 98–105.

Nadia, M., Nazrun, A., Norazlina, M., Isa, N., Norliza, M., & Ima Nirwana, S. (2012). The anti-inflammatory, phytoestrogenic, and antioxidative role of *Labisia pumila* in prevention of postmenopausal osteoporosis. *Adv Pharm Sci.*, 2012: 706905.

Nah, S.Y., Kim, D.H., & Rhim, H. (2007). Ginsenosides: Are any of them candidates for drugs acting on the central nervous system? *CNS Drug Rev.*, 13, 381–404.

Najmi, A., Nasiruddin, M., Khan, R. A., & Haque, S. F. (2008). Effect of *Nigella sativa* oil on various clinical and biochemical parameters of insulin resistance syndrome. *Int J Diabetes Dev Ctries.*, 28, 11.

Narayanaswamy, R., & Ismail, I. S. (2015). Cosmetic potential of Southeast Asian herbs: An overview. *Phytochem Rev.*, 14, 419–428.

Nash, K. M., & Shah, Z. A. (2015). Current perspectives on the beneficial role of *Ginkgo biloba* in neurological and cerebrovascular disorders. *Integr Med Insights.*, 10, 1–9.

Neag, M. A., Mocan, A., Echeverria, J., Pop, R. M., Bocsan, C. I., Crisan, G., & Buzoianu, A. D. (2018). Berberine: Botanical occurrence, traditional uses, extraction methods, and relevance in cardiovascular, metabolic, hepatic, and renal disorders. *Front Pharmacol.*, 9, 557.

Neamsuvan, O., & Ruangrit, T. (2017). A survey of herbal weeds that are used to treat gastrointestinal disorders from southern Thailand: Krabi and Songkhla provinces. *J Ethnopharmacol.*, 196, 84–93.

Neveu, V., Perez-Jiménez, J., Vos, F., Crespy, V., Du Chaffaut, L., Mennen, L., Knox, C., Eisner, R., Cruz, J., Wishart, D., & Scalbert, A. (2010). Phenol-explorer: An online comprehensive database on polyphenol contents in foods. *Database.*, 2010: bap024.

Noureddini, M., Nadali, M., & Banafshe, H. R. (2018). Effect of aqueous extract of *Quercus infectoria* gall on the basic contractility, frequency and strength of isolated virgin rat uterus smooth muscle. *Health Biotechnol Biopharma.*, 1(4), 56–65.

Oniga, I., Parvu, A. E., Toiu, A., & Benedec, D. (2007). Effects of *Salvia officinalis* L. extract on experimental acute inflammation. *Rev Med Chir Soc Med Nat Iasi.*, 111, 290–294.

Opalchenova, G., & Obreshkova, D. (2003). Comparative studies on the activity of basil—an essential oil from *Ocimum basilicum* L. against multidrug resistant clinical isolates of the genera *Staphylococcus*, *Enterococcus* and *Pseudomonas* by using different test methods. *J Microbiol Methods.*, 54, 105–110.

Oyama, Y., Hayashi, A., & Ueha, T. (1993). Ca^{2+}-induced increase in oxidative metabolism of dissociated mammalian brain neurons: effect of extract of *Ginkgo biloba* leaves. *Jpn J Pharmacol.*, 61, 367–370.

Park, J. H., Choi, J. Y., Son, D. J., Park, E. K., Song, M. J., Hellström, M., & Hong, J. T. (2017). Anti-inflammatory effect of titrated extract of *Centella asiatica* in phthalic anhydride-induced allergic dermatitis animal model. *Int J Mol Sci.*, 18, 738.

Patil, C. S., Singh, V. P., Satyanarayan, P. S., Jain, N. K., Singh, A., & Kulkarni, S. K. (2003). Protective effect of flavonoids against aging- and lipopolysaccharide-induced cognitive impairment in mice. *Pharmacology*, 69, 59–67.

Perry, N. S. L., Menzies, R., Hodgson, F., Wedgewood, P., Howes, M. R., Brooker, H. J., Wesnes, K. A., & Perry, E. K. (2018). A randomised double-blind placebo-controlled pilot trial of a combined extract of sage, rosemary and melissa, traditional herbal medicines, on the enhancement of memory in normal healthy subjects, including influence of age. *Phytomedicine*, 39, 42–48.

Piao, Y. L., & Liang, X. C. (2014). Astragalus membranaceus injection combined with conventional treatment for viral myocarditis: A systematic review of randomized controlled trials. *Chin J Integr Med.*, 20, 787–791.

Picard, F., Kurtev, M., Chung, N., Topark-Ngarm, A., Senawong, T., Machado De Oliveira, R., Leid, M., Mcburney, M. W., & Guarente, L. (2004). Sirt1 promotes fat mobilization in white adipocytes by repressing PPAR-gamma. *Nature.*, 429, 771–776.

Polyak, S. J., Morishima, C., Lohmann, V., Pal, S., Lee, D. Y., Liu, Y., Graf, T. N., & Oberlies, N. H. (2010). Identification of hepatoprotective flavonolignans from silymarin. *Proc Natl Acad Sci USA.*, 107, 5995–5999.

Pradeep, A., Dinesh, M., Govindaraj, A., Vinothkumar, D., & Ramesh Babu, N. G. (2014). Phytochemical analysis of some important medicinal plants. *Int J Biol Pharm Res.*, 5, 48–50.

Pradhan, S. C., & Girish, C. (2006). Hepatoprotective herbal drug, silymarin from experimental pharmacology to clinical medicine. *Indian J Med Res.*, 124, 491–504.

Prochazkova, K.; Sejna, I.; Skutil, J. & Hahn, A. (2018). *Ginkgo biloba* extract EGb 761⁰ versus pentoxifylline in chronic tinnitus: A randomized, double-blind clinical trial. *Int J Clin Pharm.*, 40, 1335–1341.

Prueksrisakul, T., Chantarangsu, S., & Thunyakitpisal, P. (2015). Effect of daily drinking of Aloe vera gel extract on plasma total antioxidant capacity and oral pathogenic bacteria in healthy volunteer: A short-term study. *J Complement Integr Med.*, 12, 159–164.

Prybylowski, K., & Wenthold, R. J. (2004). N-methyl-D-aspartate receptors: Subunit assembly and trafficking to the synapse. *J Biol Chem.*, 279, 9673–9676.

Puttarak, P., Dilokthornsakul, P., Surasak, S., Dhippayom, T., Kon Gkaew, C., Sruamsiri, R., Chuthaputti, A., & Chaiyakunapruk, N. (2017). Effects of Centella asiatica (L.) Urb. on cognitive function and mood related outcomes: A systematic review and meta-analysis. *Sci Rep.*, 7, 10646.

Radha, M. H., & Laxmipriya, N. P. (2015). Evaluation of biological properties and clinical effectiveness of Aloe vera: A systematic review. *J Tradit Complement Med.*, 5, 21–26.

Rahman, S. A., Ahmad, N. A., Samat, N. H. A., Zahri, S., Abdullah, A. R., & Chan, K. L. (2017). The potential of standardized quassinoid-rich extract of –Eurycoma longifolia in the regulation of the oestrous cycle of rats. *Asian Pac J Trop Biomed.*, 7, 27–31.

Rajeshkumar, N. V., Joy, K. L., Kuttan, G., Ramsewak, R. S., Nair, M. G., & Kuttan, R. (2002). Antitumour and anticarcinogenic activity of *Phyllanthus amarus* extract. *J Ethnopharmacol.*, 81, 17–22.

Ramalashmi, K., Prasanna Vengatesh, K., Magesh, K., Sanjana, R., Siril Joe, S., & Ravibalan, K. (2018). A potential surface sterilization technique and culture media for the isolation of endophytic bacteria from *Acalypha indica* and its antibacterial activity. *J Med Plants Stud.*, 6, 181–184.

Rambaldi, A., Jacobs, B. P., Iaquinto, G., & Gluud, C. (2005). Milk thistle for alcoholic and/ or hepatitis B or C liver diseases: A systematic cochrane hepato-biliary group review with meta-analyses of randomized clinical trials. *Am J Gastroenterol.*, 100, 2583–2591.

Ramezani, M., Azarabadi, M., Fallah Huseini, H., Abdi, H., Baher, G., & Huseini, M. (2008). The effects of *Silybum marianum* (L.) Gaertn. seed extract on glycemic control in Type II diabetic patient's candidate for insulin therapy visiting endocrinology clinic in Baqiyatallah Hospital in the Years of 2006. *J Med Plants.*, 2, 79–84.

Rasoolijazi, J., Azad, N., Joghataei, M. T., Kerdari, M., Nikbakht, F., & Soleimani, M. (2013). The protective role of carnosic acid against beta-amyloid toxicity in rats. *Sci World J.*, 2013, 917082.

Rehman, S. U., Choe, K., & Yoo, H. H. (2016). Review on a traditional herbal medicine, *Eurycoma longifolia* Jack (*Tongkat Ali*): Its traditional uses, chemistry, evidence-based pharmacology and toxicology. *Molecules*, 21, 331.

Ren, S., Zhang, H., Mu, Y., Sun, M., & Liu, P. (2013). Pharmacological effects of Astragaloside IV: A literature review. *J Tradit Chin Med.*, 33, 413–416.

Renzulli, C., Galvano, F., Pierdomenico, L., Speroni, E., & Guerra, M. C. (2004). Effects of rosmarinic acid against aflatoxin B1 and ochratoxin-A-induced cell damage in a human hepatoma cell line (Hep G2). *J Appl Toxicol.*, 24, 289–296.

Roberts, K. T. (2011). The potential of Fenugreek (*Trigonella foenum*-graecum) as a functional food and nutraceutical and its effects on glycemia and lipidemia. *J Med Food.*, 14, 1485–1489.

Rocher, M. N., Carre, D., Spinnewyn, B., Schulz, J., Delaflotte, S., Pignol, B., Chabrier, P. E., & Auguet, M. (2011). Long-term treatment with standardized *Ginkgo biloba* extract (EGb 761) attenuates cognitive deficits and hippocampal neuron loss in a gerbil model of vascular dementia. *Fitoterapia.*, 82, 1075–1080.

Rokaya, M. B., Münzbergová, Z., Shrestha, M. R., & Timsina, B. (2012). Distribution patterns of medicinal plants along an elevational gradient in central Himalaya, Nepal. *J Mt Sci.*, 9, 201–213.

Rokaya, M. B., Uprety, Y., Poudel, R. C., Timsina, B., Münzbergová, Z., Asselin, H., Tiwari, A., Shrestha, S. S., & Sigdel, S. R. (2014). Traditional uses of medicinal plants in gastrointestinal disorders in Nepal. *J Ethnopharmacol.*, 158, 221–229.

Ryan, E. T. (2011). The cholera pandemic, still with us after half a century: Time to rethink. *PLoS Negl Trop Dis.*, 5, e1003.

Sallam, A., Mira, A., Ashour, A., & Shimizu, K. (2016). Acetylcholine esterase inhibitors and melanin synthesis inhibitors from *Salvia officinalis*. *Phytomedicine.*, 23, 1005–1011.

Saller, R., Meier, R., & Brignoli, R. (2001). The use of silymarin in the treatment of liver diseases. *Drugs*, 61, 2035–2063.

Salous, A. K., Panchatcharam, M., Sunkara, M., Mueller, P., Dong, A., Wang, Y., Graf, G. A., Smyth, S. S., & Morris, A. J. (2013). Mechanism of rapid elimination of lysophosphatidic acid and related lipids from the circulation of mice. *J Lipid Res.*, 54, 2775–2784.

Saslis-Lagoudakis, C. H., Williamson, E. M., Savolainen, V., & Hawkins, J. A. (2011). Cross-cultural comparison of three medicinal floras and implications for bioprospecting strategies. *J Ethnopharmacol.*, 135, 476–487.

Schwager, J., Richard, N., Fowler, A., Seifert, N., & Raederstorff, D. (2016). Carnosol and related substances modulate chemokine and cytokine production in macrophages and chondrocytes. *Molecules*, 21, 465.

Selvamani, S., & Balamurugan, S. (2015). Antibacterial and antifungal activities of different organic solvent extracts of *Acalypha indica* (Linn.). *Asian J Plant Sci Res.*, 5, 52–55.

Shang, Y. H., Tian, J. F., Hou, M., & Xu, X. Y. (2013). Progress on the protective effect of compounds from natural medicines on cerebral ischemia. *Chin J Nat Med.*, 11, 588–595.

Shin, B. K., Kwon, S. W., & Park, J. H. (2015). Chemical diversity of ginseng saponins from Panax ginseng. *J Ginseng Res.*, 39, 287–298.

Shrestha, S., Kaushik, V. S., Eshwarappa, R. S. B., Subaramaihha, S. R., Ramanna, L. M., & Lakkappa, D. B. (2014). Pharmacognostic studies of insect gall of *Quercus infectoria* Olivier (Fagaceae). *Asian Pac J Trop Biomed.*, 4, 35.

Siew, Y. Y., Zareisedehizadeh, S., Seetoh, W. G., Neo, S. Y., Tan, C. H., & Koh, H. L. (2014). Ethnobotanical survey of usage of fresh medicinal plants in Singapore. *J Ethnopharmacol.*, 155, 1450–1466.

Sivaperumal, R., Ramya, S., Ravi, A. V., Rajasekaran, C., & Jayakumararaj, R. (2009). Herbal remedies practiced by Malayalis to treat skin diseases. *Environment We*, 4, 35–44.

Smith, L. A., Cornelius, V. R., Plummer, C. J., Levitt, G., Verrill, M., Canney, P., & Jones, A. (2010). Cardiotoxicity of anthracycline agents for the treatment of cancer: Systematic review and meta-analysis of randomised controlled trials. *BMC Cancer.*, 10, 337.

Smith, S. C., Jr. (2007). Multiple risk factors for cardiovascular disease and diabetes mellitus. *Am J Med.*, 120, S3–S11.

Sow, S. O., Muhsen, K., Nasrin, D., et al. (2016). The Burden of cryptosporidium diarrheal disease among children < 24 months of age in moderate/high mortality regions of Sub-Saharan Africa and South Asia, utilizing data from the Global Enteric Multicenter Study (GEMS). *PLoS Negl Trop Dis.*, 10: e0004729.

Spiegel, R., Kalla, R., Mantokoudis, G., Maire, R., Mueller, H., Hoerr, R., & Ihl, R. (2018). Ginkgo biloba extract EGb 761((R)) alleviates neurosensory symptoms in patients with dementia: A meta-analysis of treatment effects on tinnitus and dizziness in randomized, placebo-controlled trials. *Clin Interv Aging.*, 13, 1121–1127.

Srinivasan, K. (2006). Fenugreek (*Trigonella foenum*-graecum): A review of health beneficial physiological effects. *Food Rev Int.*, 22, 203–224.

Sudhakar, C.; Vankudothu, N.; Panjala, S.; Rao, N. B., & Anupalli, R. R. (2016). Phytochemical analysis, anti-oxidant and anti-microbial activity of *Acalypha indica* leaf extracts in different organic solvents. *Int J Phytomed.*, 8, 444–452.

Sugiyama, T., & Asaka, M. (2004). *Helicobacter pylori* infection and gastric cancer. *Med Electron Microsc.*, 37, 149–157.

Sulniute, V., Ragazinskiene, O., & Venskutonis, P. R. (2016). Comprehensive evaluation of antioxidant potential of 10 salvia species using high pressure methods for the isolation of lipophilic and hydrophilic plant fractions. *Plant Foods Hum Nutr.*, 71, 64–71.

Sun, C., Lai, X., Huang, X., & Zeng, Y. (2014). Protective effects of ginsenoside Rg1 on astrocytes and cerebral ischemic-reperfusion mice. *Biol Pharm Bull.*, 37, 1891–1898.

Sun, Y., Liu, Y., & Chen, K. (2016). Roles and mechanisms of ginsenoside in cardiovascular diseases: Progress and perspectives. *Sci China Life Sci.*, 59, 292–298.

Syamasundar, K. V., Singh, B., Thakur, R. S., Husain, A., Kiso, Y., & Hikino, H. (1985). Antihepatotoxic principles of *Phyllanthus niruri* herbs. *J Ethnopharmacol.*, 14, 41–44.

Tabassum, N., & Hamdani, M. (2014). Plants used to treat skin diseases. *Pharmacogn Rev.*, 8, 52–60.

Takemura, G., & Fujiwara, H. (2007). Doxorubicin-induced cardiomyopathy from the cardiotoxic mechanisms to management. *Prog Cardiovasc Dis.*, 49, 330–352.

Takuma, K., Mizoguchi, H., Funatsu, Y., Kitahara, Y., Ibi, D., Kamei, H., Matsuda, T., Koike, K., Inoue, M., Nagai, T., & Yamada, K. (2012). Placental extract improves hippocampal neuronal loss and fear memory impairment resulting from chronic restraint stress in ovariectomized mice. *J Pharmacol Sci.*, 120, 89–97.

Tanaka, K., Tomisato, W., Hoshino, T., Ishihara, T., Namba, T., Aburaya, M., Katsu, T., Suzuki, K., Tsutsumi, S., & Mizushima, T. (2005). Involvement of intracellular Ca^{2+} levels in nonsteroidal anti-inflammatory drug-induced apoptosis. *J Biol Chem.*, 280, 31059–31067.

Tanaka, M., Misawa, E., Yamauchi, K., Abe, F., & Ishizaki, C. (2015). Effects of plant sterols derived from Aloe vera gel on human dermal fibroblasts in vitro and on skin condition in Japanese women. *Clin Cosm Invest Dermatol.*, 8, 95–104.

Tanaka, T., Horiuchi, G., Matsuoka, M., Hirano, K., Tokumura, A., Koike, T., & Satouchi, K. (2009). Formation of lysophosphatidic acid, a wound-healing lipid, during digestion of cabbage leaves. *Biosci Biotech Biochem.*, 73, 1293–1300.

Tawfik, H. E., El-Remessy, A. B., Matragoon, S., Ma, G., Caldwell, R. B., & Caldwell, R. W. (2006). Simvastatin improves diabetes-induced coronary endothelial dysfunction. *J Pharmacol Exp Ther.*, 319, 386.

Teng, Y., Zhang, M. Q., Wang, W., Liu, L. T., Zhou, L. M., Miao, S. K., & Wan, L. H. (2014). Compound danshen tablet ameliorated $a\beta_{25-35}$-induced spatial memory impairment in mice via rescuing imbalance between cytokines and neurotrophins. *BMC Complement Alternat Med*, 14, 23.

Thompson, K. E., Ray, R. M., Alli, S., Ge, W., Boler, A., Shannon Mccool, W., Meena, A. S., Shukla, P. K., Rao, R., Johnson, L. R., Miller, M. A., & Tigyi, G. J. (2018). Prevention and treatment of secretory diarrhea by the lysophosphatidic acid analog Rx100. *Exp Biol Med.*, 243(13), 1056–1065.

Tokumura, A. (2011). Physiological significance of lysophospholipids that act on the lumen side of mammalian lower digestive tracts. *J Health Sci.*, 57, 115–128.

Tu, L., Pan, C. S., Wei, X. H., Yan, L., Liu, Y. Y., Fan, J. Y., Mu, H. N., Li, Q., Li, L., Zhang, Y., He, K., Mao, X. W., Sun, K., Wang, C. S., Yin, C. C., & Han, J. Y. (2013). Astragaloside IV protects heart from ischemia and reperfusion injury via energy regulation mechanisms. *Microcirculation.*, 20, 736–747.

Umukoro, S., & Ashorobi, R. (2006). Evaluation of anti-inflammatory and membrane stabilizing property of aqueous leaf extract of *Momordica charantia* in rats. *Afr J Biomed Res.*, 9, 119–124.

Uprety, Y., Asselin, H., Dhakal, A., & Julien, N. (2012). Traditional use of medicinal plants in the boreal forest of Canada: Review and perspectives. *J Ethnobiol Ethnomed.*, 8, 7.

Vajda, F. J. (2002). Neuroprotection and neurodegenerative disease. *J Clin Neurosci.*, 9, 4–8.

Von Boetticher, A. (2011). Ginkgo biloba extract in the treatment of tinnitus: A systematic review. *Neuropsychiatr Dis Treat.*, 7, 441–447.

Wahab, N., Yusoff, W., Shuib, A., Wan, N., & Khatiza, H. (2011). *Labisia pumila* has similar effects to estrogen on the reproductive hormones of ovariectomized rats. *Internet J Herbal Plant Med.*, 2, 1–6.

Walsh, K. (2006). Akt signaling and growth of the heart. *Circulation*, 113, 2032–2034.

Wang, C., Li, J., Lv, X., Zhang, M., Song, Y., Chen, L., & Liu, Y. (2009). Ameliorative effect of berberine on endothelial dysfunction in diabetic rats induced by high-fat diet and streptozotocin. *Eur J Pharmacol.*, 620, 131–137.

Wang, K., Zhang, D., Wu, J., Liu, S., Zhang, X., & Zhang, B. (2017). A comparative study of Danhong injection and *Salvia miltiorrhiza* injection in the treatment of cerebral infarction: A systematic review and meta-analysis. *Medicine (Baltimore)*, 96: e7079.

Wang, L. S., Lee, C. T., Su, W. L., Huang, S. C., & Wang, S. C. (2016). *Delonix regia* Leaf extract (DRLE): A potential therapeutic agent for cardioprotection. *PLoS One.*, 11: e0167768.

Wang, N., Chen, X., Geng, D., Huang, H., & Zhou, H. (2013). *Ginkgo biloba* leaf extract improves the cognitive abilities of rats with D-galactose induced dementia. *J Biomed Res.*, 27, 29–36.

Wang, S. G., Xu, Y., Xie, H., Wang, W., & Chen, X. H. (2015). Astragaloside IV prevents lipopolysaccharide-induced injury in H9C2 cardiomyocytes. *Chin J Nat Med.*, 13, 127–132.

Wang, X., Morris-Natschke, S. L., & Lee, K. H. (2007). New developments in the chemistry and biology of the bioactive constituents of Tanshen. *Med Res Rev.*, 27, 133–148.

Wang, Y., Yi, X., Ghanam, K., Zhang, S., Zhao, T., & Zhu, X. (2014). Berberine decreases cholesterol levels in rats through multiple mechanisms, including inhibition of cholesterol absorption. *Metabolism.*, 63, 1167–1177.

Wani, S. A., & Kumar, P. (2018). Fenugreek: A review on its nutraceutical properties and utilization in various food products. *J Saudi Soc Agric Sci.*, 17, 97–106.

Williams, B., Watanabe, C. M., Schultz, P. G., Rimbach, G., & Krucker, T. (2004). Age-related effects of *Ginkgo biloba* extract on synaptic plasticity and excitability. *Neurobiol Aging.*, 25, 955–962.

World Health Organization. 2018. Cardiovascular disease [Online]. Available: http://www.who. int/cardiovascular_diseases/en/.

Wu, G. B., Zhou, E. X., & Qing, D. X. (2009). Tanshinone II(A) elicited vasodilation in rat coronary arteriole: roles of nitric oxide and potassium channels. *Eur J Pharmacol.*, 617, 102–107.

Wu, Y., Wu, Z., Butko, P., Christen, Y., Lambert, M. P., Klein, W. L., Link, C. D., & Luo, Y. (2006). Amyloid-beta-induced pathological behaviors are suppressed by *Ginkgo biloba* extract EGb 761 and ginkgolides in transgenic *Caenorhabditis elegans*. *J Neurosci.*, 26, 13102–13113.

Xie, Z., Shi, M., Zhang, C., Zhao, H., Hui, H., & Zhao, G. (2016). Ginsenoside Rd protects against cerebral ischemia-reperfusion injury via decreasing the expression of the NMDA receptor 2B subunit and its phosphorylated product. *Neurochem Res.*, 41, 2149–2159.

Xu, Q. Q., Xu, Y. J., Yang, C., Tang, Y., Li, L., Cai, H. B., Hou, B. N., Chen, H. F., Wang, Q., Shi, X. G., & Zhang, S. J. (2016). Sodium tanshinone IIA sulfonate attenuates scopolamine-induced cognitive dysfunctions via improving cholinergic system. *Biomed Res Int.*, 2016, 9852536.

Xu, S. L., Bi, C. W., Choi, R. C., Zhu, K. Y., Miernisha, A., Dong, T. T., & Tsim, K. W. (2013). Flavonoids induce the synthesis and secretion of neurotrophic factors in cultured rat astrocytes: A signaling response mediated by estrogen receptor. *Evid Based Complement Alternat Med.*, 2013, 127075.

Yang, Q. Y., Lu, S., & Sun, H. R. (2010). Effects of astragalus on cardiac function and serum tumor necrosis factor-α level in patients with chronic heart failure. *Zhongguo Zhong Xi Yi Jie He Za Zhi.*, 30, 699–701.

Yang, T., Sun, Y., Lu, Z., Leak, R. K., & Zhang, F. (2017). The impact of cerebrovascular aging on vascular cognitive impairment and dementia. *Ageing Res Rev.*, 34, 15–29.

Yang, T. L., Lin, F. Y., Chen, Y. H., Chiu, J. J., Shiao, M. S., Tsai, C. S., Lin, S. J., & Chen, Y. L. (2011). Salvianolic acid B inhibits low-density lipoprotein oxidation and neointimal hyperplasia in endothelium-denuded hypercholesterolaemic rabbits. *J Sci Food Agric.*, 91, 134–141.

Yang, T. Y., Wei, J. C., Lee, M. Y., Chen, C. M., & Ueng, K. C. (2012). A randomized, double-blind, placebo-controlled study to evaluate the efficacy and tolerability of Fufang Danshen (Salvia miltiorrhiza) as add-on antihypertensive therapy in Taiwanese patients with uncontrolled hypertension. *Phytother Res.*, 26, 291–298.

Yang, Y., Cai, F., Li, P. Y., Li, M. L., Chen, J., Chen, G. L., Liu, Z. F., & Zeng, X. R. (2008). Activation of high conductance Ca^{2+}-activated K^+ channels by sodium tanshinone II-A sulfonate (DS-201) in porcine coronary artery smooth muscle cells. *Eur J Pharmacol.*, 598, 9–15.

Yang, Y., Li, X., Zhang, L., Liu, L., Jing, G., & Cai, H. (2015). Ginsenoside Rg1 suppressed inflammation and neuron apoptosis by activating PPARγ/HO-1 in hippocampus in rat model of cerebral ischemia-reperfusion injury. *Int J Clin Exp Pathol.*, 8, 2484–2494.

Yao, R. Q., Qi, D. S., Yu, H. L., Liu, J., Yang, L. H., & Wu, X. X. (2012). Quercetin attenuates cell apoptosis in focal cerebral ischemia rat brain via activation of BDNF-TrkB-PI3K/Akt signaling pathway. *Neurochem Res.*, 37, 2777–2786.

Yao, Z., Drieu, K., & Papadopoulos, V. (2001). The *Ginkgo biloba* extract EGb 761 rescues the PC12 neuronal cells from beta-amyloid-induced cell death by inhibiting the formation of beta-amyloid-derived diffusible neurotoxic ligands. *Brain Res.*, 889, 181–190.

Ye, R., Zhao, G., & Liu, X. (2013). Ginsenoside Rd for acute ischemic stroke: Translating from bench to bedside. *Expert Rev Neurother.*, 13, 603–613.

Yin, B., Liang, H., Chen, Y., Chu, K., Huang, L., Fang, L., Matro, E., Jiang, W., & Luo, B. (2013). EGB1212 post-treatment ameliorates hippocampal CA1 neuronal death and memory impairment induced by transient global cerebral ischemia/reperfusion. *Am J Chin Med*, 41, 1329–1341.

Yin, B., Xu, Y., Wei, R., & Luo, B. (2014a). *Ginkgo biloba* on focal cerebral ischemia: A systematic review and meta-analysis. *Am J Chin Med*, 42, 769–783.

Yin, Y., Qi, F., Song, Z., Zhang, B., & Teng, J. (2014b). Ferulic acid combined with astragaloside IV protects against vascular endothelial dysfunction in diabetic rats. *BioScience Trends.*, 8, 217–226.

Yu, Q. T., Qi, L. W., Li, P., Yi, L., Zhao, J., & Bi, Z. (2007). Determination of seventeen main flavonoids and saponins in the medicinal plant Huang-qi (*Radix astragali*) by HPLC-DAD-ELSD. *J Sep Sci.*, 30, 1292–1299.

Yulianti, L., Bramono, K., Mardliyati, E., & Freisleben, H. J. (2016). Effects of *Centella asiatica* ethanolic extract encapsulated in chitosan nanoparticles on proliferation activity of skin fibroblasts and keratinocytes, type I and III collagen synthesis and aquaporin 3 expression in vitro. *J Pharm Biomed Sci.*, 6, 315–327.

Yun, C. C., & Kumar, A. (2015). Diverse roles of LPA signaling in the intestinal epithelium. *Exp Cell Res.*, 333, 201–207.

Yung, Y. C., Stoddard, N. C., & Chun, J. (2014). LPA receptor signaling: Pharmacology, physiology, and pathophysiology. *J Lipid Res.*, 55, 1192–1214.

Zaha, V. G.; Qi, D.; Su, K. N.; Palmeri, M.; Lee, H. Y.; Hu, X.; Wu, X.; Shulman, G. I.; Rabinovitch, P. S.; Russell, R. R., 3rd; & Young, L. H. (2016). AMPK is critical for mitochondrial function during reperfusion after myocardial ischemia. *J Mol Cell Cardiol.*, 91, 104–113.

Zahidin, N. S., Saidin, S., Zulkifli, R. M., Muhamad, I.I., Ya'akob, H., & Nur, H. (2017). A review of *Acalypha indica* L. (Euphorbiaceae) as traditional medicinal plant and its therapeutic potential. *J Ethnopharmacol.*, 207, 146–173.

Zaidi, S. F. H., Yamada, K., Kadowaki, M., Usmanghani, K., & Sugiyama, T. (2009). Bactericidal activity of medicinal plants, employed for the treatment of gastrointestinal ailments, against *Helicobacter pylori*. *J Ethnopharmacol.*, 121, 286–291.

Zeng, X.H., Zeng, X.J., & Li, Y.Y. (2003). Efficacy and safety of berberine for congestive heart failure secondary to ischemic or idiopathic dilated cardiomyopathy. *Am J Cardiol.*, 92, 173–176.

Zeng, X., & Zeng, X. (1999). Relationship between the clinical effects of berberine on severe congestive heart failure and its concentration in plasma studied by HPLC. *Biomed Chromatogr.*, 13, 442–444.

Zhang, G., Liu, A., Zhou, Y., San, X., Jin, T., & Jin, Y. (2008a). Panax ginseng ginsenoside-Rg2 protects memory impairment via anti-apoptosis in a rat model with vascular dementia. *J Ethnopharmacol.*, 115, 441–448.

Zhang, Q. J., Wang, Z., Chen, H. Z., Zhou, S., Zheng, W., Liu, G., Wei, Y. S., Cai, H., Liu, D. P., & Liang, C. C. (2008b). Endothelium-specific overexpression of class III deacetylase SIRT1 decreases atherosclerosis in apolipoprotein E-deficient mice. *Cardiovasc Res.*, 80, 191–199.

Zhang, L., Eslick, G. D., Xia, H. H. X., Wu, C., Phung, N., & Talley, N. J. (2010). Relationship between alcohol consumption and active *Helicobacter pylori* infection. *Alcohol Alcohol.*, 45, 89–94.

Zhang, Z., Peng, D., Zhu, H., & Wang, X. (2012). Experimental evidence of *Ginkgo biloba* extract EGB as a neuroprotective agent in ischemia stroke rats. *Brain Res Bull.*, 87, 193–198.

Zhao, Y., Xu, P., Hu, S., Du, L., Xu, Z., Zhang, H., Cui, W., Mak, S., Xu, D., Shen, J., Han, Y., Liu, Y., & Xue, M., (2015). Tanshinone II A, a multiple target neuroprotectant, promotes caveolae-dependent neuronal differentiation. *Eur J Pharmacol.*, 765, 437–446.

Zheng, M.; Xin, Y.; Li, Y.; Xu, F.; Xi, X.; Guo, H.; Cui, X.; Cao, H.; Zhang, X. & Han, C. (2018). Ginsenosides: A potential neuroprotective agent. *BioMed Res. Int.*, 2018, 8174345.

Zhou, L. J., & Zhu, X. Z. (2000). Reactive oxygen species-induced apoptosis in PC_{12} cells and protective effect of bilobalide. *J Pharmacol Exp Ther.*, 293, 982–988.

Zhou, Y., Li, W., Xu, L., & Chen, L. (2011). In *Salvia miltiorrhiza*, phenolic acids possess protective properties against amyloid beta-induced cytotoxicity, and tanshinones act as acetylcholinesterase inhibitors. *Environ Toxicol Pharmacol.*, 31, 443–452.

Zhou, Z., Liu, Y., Miao, A. D., & Wang, S. Q. (2005). Protocatechuic aldehyde suppresses TNF-α-induced ICAM-1 and VCAM-1 expression in human umbilical vein endothelial cells. *Eur J Pharmacol.*, 513, 1–8.

Zhu, Y. P., Shen, T., Lin, Y. J., Chen, B. D., Ruan, Y., Cao, Y., Qiao, Y., Man, Y., Wang, S., & Li, J. (2013). Astragalus polysaccharides suppress ICAM-1 and VCAM-1 expression in TNF-α-treated human vascular endothelial cells by blocking NF-kB activation. *Acta Pharmacol Sin.*, 34, 1036–1042.

Zuo, W., Yan, F., Zhang, B., Li, J., & Mei, D. (2017). Advances in the studies of *Ginkgo biloba* leaves extract on aging-related diseases. *Aging Dis.*, 8, 812–826.

CHAPTER 5

Herbs in Cancer Therapy

ANNUM MALIK[1], SHAHZADI SIDRA SALEEM[1], KIFAYAT ULLAH SHAH[1*], LEARN-HAN LEE[2], BEY HING GOH[2], and TAHIR MEHMOOD KHAN[2,3]

[1]*Department of Pharmacy, Quaid-i-Azam University, Islamabad, Pakistan*

[2]*School of Pharmacy, Monash University, Bandar Sunway 45700, Selangor, Malaysia*

[3]*Institute of Pharmaceutical Sciences, University of Veterinary and animal sciences, Outfall Road, Lahore, Pakistan*

*Corresponding author. E-mail: tahir.mehmood@monash.edu

ABSTRACT

Herbs are used to treat oncology-related disorder from ancient times. Various folklore literatures approve the use of variety of herbal supplements that has been observed to be effective in the management of cancer. However, today, there is an urgent need of high-quality trials and experimental studies to investigate and report the clinical efficacy and safety of herbs in treatment of cancer. More information is needed about the doses of herbal compounds to bring a specific response without causing undesirable effects.

5.1 INTRODUCTION

Amongst the myriad number of diseases, cancer is responsible for contributing to the second highest death toll on earth (Tunstall-Pedoe 2006). It is characterized by the ability of cells to continuously proliferate, without being controlled. The most commonly used treatment for cancer is chemotherapy, in addition to radiotherapy and drugs of chemical nature.

Chemotherapeutic agents target cancer cells only as they lose most of the regulatory functions exhibited by the normal cells. Chemotherapy, however, comes with deleterious side effects. For example, a frequently used chemotherapeutic agent, 5-fluorouracil may cause myelotoxicity and cardiotoxicity (Rexroth and Scotland 1994) in cancer patients. Such toxicities create a hindrance in the treatment of cancer with the allopathic treatment. Henceforth, various naturally occurring compounds have been suggested to combat cancer as an alternative means of therapy, since they have fewer side effects. Although there are numerous plants possessing the anticancerous activity, only four classes of plant derivatives are being marketed today. These include *vinca* alkaloids such as vincristine, vinblastine and vindesine; epipodophyllotoxins such as etoposide and teniposid; the taxanes such as paclitaxel and docetaxel; and the camptothecin derivatives, which mainly include camptothecin and irinotecan (Desai et al. 2008).

Medicinal herbs and their phyto-derivatives are increasingly gaining popularity in terms of their utility in anticancer therapy. When used with the conventional treatment, these herbs showed immense improvement in survival, modulation of immune system, and quality of life of patients suffering from cancer, in different clinical studies (Yin et al. 2013). Different herbs show specificity for different types of cancers. For example, 200 mg of Vitamin A, fenretinide, lowers the chances of recurrence of breast cancer in women (Gerber et al. 2006). Lycopene lowers the risk of prostate cancer in humans (Giovannucci et al. 2002), whereas *Prunus armeniaca*, from the Rosaceae family, fights against lung cancer (Bröker and Giaccone 2002). Curcumin is known to inhibit the progression of liver fibrogenesis and carcinogenesis (Luk et al. 2007). New technologies are now being developed that can formulate these herbs as nanoparticles in the nanomedicine. This advancement in nanotechnology can not only control the release of the drug but also improve the efficacy of the plant-based product, as well as investigate new ways of administration (Greenwell et al. 2015).

The graph in Figure 5.1 shows a tremendous increase in the changes in the use of US herbs and spice. A plethora of herbs and spices, which were initially used as constituents of diet, have been shown to have anticancer properties (Kaefer and Milner 2008). The antimicrobial, antitumorigenic, and antioxidant activities possessed by these culinary herbs and spices majorly contribute to the antineoplastic behavior.

5.2 MEDICINAL HERBS AND CANCER

Medicinal herbs may include plants, plant preparations, herbal complexes, or products. The Plant kingdom has been known to possess anticancerous activity for a very long time now (Table 5.1). *Podophyllum peltatum* (mayapple) was used for the extraction of lignans and podophyllotoxins that led to the development of drugs for curing testicular and lung cancer (Pettit et al. 1995). It is rarely possible to develop a chemically derived drug that is not toxic to healthy cells and has high specificity to the cytotoxic nature of cancerous cells. This hurdle in cancer treatment has placed naturally occurring products into limelight. Active compounds derived from plant species can induce apoptosis in cancer cells, thereby suppressing their proliferation and causing cellular death (Kumar et al. 2013). The major plant-derived compounds responsible for anticancerous activity are polyphenols, flavonoids, taxols, and brassinosteriods. These have been discussed as follows.

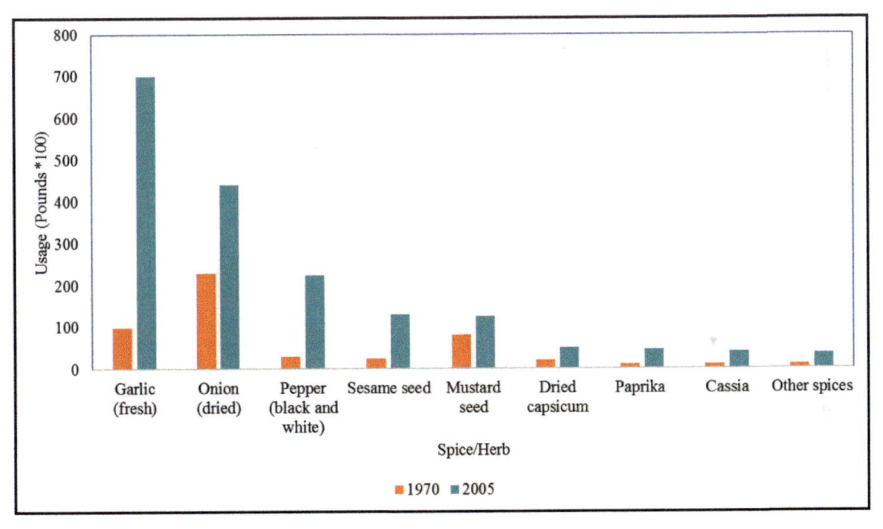

FIGURE 5.1   From 1970 to 2005, tremendous changes in the application of herbs and spices took place as depicted by the bar chart.

5.2.1   POLYPHENOLS

Polyphenols that exhibit anticancerous activity are namely tannins, resveratrol, gallacatechins, curcumin, and flavonoids. Foods such as peanuts and grapes are rich in resveratrol which is a phytoalexin, whereas gallacatechins are found in green tea (Azmi et al. 2006). Antioxidants prevent free radicals from causing cancer in humans; hence, the consumption of such food improves overall health and well-being. The alkaloid taxol was first extracted from *Taxus brevifolia* (*T. brevifolia*). Polyphenols act by inducing apoptotic cellular

TABLE 5.1 Compounds Along with Their Mechanism of Action (Millimouno et al., 2014)

Compounds	Natural Source	Mechanism of Action	Types of Cancers
Flavonoids			
Honokiol	*Magnolia officinalis,*	Antioxidant, antiproliferation, anti-inflamation, immunomodulation, antiangiogenesis	Glioblastoma, leukemia, breast, ovarian, prostate and squamous cell carcinoma
Magnolol	*Magnolia grandiflora Magnolia officinalis, Magnolia obovata*	Antiproliferation, immunomodulation, anticancer, antioxidant and hepatoprotective effects	Glioblastoma, fibrosarcoma, liver cancer, thyroid cancer, neuroblastoma
Sesqueterpenes			
Parthenolide	*Tanacetum parthenium, Tanacetum vulgare*	Antioxidant, autophagy, anti-inflammation, cytotoxic effects	Malignant glioma, Epidermal tumorigenesis, burkitt lymphoma, osteosarcoma
Alantolactone	*Inula helenium,*	Antimicrobial, anticancer, cytotoxicity, oxidoreductase and antiproliferative	Prostate, colon, glioblastoma, liver,lung, leukemia
Diterpenoids			
Oridonin	*Isodons rubescens*	Antioxidant, autophagy, anti-inflammation, immunomodulation	Breast, hepatoma, astrocytoma, leukemia, multiple myloma, murine fibrosarcoma, osteosarcoma
Pseudolaric Acid B	*Pseudolarix kaempferi*	Antiproliferation, immunomodulation, anticancer, antiangiogenesis effects	Microvessel endothelial, glioblastoma, breast, umbilical vein endothelial, murine fibrosarcoma

TABLE 5.1 *(Continued)*

Compounds	Natural Source	Mechanism of Action	Types of Cancers
Polyphenolic			
Wedelolactone	*Eclipta alba, Eclipta prostrata*	Antiproliferation, antioxidant, hepatoprotective effects	Breast, prostate, neuroblastoma, mammary carcinosarcoma, leukemia, adenoma
Alkaloids			
Evodiamine	*Evodia rutaecarpa*	Antioxidant, antiproliferation, antimicrobial, antimetastatic, anticarcinogenesis	Hepatocellular, leukemia, human thyroid cancer, colorectal, cervix carcinoma, breast cancer

death of cancer cells. Copper ions are bound to chromatin. Polyphenols cause the regulation of their mobilization, leading to DNA fragmentation, and hence initiating the process of apoptosis (Azmi et al. 2006).

Another mechanism of action of polyphenols is their ability to promote the growth of proteins of cancer cells. They perform this action by forming different types of bonds such as methylation, phosphorylation, or acetylation. Studies on curcumin have shown to suppress the expression of the tumor necrosis factor (TNF) in cancer cells, thereby causing cytotoxicity (Gupta et al. 2014).

Ginger, also officially known as *Zingiber officinale*, is famous for the usage of its rhizome in tradition therapy. Phenolic compounds of ginger possess anti-cancerous activity by inducing apoptosis in cancer cells (Taraphdar et al. 2001).

5.2.2   FLAVONOIDS

Flavonoids, or bioflavonoids, include about 3000 natural phenolic structures. They commonly occur in almost every vegetable, fruit, and herb. They are also found in tea and coffee. Flavonoids constitute a considerable part of our daily dietary value, mostly in the form of quercetin (Kühnau 1976). They act as anti-inflammatory, enzyme inhibitors, that can potentially improve capillary resistance and battle free radicals. Flavonoids are classified into flavanones, flavones, isoflavones, flavonols, flavanols, and anthocyanidins. Flavanones have limited distribution, found mainly in citrus fruits such as lemons and oranges (Hollman et al. 1997). Flavones are widely distributed, such as luteolin and apigenin. Isoflavones include genistein, which can potentially inhibit human prostate cancer cells, and daidzein. Food rich in isoflavones includes legumes such as soy. Flavonols are found as naturally occurring glycosides, and the major ones include kaempferol and quercetin. Flavanols, or flavan-3-ols, have limited distribution and are found in tea, apples, broccoli, etc. Catechins such as epigallocatechin gallate (EGCG) are an example of flavanols. Anthrocyanidins are red-blue pigments found in berries. They are responsible for pigmentation in fruits.

5.2.2.1   IN VIVO STUDIES

Studies performed on animals to investigate antitumor effects of quercetin concluded that injecting 20-800 mg/kg of the drug into intraperitoneal fluids drastically controlled the growth of human head and neck squamous cell carcinoma in rats (Castillo et al. 1989). Growth of melanoma cells transplanted in mice was inhibited by the intraperitoneal administration of 25 and 50 mg/kg of drug (Caltagirone et al. 2000).

Similarly, Genistein has been shown to stop the growth of melanoma cells in mice at doses of 40 mg/kg (Record et al. 1997) and metastasis of melanoma cells and tumor growth at doses of 120 and 240 mg/kg in mice, respectively (Li et al. 1999).

About 110 mg/kg of EGCG (from tea) prevented the metastasis of transplanted lung cancer in mice (Sazuka et al. 1995).

5.2.3 BRASSINOSTEROIDS

Brassinosteroids are polyhydroxysteroids that belong to a class of phytohormones. They play various regulatory roles in the physiological processes of plants, such as cell division and stem and root elongation. They are referred to as steroidal, owing to their mechanism of action. They bind to the receptors inside the cell membrane, initiating a cascade of regulatory effects involving a large number of genes. Over 40 naturally occurring brassinosteroids have been identified, out of which two were used as anticancerous agents on different cell lines: 28-homocastasterone and 24-epibrassinolide. Brassinosteroids interact with the cell cycle and initiate regulatory responses that cause growth inhibition and apoptotic cell death of the cancerous cells (Malíková et al. 2008).

Brassinosteroids are effective against many types of cell lines, such as lung carcinoma A-549, osteosarcoma HOS cell lines, and cell lines in prostate and breast cancer (Malíková et al. 2008). In breast cancer, the proteins that are targeted includes human epidermal growth factor receptor-2 (HER2) and the estrogen receptor (Steigerová et al. 2010). The androgen receptor is the major protein responsible for the development of prostate cancer and is structurally similar to the estrogen receptor. Brassinosteroids bind with these receptors and cause growth inhibition of the cancer cells (Steigerová et al. 2012).

5.2.4 ALKALOIDS

Alkaloids have potent anticancerous activity. They have heterocyclic ring structures with a nitrogen atom. They are widely distributed in the Plant kingdom and belong to families such as Loganiaceae, Leguminosae, Papaveraceae, Menispermaceae, Solanaceae, and Ranunculaceae (Wang et al. 2009). Following alkaloids are known to possess activity against cancer cell lines:

a) Berberine: It belongs to the isoquinoline class of compounds, found in natural herbs such as Rhizoma coptidisa widely used Chinese herb (Chen et al. 2008). In cancer cells, cell cycle arrest at G_1 or G_2/M phases, as well as cellular death by apoptosis is caused by Berberine (Sun et al. 2009, Eom et al. 2010, Burgeiro et al. 2011).

b) Sanguinarine: It is a benzophenanthridine alkaloid, belonging to the papaveraceae family. It is extracted from *Sanguinaria canadensis* L. and Chelidonium majus L. (Mahady and Beecher 1994). Sanguinarine causes cancerous cells death by inducing apoptosis which ultimately inhibits their proliferation. It also makes the breast cancer cells vulnerable to apoptosis caused by TNF (Kim et al. 2008).

c) Evodiamine: It belongs to the quinolone class of compounds and is extracted from a Chinese herb called *Evodia rutaecarpa* (Table 5.2). Several studies performed to investigate the mechanism of action of evodiamine in cancer prevention illustrated that the herb induces cell cycle arrest, which ultimately results in the control of angiogenesis and metastasis in a number of cancer cell lines (Shyu et al. 2006, Kim et al. 2008).

Other alkaloids that are known to inhibit the proliferation of cancer cell lines include matrine, piperine, tetrandrine, chelerythrine, lycorine (Lu et al. 2012), taxol, and *vinca* alkaloids.

5.2.5 LECTINES

Viscum album L. has been used as an anti-cancerous agent, in adjuvant treatment since a very long time. It boosts up the immune system, that is, it increases both the quantity and the efficiency of neutrophils and natural killer cells. According to an investigation, the antineoplastic effects of lectin were observed by acquiring an extract from *Viscum album* C., which caused apoptosis-induced cellular death in tumor cell lines (Taraphdar et al. 2001).

5.2.6 TERPENOIDS

Xanthorrhizol is the main antineoplastic agent that is a sesquiterpeoid in nature. It is isolated from the rhizome of Curcuma xanthorrhiza Roxb. Several in vivo studies have been performed on the clinical usefullness of Xanthorrhizol, which have shown that the terpenoid prevents the onset and progression of tumors. Enzymes such as ornithine decarboxylase and cyclooxgenase-2

(COX-2) are inhibited by terpenoids resulting in reduced nuclear factor-kB (NF-kB) signaling in an in vivo lung metastasis of a mouse (Cheah et al. 2008).

TABLE 5.2 Compounds Causing Cell Cycle Arrest (Millimouno et al. 2014)

Compounds	Cell-Cycle Arrest (Phase)
Flavonoids	
Magnolol	G_0–G_1
Casticin	G_2–M
Honokiol	G_2–M or G_0-G_1
Jaceosidin	G_2–M
Sesquiterpenes	
Costunolide	G_2–M
Parthenolide	G_2–M
Alantolactone	G_2–M
Isoalantolactone	G_2–M
Diterpenoids	
Pseudolaric acid B	G_2–M
Oridonin	G_2–M
Polyphenolic	
Wedelolactone	S and G_2–M
Alkaloids	
Evodiamine	G_2–M

5.3 FDA APPROVED PLANT-DERIVED MEDICINES

As discussed earlier, despite the large number of herbs possessing anticancerous activities, only four major categories have been approved by the Food and Drug Administration (FDA) so far. They have been discussed in the following sections.

5.3.1 VINCA ALKALOIDS

Vinca alkaloids are extracted from Catharanthus roseus G. Don, the pink periwinkle plant. Vinblastine, vinorelbine, vincristine, and vindesine have currently been approved for their clinical use as anticancerous agents (Moudi et al. 2013). They exhibit cytotoxic effects by interacting with tubulin and interrupting the microtubule function, involving the spindle fiber assembly. This leads to metaphase arrest, causing cellular death (Himes and therapeutics 1991). They are used in combination chemotherapy regimens of testicular carcinoma, Hodgkin and non-Hodgkin lymphomas, breast cancer, osteosarcoma, advanced lung cancer, acute leukemia, rhabdomyosarcoma, neuroblastoma and Wilm's tumor (Rowinsky and Donehower 1995).

5.3.2   TAXANES

Taxanes are one of the most effective up-to-date antitumour agents. They are used for the treatment of several cancers including lung cancer, ovarian cancer, breast cancer, Kaposi's sarcoma, and several other metastatic cancers. Taxol or paclitaxel and taxotere or docetaxel are derived from a specie called *T. brevifolia* (Ireson et al. 2002). Taxanes act by inhibiting depolymerization of microtubules, that is, they target mitosis during metaphase and anaphase, thereby causing the apoptotic cellular death of the cancer cell lines. With reference to their mechanism of action, taxanes are often refered to as "spindle poisons" (Hagiwara and Sunada 2004).

5.3.3   PODOPHYLLOTOXINS

An isomer of podophyllotoxin, called epipodophyllotoxin, is isolated from *Podophyllum peltatum*'s roots. This isomer comrpises two active compounds called teniposide and etoposide. Their mechanism of action is inhibition of enzyme, topoisomerase II. Teniposide is used for the treatment of bladder cancer, lymphoma, and central nervous system tumors. Etoposide is FDA approved for the treatment of acute myeloid leukemia, lymphoma, lung, ovarian and testicular cancers, and chorio carcinoma (Ireson et al. 2002, Shoeb 2006).

5.3.4   CAMPTOTHECIN

Camptothecin is a phytoalkaloid—discovered in a Chinese tree called Camptotheca acuminate. Topotecan and irinotecan are chemical analogs of camptothecin, currently under clinical trials for use in combination therapy or alone in a variety of cancers. They are FDA approved and are used for the treatment of ovarian cancer, small cell lung, and metastatic colorectal cancer (Chauhan et al. 2015).

5.4 VARIOUS MECHANISMS OF ACTION OF HERBAL MEDICINES

A major role in cancer prevention and treatment is played by the natural compounds, including the culinary herbs and spices. Most of the antitumor agents currently used are of natural origin (Efferth et al. 2007). These plants come at the forefront for being the chief source of natural medicine. The most rational way of affecting carcinogenesis is to interfere with the modulation steps involved in cancer. These modulation steps involved in carcinogenesis include initiation, promotion, and progression which may be affected by interfering with various signal transduction pathways (Fresco et al. 2006). Drugs obtained from herbal sources act through various mechanisms. These mechanisms of action of herbal drugs for cancer prevention and treatment can be summarized as follows.

5.4.1 BLOCKING OF SIGNAL TRANSDUCTION PATHWAYS OF CELLS

In new clinical procedures, emphasis is laid on the blocking of specific cell signaling transduction pathways involved in converting normal cells into cancerous cells. Herbal drugs play their anticancerous role by blocking some of these pathways as follows:

1. *Pathway involving the activation of activator protein-1 (AP-1) and NF-kB*: Pathogenesis of various inflammatory diseases including cancer, diabetes, rheumatoid arthritis, and atherosclerosis depends upon the activation of the NF-kB activation pathway.

   NF-kB along with other transcription factors including AP-1 regulates genes which play their role in angiogenesis and aggressive growth leading to cancer (Hemalswarya and Doble 2006). Mountain ginseng extract has shown to inhibit the lungs cancer cell growth through attenuating cancerous cells proliferation and by inducing apoptosis which is resulted through the controlling of NF-kB signaling transduction pathway (Hwang et al. 2012).

2. *Signal transduction through PTK*: Polypeptide growth factors includes platelet derived growth factor (PGDF). These binds to their specific receptors which are called tyrosine kinesis. This binding of peptide growth factors with their receptors results in increased signal transduction to cancerous cells. Overexpression has been reported of these growth factors and their associated tyrosine kinesis receptors (Hemalswarya and Doble 2006).

3. *MAPK pathway*: Activation of mitogen activated protein kinase (MAPK) cause proliferation of the cells. Variety of tumors are owing to the dysregulation in this MAPK pathway. Therefore an effective chemotherapy involves targeting of this MAPK pathway (Hemalswarya and Doble 2006). Zengshenping (zsp) is applied because it increases Capase-3 expression thus inducing tumor cell apoptosis (Guan et al. 2012).

4. *Cyclooxygenase-2 (COX-2) pathway*: Proliferation as well as angiogenesis, either can be inhibited or stopped by the inhibition of cyclooxygenase (COX), mainly COX-2. This inhibition results in termination of the prostaglandin (PG) cascade ultimately effecting the growth of cancerous cells (Cragg and Newman 2005).

5.4.2   CANCER DUE TO CELL CYCLE ALTERATIONS

Normal proliferation and growth of cells in natural and regular way including the continuation of cell cycle is ensured by the presence of checkpoints in G1, G2 phases. While in case of cancerous cells, proliferation becomes uncontrolled and is not paused at these checkpoints (G1/S and G2/M) resulting in uncontrolled cancerous cells growth (Hemalswarya and Doble 2006). Available evidence proved that natural compounds possess the capability of inducing cell-cycle arrest at either G_2-M, or S or G_0-G_1 phase.

5.4.3   INTERFERENCE WITH MICROTUBULES

Two important classes of the tubulin-binding agents include *Vinca* alkaloids and taxanes. *Vinca* alkaloids are microtubule-targeting drugs which binds with α/β-tubulin dimers and as a result destabilized the microtubules. (Hassan and MA). Recent evidence indicates that microtubule function is disrupted by both kinds of drugs including taxanes that are known to be microtubule stabilizing drugs and *vinca* alkaloids that are considered as microtubule destabilizing drugs (Escuin et al. 2005). Hence, they prevent the proper alignment of daughter chromosomes to the microtubules. This disruption in proper alignment of daughter chromosomes on microtubules results in the inhibition of phases of mitotic cell division, particularly metaphase and anaphase. Such inhibition of cell division or particularly called cell cycle arrest may ultimately be followed by apoptosis (Nobili et al. 2009). For example, *vinca* alkaloids prevents cell cycle progression and induce mitotic block and apoptosis. Whatever the concentration of *vinca* alkaloids, mitochondria appear to be at the point of convergence for the apoptotic signals (Pourroy et al. 2004). Podophyllotoxin binds to microtubules, thus preventing their formation and destabilizing them (Darwiche et al. 2007).

5.4.4 TOPOISOMERASE INHIBITORS

Inhibition of topoisomerase plays a pivotal role in the treatment of cancer. Several herbal drugs are shown to be efficacious in controlling the activity of topoisomerase. Some of these herbal drugs includes Camptothecins inhibiting topoisomerase I while epopodo phylotoxins show anticancer activity by inhibiting topoisomerase II (Nobili et al. 2009). Etoposide inhibit topoisomerase II (Topo II), enzyme that produce transient double strand breaks in the DNA. Etoposide stabilizes a complex of the Topo II enzyme with cleaved DNA, inducing permanent breaks in the DNA and triggering cell death (Baldwin and Osheroff 2005).

5.4.5 APOPTOSIS

Vinorelbine and Vinblastine by phosphorylation cause Bcl-2 inactivation. This inactivation of Bcl-2 cause the basis for death of cancerous cells by apoptosis (Fan et al. 2000, Liu et al. 2001). Recently mitotic catastrophe has been described in which Vincristine induces cell death. This mode of cell death involves aberrant mitosis forming tetraploid non-viable cells containing multiple micronuclei (Darwiche et al. 2007). In response to Vincristine, the apoptosis resistant HL60-derived HCW-2 cell line underwent mitotic catastrophe, which led to the capase-3 activation and oligonucleosomal DNA degradation (Magalska et al. 2006). Campothecins induce cell death by upregulating proapoptotic genes and downregulating antiapoptotic genes (Darwiche et al. 2007). Apoptosis by Homoharringtonine is confirmed in several tumor types including leukemia, lymphoma, neuroblastoma and carcinoma (Borgne et al. 2006). Homoharringtonine (proliferation and protein inhibitor) may be more effective in resistant cancer when used in combination treatments with nucleoside analogs or mitosis inhibitors (Darwiche et al. 2007).

5.4.6 ANTIOXIDANT, ANTI-INFLAMMATORY AND ANTICANCER ACTIVITIES

Flavonoids possess powerful antioxidant properties due to which they reduce the risk of major chronic diseases including cancer. Flavonoids have the ability to scavenge a wide range of reactive species including hydroxyl, peroxyl and superoxide radicals as well as hypochlorous acid (Hollman and Arts 2000). Quercetin is a flavonoid possessing not only antioxidant property but also has antiproliferative effects and is also capable of inducing apoptosis (Huang et al. 2009). Tannins are also powerful antioxidant agents. These tannins include both hydrolysable as well as condensed tannins which also possess various other anticancer activities including anti-inflammatory, antimutagenic, blocking of transduction and apoptotic activities (Huang et al. 2009).

5.5 KINDS OF HERBAL MEDICINES

Herbal medicines are categorized into two categories depending exclusively upon the mode of actions as either of immunomodulation or direct chemo-prevention (Safarzadeh et al. 2014). These categories of herbal medicines are described as:

5.5.1 HERBS POSSESSING IMMUNOMODULATION PROPERTIES

Vital role herbal medicines play in cancer treatment is of immunomodulation. Immune system surveillance is evaded by the cancer cells in cancer patients. As a result of this patients innate and cellular immunity becomes attenuated. So, the role played by various herbal medicines and herbal complexes includes not only the stimulation of the innate immunity but also the protection of the bone marrow against the myelosuppressive effects of chemotherapy. *Ganoderma lucidum, Scutellaria baicalensia, Sophora flavescent,* and *Isatis tinctoria* are some of the herbal drugs included in this class of herbal drugs possessing such immunomodulation properties (Safarzadeh et al. 2014).

G. lucidum is medicinal fungus having effects on immune system. It stimulates the immune system by activating the macrophages, t-cells and NK cells (natural killer cells). It also causes an increase in TNF, interleukins and interferon's by stimulating their production (Smith-Hall et al. 2012). Leukocytes increase and improved immune system was shown by the use of *Sophora flavescens* (Yang et al. 2006). Heat and toxin clearing plant group includes *S. baicalensis*. This plant also has anti-tumor and immune system stimulatory effects which includes inhibition of platelet aggregation and induction of apoptosis.

Similarly, *I. tinctoria* involves the herbal compounds inhibiting DNA synthesis in neoplastic cells simultaneously resulting in the stimulation of immune system (Tavakoli et al. 2012). Many phenolic acids have been shown to exhibit anti-inflammatory effects thus enhancing immunity (Fresco et al. 2006).

5.5.2 HERBS POSSESSING CHEMOPREVENTIVE PROPERTIES

Various strategies used to interfere with or prevent cancer progression or development are involved in chemoprevention. These strategies may use natural complexes or their synthetic analogs (Safarzadeh et al. 2014). One of the main activities of chemopreventive plants is to induce apoptosis in the malignant cells (Pal and Shukla 2003, Yang et al. 2013). In vivo studies of Saffron shows ability of inhibiting tumorigenesis through different mechanisms. Some of these mechanisms involve inhibition of RNA, DNA synthesis, free radical chain or topoisomerase II inhibition (Abdullaev, 2002). Curcumin inhibits cycloeskinase-2 production in epithelial colon cells and also induce apoptosis in cancer cells by blocking NF-kB signaling pathway. This blocking is caused by the control of IKB enzyme phosphorylation (Plummer et al. 1999, Ireson et al. 2002).

5.6 PHARMACEUTICAL INTERACTIONS OF HERBAL MEDICINES AND CONCOMITANTLY ADMINISTERED OTHER MEDICINAL AGENTS

A condition in which a substance may be safe or dangerous is termed as safety. It may also be defined as considering beneficial effects against side effects (Tavakoli et al. 2012). Important source comprising the data about the safety and efficacy of herbal compounds is the German Commission E Monographs (Blumenthal 1998). Plant products being natural are generally considered to be safe but various side effects has also been reported after these have been administered. Different mechanisms accounting for such adverse effects may include toxic effects of these administered herbal agents, their possible allergenic reactions and also their pharmaceutical interactions with concomitantly administered chemical medicines (Bent and Ko 2004). Active compounds of some plants have dangerous side effects like cardiotoxicity and

hepatotoxicity and some of these can produce metabolites having carcinogenic activity. As a result, cellular metabolism may be altered by such active
metabolites resulting in cancer. *Aristolochia fangchi* is not only a potent
carcinogen but also is notorious of causing kidney failure (Safarzadeh et al.
2014). Polyphenol when used for chemoprevention shows side effects such
as liver failure, dermatitis and anemia (Galati and O'brien 2004). Several
contaminants present in the herbal products may also contribute to these side
effects. For example, a study conducted showed that 25% of 260 medicines
studied contained high level of non-organic pollutants. These nonorganic
pollutants include lead (Pb), mercury (Hg), and arsenic (Ar) (Bent and Ko,
2004). Similarly, allergic reactions are caused by salicylic glycosides and
lactonic sesquiterpenes (Firenzuoli and Gori 2007).

Essential oil of basil contains estragole which possess potent mutagenic
effects (Kaefer and Milner 2008). When high dose of Nutmeg is taken, psychic
disturbances can occur. Similarly, if nine teaspoons of ground nutmeg are taken
per day, an atropine like effect can be produced (Kaefer and Milner 2008).

5.6.1   FLAVONOID/PHENOLIC TOXICITY

Reduction of medicinal products containing phenol groups can be caused
by transition metals such as copper (Cu) and iron (Fe) in the presence of
oxygen. As a result, reactive oxygen species and phenoxyl radicals are formed
which ultimately results in the damaged to DNA, lipids and other biological
molecules (Li and Trush 1994). These copper ions may be released from sites
such as tissue injury, for example, atherosclerotic lesions (Smith et al. 1992).
In the initiation stage of atherosclerosis, phenoxyl radicals have shown to be
involved (Heinecke et al. 1993). Flavonoid consumption in large amount can
only be approved if it is in vivo potential for oxidative stress is evaluated and
declared safe (Galati and O'brien 2004).

Liver necrosis has been reported as a result of tannic acid use. Formation of
intermediates possessing pro-oxidant activities and inhibition of functioning
of antioxidant enzymes are the possible reported causes of this side effect
(Chung et al. 1998).

Cyanidanol (flavonoid catechin) produced both hemolytic anemia and
thrombocytopenia (Galati and O'brien 2004). Several case reports presented
with liver failure following the ingestion of herbal preparations of Kava
(*Piper methysticum*) due to which it has been banned in Switzerland and
Germany (Mathews et al. 2002).

5.6.2 PHYTOESTROGEN AND POSSIBLE HUMAN TOXICITY

Studies involving the use of phytoestrogen in experimental animals and human populations have shown a decrease in fertility and sexual dysfunction at high doses due to which decision about its use is controversial (You et al. 2002).

5.7 TYPES OF VARIOUS HERB-DRUG INTERACTIONS

Concomitant administration of herbal medicines with other chemical drugs can induce herb-drug interactions, resulting in the change in the intended action of the prescribed drug or toxic or side effects (Table 5.3).

TABLE 5.3 Published Interactions of Herbs with Concomitantly Administered Drugs (Bent and Ko 2004)

Herb	Drug	Interaction
Garlic	Warfarin	Bleeding, raised INR
	Chlorpropamide	
Ginkgo biloba	Warfarin	Bleeding
	Aspirin	Bleeding
	Thiazide diuretic	Blood pressure
	Trazodone	Increased sedation
Ginseng	Warfarin	Decreased INR
	Phenelzine	Insomnia, headache, tremulousness
Kava	Alprazolam	Sedation
St. John's wort	Amitriptyline	Reduction in plasma concentrations
	Cyclosporine	Reduction in plasma concentrations
	Digoxin	Reduction in plasma concentrations
	Oral contraceptives	Altered menstrual bleeding
	Paroxetine	Central serotonin excess
	Sertraline	Central serotonin excess
	Theophylline	Reduction in plasma concentrations
	Warfarin	Decreased INR

5.7.1 PHARMACODYNAMIC INTERACTIONS EXAMPLES

Pharmacodynamics involves the mechanism in which a body organ or tissue is affected by the drug (Safarzadeh et al. 2014). Due to such interactions the effect of concomitantly administered drug with herb may be increased (synergistic property) leading to toxic effects or the drugs effect may be decreased (antagonizing property). For example, when genistein is used in combination with β-lapachone, more intense apoptosis is produced (Safarzadeh et al. 2014). Valerian, a herbal complex used as a pain killer, shows a decrease in its effects with the concomitant administration of benzodiazepine (Hemalswarya and Doble 2006, Wang et al. 2012).

5.7.2   *PHARMACOKINETIC INTERACTIONS EXAMPLES*

Pharmacokinetic parameters include absorption, distribution, metabolism and excretion, such parameters of administered drugs can be affected by the concomitant administration of herbal drugs. Hepatic enzymes are mainly involved in the drug metabolism, so when such enzymes are affected by the herbal constituents then such interactions become clinically important (Islam and Iskander 2004). For example, a clinical trial research conducted for studying the interaction between ginseng and warfarin showed that anticoagulant effects of warfarin were reduced because of concomitant use of ginseng (Abdullaev 2002). As cytochrome P450 enzyme system was involved in the metabolism of warfarin, so it appears that ginseng increase warfarin metabolism by increasing P450 enzyme function

5.8   WHO ROLE

Herbal medicine has been considered as a holistic treatment approach by WHO. This hypothesis states that apart from external pathologic elements, immune system imbalance can also contribute to the development of diseases (Safarzadeh et al. 2014). These disorders can be improved by the endogen herbal systems. WHO introduced some guidelines to evaluate herbal medicines such as quality control analysis of active constituents, safety assessment, documentation and efficacy evaluation (Kamboj, 2000).

5.9   CONCLUSION

Promising anticancer drug approach focuses on discovery of natural molecules that specifically attack genetic alterations and deregulates biochemical pathways involved in causing cancer while sparing normal cells. Dietary plants contain natural bioactive compounds, that apart from providing nutrition may also provide other health benefits such as reducing the risk of many chronic diseases including cancer (Liu 2003). These dietary compounds can cause cell cycle arrest/cell death or either inhibit various signaling pathways involved in cancer (Kwon et al. 2007).

Owing to possessing the potential for immunomodulation and chemoprevention, natural sources can be investigated as novel treatment options in cancer. But extracting effective constituents from natural sources especially plants was a time-consuming process in the past but nowadays this process of extracting active constituents of plants has been accelerated by using modern techniques. Plant derived anticancer agents have been used widely and have shown efficacy against a wide spectrum of human cancers and also provided hope for the treatment of cancers that have shown resistance to conventional anticancer therapies and treatments. Most plant derived anticancer agents induce intrinsic apoptotic pathways.

As little is known about the efficacy and safety of herbal drugs, this area require further research (Hemalswarya and Doble 2006). For ensuring herbal medicine safety, more comprehensive information about the health benefits and possible risks accompanying the use of herbal medicines is needed. More information is needed about the doses of herbal compounds to bring a specific response without causing undesirable effects.

KEYWORDS

- **herbs**
- **cancer**
- **safety**
- **efficacy**

REFERENCES

Abdullaev, F. I. (2002). "Cancer chemopreventive and tumoricidal properties of saffron (*Crocus sativus* L.)." *Exp Bio Med* 227(1): 20–25.

Azmi, A. S., et al. (2006). "Plant polyphenols mobilize endogenous copper in human peripheral lymphocytes leading to oxidative DNA breakage: A putative mechanism for anticancer properties." *FEBS Lett* 580(2): 533–538.

Baldwin, E. and Osheroff, N. (2005). "Etoposide, topoisomerase II and cancer." *Cur Med Chem Anticancer Agents* 5(4): 363–372.

Bent, S. and Ko, R. (2004). "Commonly used herbal medicines in the United States: A review." *Am J Med* 116(7): 478–485.

Blumenthal, M., et al. (1998). *The Complete German Commission E Monographs—Therapeutic Guide to Herbal Medicines.* Boston, MA: Integrative Medicine Communications.

Borgne, A., et al. (2006). "Analysis of cyclin B1 and CDK activity during apoptosis induced by camptothecin treatment." *Oncogene* 25(56): 7361.

Bröker, L. E. and Giaccone G. E. (2002). "The role of new agents in the treatment of non-small cell lung cancer." *Eur J Cancer* 38(18): 2347–2361.

Burgeiro, A., et al. (2011). "Involvement of mitochondrial and B-RAF/ERK signaling pathways in berberine-induced apoptosis in human melanoma cells." *Anticancer Drugs* 22(6): 507–518.

Caltagirone, S., et al. (2000). "Flavonoids apigenin and quercetin inhibit melanoma growth and metastatic potential." *Int J Cancer* 87(4): 595–600.

Castillo, M. H., et al. (1989). "The effects of the bioflavonoid quercetin on squamous cell carcinoma of head and neck origin." *Am J Surg* 158(4): 351–355.

Chauhan, A, Tyagi, R. (2015). Herbal anti-cancer drugs: A better way to cure the disease. *Pharm Anal Acta* 6: e176.

Cheah, Y. H., et al. (2008). "Antiproliferative property and apoptotic effect of xanthorrhizol on MDA-MB-231 breast cancer cells." *Anticancer Res* 28(6A): 3677–3689.

Chen, J., et al. (2008). "Analysis of major alkaloids in *Rhizoma coptidis* by capillary electrophoresis-electrospray-time of flight mass spectrometry with different background electrolytes." *Electrophoresis* 29(10): 2135–2147.

Chung, K.-T., et al. (1998). "Tannins and human health: A review." *Crit Rev Food Sci Nutr* 38(6): 421–464.

Cragg, G. M. and Newman, D. J. (2005). "Plants as a source of anti-cancer agents." *J Ethnopharmacology* 100(1-2): 72–79.

Crane, C.H., Skibber, J.M., Feig, B.W., Vauthey, J.N., Thames, H.D., Curley, S.A., Rodriguez-Bigas, M.A., Wolff, R.A., Ellis, L.M., Delclos, M.E. and Lin, E.H., 2003. Response to preoperative chemoradiation increases the use of sphincter-preserving surgery in patients with locally advanced low rectal carcinoma. Cancer. *97*(2): 517–524

Darwiche, N., et al. (2007). "Cell cycle modulatory and apoptotic effects of plant-derived anticancer drugs in clinical use or development." *Exp Opin Drug Discov* 2(3): 361–379.

Desai, A. G., et al. (2008). "Medicinal plants and cancer chemoprevention." *Curr Drug Metab* 9(7): 581–591.

Hemalswarya, S. and Doble, M. (2006). "Potential synergism of natural products in the treatment of cancer." *Phtother Res* 20(4): 239–249.

Efferth, T., et al. (2007). "From traditional Chinese medicine to rational cancer therapy." *Trends Mol Med* 13(8): 353–361.

Eom, K. S., et al. (2010). "Berberine-induced apoptosis in human glioblastoma T98G cells is mediated by endoplasmic reticulum stress accompanying reactive oxygen species and mitochondrial dysfunction." *Biol Pharm Bull* 33(10): 1644–1649.

Escuin, D., et al. (2005). "Both microtubule-stabilizing and microtubule-destabilizing drugs inhibit hypoxia-inducible factor-1α accumulation and activity by disrupting microtubule function." *Cancer Res* 65(19): 9021–9028.

Fan, M., et al. (2000). "Modulation of mitogen-activated protein kinases and phosphorylation of Bcl-2 by vinblastine represent persistent forms of normal fluctuations at G2-M." *Cancer Res* 60(22): 6403–6407.

Firenzuoli, F. and Gori, L. (2007). "Herbal medicine today: Clinical and research issues." *Evid Based Complement Alternat Med* 4(S1): 37–40.

Fresco, P., et al. (2006). "New insights on the anticancer properties of dietary polyphenols." *Med Res Rev* 26(6): 747–766.

Galati, G. and O'brien, P. J. (2004). "Potential toxicity of flavonoids and other dietary phenolics: Significance for their chemopreventive and anticancer properties." *Free Radic Bio Med* 37(3): 287–303.

Gerber, B., et al. (2006). "Complementary and alternative therapeutic approaches in patients with early breast cancer: A systematic review." *Breast Cancer Res Treat* 95(3): 199-209.

Giovannucci, E., et al. (2002). "A prospective study of tomato products, lycopene, and prostate cancer risk." *J Natl Cancer Inst* 94(5): 391–398.

Greenwell, M., et al. (2015). "Medicinal plants: Their use in anticancer treatment." *Int J Sci Res* 6(10): 4103.

Guan, X., et al. (2012). "Inhibitory effects of Zengshengping fractions on DMBA-induced buccal pouch carcinogenesis in hamsters." *Chinese Med J* 125(2): 332–337.

Gupta, S. C., et al. (2014). "Downregulation of tumor necrosis factor and other proinflammatory biomarkers by polyphenols." *Arch Biochem Biophys* 559: 91–99.

Hagiwara, H. and Sunada, Y. (2004). "Mechanism of taxane neurotoxicity." *Breast Cancer* 11(1): 82–85.

Hassan, M. A. I. A., et al. (2006). Paclitaxel and vincristine potentiate adenoviral oncolysis that is associated with cell cycle and apoptosis modulation, whereas they differentially affect the viral life cycle in non-small-cell lung cancer cells." *Cancer Gene Ther* 13: 1105–1114.

Heinecke, J. W., et al. (1993). "Tyrosyl radical generated by myeloperoxidase catalyzes the oxidative cross-linking of proteins." *J Clin Invest* 91(6): 2866–2872.

Himes, R. H. (1991). "Interactions of the catharanthus (Vinca) alkaloids with tubulin and microtubules." *Pharmacol Ther* 51(2): 257–267.

Hollman, P. C., et al. (1997). "Bioavailability of flavonoids from tea." *Crit Rev Food Sci Nutr* 37(8): 719–738.

Hollman, P. C. H. and Arts, I. C. W. (2000). "Flavonols, flavones and flavanols–nature, occurrence and dietary burden." *J Sci Food Agri* 80(7): 1081–1093.

Huang, W.-Y., et al. (2009). "Natural phenolic compounds from medicinal herbs and dietary plants: potential use for cancer prevention." *Nutr Cancer* 62(1): 1–20.

Hwang, J. W., et al. (2012). "Mountain ginseng extract exhibits anti-lung cancer activity by inhibiting the nuclear translocation of NF-κB." *Am J Chin Med* 40(01): 187–202.

Ireson, C. R., et al. (2002). "Metabolism of the cancer chemopreventive agent curcumin in human and rat intestine." *Cancer Epidemiol Biomarkers Prev* 11(1): 105–111.

Islam, M. and Iskander, M. N. (2004). "Microtubulin binding sites as target for developing anticancer agents." *Mini Rev Med Chem* 4(10): 1077–1104.

Kaefer, C. M. and Milner, J. A. (2008). "The role of herbs and spices in cancer prevention." *J Nutr Biochem* 19(6): 347–361.

Kamboj, V. P. (2000). "Herbal medicine." *Curr Sci* 78(1): 35–39.

Kim, S., et al. (2008). "Sanguinarine-induced apoptosis: Generation of ROS, down-regulation of Bcl-2, c-FLIP, and synergy with TRAIL." *J Cell Biochem* 104(3): 895–907.

Kühnau, J. (1976). The flavonoids. A class of semi-essential food components: Their role in human nutrition. World Rev Nutr Diet 24: 117–191.

Kumar, S., et al. (2013). "The anticancer potential of flavonoids isolated from the stem bark of *Erythrina suberosa* through induction of apoptosis and inhibition of STAT signaling pathway in human leukemia HL-60 cells." *Chem Biol Interact* 205(2): 128–137.

Kwon, K. H., et al. (2007). "Cancer chemoprevention by phytochemicals: potential molecular targets, biomarkers and animal models 1." *Acta Pharma Sinica* 28(9): 1409–1421.

Li, D., et al. (1999). "Soybean isoflavones reduce experimental metastasis in mice." *J Nutrit* 129(5): 1075–1078.

Li, Y. and Trush, M. A. (1994). "Reactive oxygen-dependent DNA damage resulting from the oxidation of phenolic compounds by a copper-redox cycle mechanism." *Cancer Res* 54(Suppl. 7): 1895s–1898s.

Liu, R. H. (2003). "Health benefits of fruit and vegetables are from additive and synergistic combinations of phytochemicals." *Am J Clin Nutr* 78(3): 517S–520S.

Liu, X., et al. (2001). "Differential effect of vinorelbine versus paclitaxel on ERK2 kinase activity during apoptosis in MCF-7 cells." *Br J Cancer* 85(9): 1403.

Lu, J.-J., et al. (2012). "Alkaloids isolated from natural herbs as the anticancer agents." *Evid Based Complement Alternat Med* 2012: 485042.

Luk, J. M., et al. (2007). "Traditional Chinese herbal medicines for treatment of liver fibrosis and cancer: From laboratory discovery to clinical evaluation." *Liver Int* 27(7): 879–890.

Magalska, A., et al. (2006). "Resistance to apoptosis of HCW-2 cells can be overcome by curcumin-or vincristine-induced mitotic catastrophe." *Int J Cancer* 119(8): 1811–1818.

Mahady, G. B. and Beecher, C. W. (1994). "Quercetin-induced benzophenanthridine alkaloid production in suspension cell cultures of *Sanguinaria canadensis*." *Planta Med* 60(06): 553–557.

Malíková, J., et al. (2008). "Anticancer and antiproliferative activity of natural brassinosteroids." *Phytochemistry* 69(2): 418–426.

Mathews, J. M., et al. (2002). "Inhibition of human cytochrome P450 activities by kava extract and kavalactones." *Drug Metab Dispos* 30(11): 1153–1157.

Millimouno, F. M., et al. (2014). "Targeting apoptosis pathways in cancer and perspectives with natural compounds from mother nature." *Cancer* Prev Res 7(11): 1081–1107.

Moudi, M., et al. (2013). "Vinca alkaloids." *Int J Prev Med* 4(11): 1231–1235.

Nobili, S., et al. (2009). "Natural compounds for cancer treatment and prevention." *Pharm Res* 59(6): 365–378.

Pal, S. K. and Shukla, Y. (2003). "Herbal medicine: Current status and the future." *Asian Pac J Cancer Prev* 4(4): 281–288.

Pettit, G. R., et al. (1995). "Antineoplastic agents, 325. Isolation and structure of the human cancer cell growth inhibitory cyclic octapeptides phakellistatin 10 and 11 from Phakellia sp." *J Nat Prod* 58(6): 961–965.

Plummer, S. M., et al. (1999). "Inhibition of cyclo-oxygenase 2 expression in colon cells by the chemopreventive agent curcumin involves inhibition of NF-κB activation via the NIK/ IKK signalling complex." *Oncogene* 18(44): 6013.

Pourroy, B., et al. (2004). "Low concentrations of vinflunine induce apoptosis in human SK-N-SH neuroblastoma cells through a postmitotic G1 arrest and a mitochondrial pathway." *Mole Pharm* 66(3): 580–591.

Qi, F., et al. (2010). "Chinese herbal medicines as adjuvant treatment during chemo- or radio-therapy for cancer." *Biosci Trends* 4(6): 297–307.

Record, I. R., et al. (1997). "Genistein inhibits growth of B16 melanoma cells in vivo and in vitro and promotes differentiation in vitro." *Int J Cancer* 72(5): 860–864.

Rexroth, G. and Scotland, V. J. M. K. (1994). "Cardiac toxicity of 5-fluorouracil." *Medizinische Klinik (Munich, Germany)* 89(12): 680–688.

Rowinsky, E. and Donehower, R. J. N. (1995). "Paclitaxel (taxol)." *New Engl J Med* 332: 1004–1014.

Safarzadeh, E., et al. (2014). "Herbal medicine as inducers of apoptosis in cancer treatment." *Adv Pharm Bull* 4(Suppl. 1): 421.

Sazuka, M., et al. (1995). "Inhibitory effects of green tea infusion on in vitro invasion and in vivo metastasis of mouse lung carcinoma cells." *Cancer Lett* 98(1): 27–31.

Shoeb, M. (2006). "Anticancer agents from medicinal plants." *Bangladesh J Pharmacol* 1(2): 35–41.

Shyu, K.-G., et al. (2006). "Evodiamine inhibits in vitro angiogenesis: Implication for antitumorgenicity." *Life Sci* 78(19): 2234–2243.

Smith-Hall, C., et al. (2012). "People, plants and health: A conceptual framework for assessing changes in medicinal plant consumption." *J Ethnobiol Ethnomed* 8(1): 43.

Smith, C., et al. (1992). "Stimulation of lipid peroxidation and hydroxyl-radical generation by the contents of human atherosclerotic lesions." *Biochem J* 286(3): 901–905.

Steigerová, J., et al. (2010). "Brassinosteroids cause cell cycle arrest and apoptosis of human breast cancer cells." *Chem Biol Interact* 188(3): 487–496.

Steigerová, J., et al. (2012). "Mechanisms of natural brassinosteroid-induced apoptosis of prostate cancer cells." *Food Chem Toxicol* 50(11): 4068–4076.

Sun, Y., et al. (2009). "A systematic review of the anticancer properties of berberine, a natural product from Chinese herbs." *Anticancer Drugs* 20(9): 757–769.

Taraphdar, A. K., et al. (2001). "Natural products as inducers of apoptosis: Implication for cancer therapy and prevention." *Curr Sci* 80(11): 1387–1396.

Tavakoli, J., et al. (2012). "Evaluation of effectiveness of herbal medication in cancer care: A review study." *Iran J Cancer Prev* 5(3): 144–156.

Tunstall-Pedoe, H. (2006). Preventing chronic diseases: A vital investment. WHO Global Report. Geneva: World Health Organization. http://www.who.int/chp/chronic_disease_report/en.

Wang, C.-Z., et al. (2012). "Herbal medicines as adjuvants for cancer therapeutics." *Am J Chin Med* 40(04): 657–669.

Wang, Z., et al. (2009). "microRNA-21 negatively regulates Cdc25A and cell cycle progression in colon cancer cells." *Cancer Res.* 69(20), 8157–8165.

Yang, H.-L., et al. (2006). "Growth inhibition and induction of apoptosis in MCF-7 breast cancer cells by Antrodia camphorata." *Cancer Lett* 231(2): 215–227.

Yang, S., et al. (2013). "Antiproliferative activity and apoptosis-inducing mechanism of constituents from *Toona sinensis* on human cancer cells." *Cancer Cell Int* 13(1): 12.

Yin, S.-Y., et al. (2013). "Therapeutic applications of herbal medicines for cancer patients." 2013.

You, L., et al. (2002). "Modulation of mammary gland development in prepubertal male rats exposed to genistein and methoxychlor." *Toxic Sci* 66(2): 216–225.

CHAPTER 6

Herbs in Cancer Therapy: A Preamble

ANDREW G. MTEWA[*,1,2], FANUEL LAMPIAO[3], KENNEDY NGWIRA[4], PATRICK E, OGWANG[2], and DUNCAN C. SESAAZI[2,5]

[1]*Department of Chemistry, Malawi Institute of Technology, Malawi University of Science and Technology, P.O. Box 5196, Limbe, Malawi*

[2]*Pharmbiotechnology and Traditional Medicine Center of Excellence, School of Medicine, Mbarara University of Science and Technology, P.O. Box 1410, Mbarara, Uganda*

[3]*Africa Centre of Excellency in Public health and Herbal Medicine, College of Medicine, University of Malawi, P/Bag 360, Chichiri, Blantyre 3, Malawi*

[4]*School of Chemistry, University of Witwatersrand, P/Bag 3, Wits 2050, Johannesburg, South Africa*

[5]*Department of Pharmaceutical Sciences, School of Medicine, Mbarara University of Science and Technology, P.O. Box 1410, Mbarara, Uganda*

Corresponding author. E-mails: amtewa@must.ac.mw; andrewmtewa@yahoo.com

ABSTRACT

Cancer refers to the uncontrolled growth of cells that are capable of proliferating to other parts of the body. It is a disease that has strongly weighed heavily on societies for many years and still is. It affects the economy in a way of using budgetary allocations for drug importation as well as by paralyzing productivity of people suffering from the disease or their caretakers. Herbal medicines present a cheaper, more accessible, and loosely safer alternative to cancer treatment. Various studies have revealed anticancer potential that most herbs have in different parts of the world. This chapter aims at collecting some herbs that are known to be active against cancers and discusses their specific targets and, wherever possible, mechanisms of action. The plants that have already been explored have a great potential that they still have more drug compounds that are yet to be identified. If the discovery of drugs against cancers is given much attention and funding, more drug lead molecules can be found which can later be modulated to form better and cheaper anticancer drugs. Collaboration among the academia, funders, and the industry is the only way to make meaningful strides in the advancement of herbal therapy against cancers.

6.1 INTRODUCTION

Herbal medicine use is on the rise and is going towards fusion with conventional medicines. Generally, they are currently being used less methodically, largely with less or no quality control principles but promising a huge potential to be used against cancers.

Cancer is loosely defined as a collection of uncontrolled growth of cells in or on the body that are capable of spreading to other areas. Their spread can be intracellular, actual cell migration through the lymphatic and the circulatory systems. Benign cells are early cancerous cells, that are at a curable or reversible stage and malignant cells are fully grown cancers, usually very difficult to cure. They are normally described by the areas or organs they are originating from.

As therapy against cancers, herbal medicines have been used for years. There are various mechanisms in which herbal medicines work against cancers. They work as suppressive to the growth of cancerous cells and sometimes they work by killing such cells. These include the following specific mechanisms of action: cell cycle modification, protein tyrosine kinase pathways, topoisomerase inhibition, cyclooxygenase pathways, nuclear factor pathways with activator protein, cell signal transduction disruption pathways, and mitogen-activated protein kinases signal pathways. The nature of cancerous cells is such that their dose response to drugs is generally a little faster than normal cells, which make it a bit easier to control drugs under study or designing from injuring normal cells.

6.2 HERBAL PRODUCTS IN CANCER TREATMENT

Conventional peptide drugs, radiotherapy, gene therapy, and other therapies of the same nature have proven to be effective to some extent but having some serious short comings which include high costs of treatment, significantly denying access to average patients, serious adverse effects, and low therapeutic indices among others. Lately, herbal products present a potential pool of alternative medications that can be used directly or prepared in some ways. More workable options are isolates from the botanicals that can easily be modified to improve drug properties for effective outcomes. Some of the reliably effective conventional drugs today are derived from botanical species. These include the Vinca alkaloids (Vinblastine and Vincristine) obtained from *Catharanthus roseus* which are used to treat bladder, leukemia, and testicular cancers and Paclitaxol, originally isolated from *Taxus brevifolia*, used to treat breast and ovarian cancers (Shaikh et al., 2016). Generally, in the year 2000, 40% of top drug sales were natural product medicines or their derivatives (Butlet, 2004) while camptothecin and Paclitaxol, anticancer drugs from herbs, amounted to an estimated third of global sales ($3.0bn of $9.0bn) in 2002 (Oberlines and Kroll, 2004).

Herbal medicines are relatively more accessible in terms of quantity and expenses. The discovery of herbs as therapy against cancers brought in the need for further research on a lot more potential plants to diversify possible sources of the drugs. More research came into play with the aim of perfecting safety and therapeutic aspects of the same. Various plant parts or their combinations have been prepared in different dosage forms from powder, liquid extracts, whole plant or their parts, tincture, and paste among others. Without robust research on herbal plants as alternative cancer therapies or potential sources of better conventional medicines, the medical science community can lose a great opportunity of getting a break through against cancers that presently remain a menace in the whole world. It is not easy to come across plants with known anticancer properties due to isolated publications of data on the topic. This chapter aims at putting together herbal plants known to have potential and proven bio activities against various forms of cancers as reported by various researchers from different regions of the world.

6.3 SOME KNOWN HERBS IN CANCER THERAPY

Some commonly known herbs used against cancers are summarized in Table 6.1 of this chapter. This shows an extensive potential of herbal medicines to develop better and more accessible drugs against cancers. As limited by space, this work is not exhaustive. More work can be done country by country to reveal more plants that are currently being used against cancers but are not yet well reported. A classic example is *Eichhornia crassipes* which is usually known for its invasive characteristics over water bodies and normally removed and dumped off. For example, in Malawi, this plant is blamed for excessive disruption of power generation turbines year by year. The electricity company spends a lot of money every year to clear the shire river up of these plants and then dumps them away as wastes. If more research were to be conducted on the plant, more molecules can be found in addition to the nine isolated by Aboul-Enein et al. (2014) in egypt. Compounds from the plants, as some are known and many others yet unkown, can save countries a lot of budgetary allocations for drugs against cancers and also recover the money used in the clearing of water bodies by investing in drug development research and activities with ptential to bring out anticancer drugs. Only a few people in Malawi and Northern Uganda claim that this plant has anticancer and wound healing propetries, an indication that there are a lot more drug-like compounds in many other plants that can help on the management of diseases and health but still remain unknown due to lack of research on them.

6.4 RECOMMENDATIONS FOR THE FUTURE

Having reported a list of herbs with anticancer properties, it is being made clear that it is not exhaustive. More herbs can still be found in literature and also from new research to keep the list longer. One thing that comes out clearly is that with the current rate of deforestation in many countries, due to agricultural activities, industrial and home-stead constructions, and art among others, there may not be enough plants left for medicinal use in the near future. There is an urgent need for sound afforestation programs in all countries, particularly developing countries so that these anticancer herbal species, together with other medicinal species as well, should be maintained and availed for continuous use. Challenged with population growth, the demand for land will keep increasing, taking away even the small pieces of land that could serve as forests. This forces us to consider biotechnological approaches of growing plant species unconventionally. This approach is way expensive compared to afforestation but with limited choices available to mankind, there will be a need for community, state, or entrepreneurial investments in the near future to save some species from going extinct.

TABLE 6.1 Some Known Herbs in Cancer Therapy

Sl.N.	Name of Plant	English Name	Demonstrated Anticancer Activities	References
1	*Acacia seyal*	Acacia	Reported to be cancer preventive with cytotoxicity on: A549 (lung carcinoma), MCF-7 (breast adeno carcinoma), HepG2 (hepatocellular carcinoma), and HCT-116 (colorectal carcinoma).	El-Hallouty et al., 2015; Patel et al., 2014
2	*Cynodon dactylon*		Hepatoprotective properties were reported.	Venkateswaralu et al., 2015
3	*Eichhornia crassipes*	Water hyacinth	Hexane-ethyl acetate fractions have shown cytotoxic activities against HeLa, MCF-7, lung, HepG2, and other carcinoma. At least nine known active compounds have been isolated and all shown to be active.	Aboul-Enein et al., 2014
3	*Agaricus campestris*	Mushroom	Reported reduction in glycemia levels in colorectal cancer patients. Also reported cytotoxicity on cancerous human cell lines HO-8910 & 7721.	Elbatrawy et al., 2015
4	*Allium cepa*	Onion	Reduced risks of breast, esophageal, lung, prostate, colorectal, distal colon and stomach cancers were reported.	Chan et al., 2015; Colli & Amling, 2009; Kim & Kwon, 2009; Takezaki et al., 1999;
5	*Cucurbita maxima*	Pumpkin	Leaf ethyl acetate fractions showed significant antitumor activities on HepG2 cell lines.	Murganatham et al., 2015
6	*Allium sativum*	Garlic	Reduced risks of esophageal, breast, stomach, prostate, colorectal, distal colon, and lung cancers were reported.	Chan et al., 2015; Colli & Amling, 2009; Kim & Kwon, 2009
7	*Aloe vera*	Aloe	Suppression of the proliferation of cancer in human neuroblastoma cell lines (TGW, IMR-32, NBL-S, and CHP-126) was reported. Aloe emodin, a hydroxyanthraquinone compound, showed antineoplastic activities in primary stem-like cells (melanospheres) and in metastatic human melanoma cell lines. Adverse effects (dry skin desquamation and pain) of cancer radiotherapy were contained using aloe soap product, gel, and aqueous extracts in patients.	Tabolacci et al., 2015; Yonehara et al., 2015

TABLE 6.1 *(Continued)*

Sl.N.	Name of Plant	English Name	Demonstrated Anticancer Activities	References
8	*Apium graveolens*	Celery	Tumor growth inhibiting compounds; sedanolide and 3-n-butylphthalide were isolated from the plant leaves. Seed extracts were reported to have antiproliferative properties on BGC-823, a cancerous human stomach cell line though apoptotic mechanism of action.	Gao et al., 2011; Subhadradevi et al., 2011
9	*Cassia auriculata*		Ethanolic extracts showed anticancer activities. Some studies have also shown hepatoprotective effects.	Muruganantham et al., 2015
10	*Artemisia absinthium*	Wormwood	Antiproliferative activities on cancer of the breast through apoptosis were reported which were effective through the MEK/ERK pathway and the Bcl-2 protein group modulation.	Zaid et al., 2010a
11	*Arum palaestinum*	Palestine Arum	An isolated alkylated piperazine from the plant showed significant cytotoxicity against tumor cells in vitro.	El-Desouky et al., 2007
			Ethyl acetate fractions showed suppression of proliferation of and lymphoblastic leukemia cells (1301) and breast carcinoma cells (MCF-7). Other studies reported prostate cancer growth inhibition in mice.	
12	*Astoma seselifolium*	Astoma	The plant is reported to have anticancer prophylactic properties	Zaid et al., 2010a
13	*Terminalia chebula*		Silver nanoparticles from methanolic extracts of the plant showed anticancer activities against colon cancer cell lines.	Bupesh et al., 2016
14	*Beta vulgaris*	Beet Root	Tumor growth inhibitory activities were reported on mice lung and skin bioassays. Cytotoxicity was reported on breast cancer cell lines (MCF-7) and androgen-free human prostate cancer cell lines (PC-3).	Georgiev et al., 2010 ;Nowacki et al., 2015
15	*Boswellia carterii*	Olibanum	An isolated compound from the plant, acetyl-11-keto-β-boswellic acid, was reported to inhibit prostate cancer promoting pathways. Crude extracts were reported to have cytotoxic activities on HCT 116 and HepG2 cell lines.	Ahmed et al., 2015; Yuan et al., 2008

TABLE 6.1 *(Continued)*

Sl.N.	Name of Plant	English Name	Demonstrated Anticancer Activities	References
16	*Brassica nigra*	Mustard	An essential oil from the plant, allylisothiocyanate (AITC) was reported to have antineoplastic activities on cancer of the bladder with mutated (T24) TP53 or wild type RT4 genes. Extracts from the plant were demonstrated to work antagonistically against adverse effects of benzo(α)pyene metabolites mutagenicity. Tumoricidal activities were reported on human promyelocytic leukemia cell lines (HL60) and drosophila melanogaster (SMART).	Savio et al., 2015
17	*Brassica oleracea*	Wild Cabbage	The growth of estrogen receptor negative (BT20 and MDA-MB-231) and estrogen receptor positive (BT474 and MCF-7) were reported to be inhibited by extracts from the plant. The plant's activities are reported to warrant cancer preventive effects.	Brandi et al., 2005
18	*Bryonia syriaca*	Syrian Bryony	Leaf extracts were reported to have anticancer potential making the plant useful in cancer treatment.	Zaid et al., 2010a
19	*Capparis spinosa*	Caper	Isolated proteins from the plant showed the inhibition of breast cancer cell lines (MCF-7), colon cancer cell lines (HT29), and hepatoma cell lines (HepG2) proliferation. An infusion in water and the essential oils from the plant separately demonstrated significant inhibition of the proliferation of colon cancer (HT29). Extracts from the plant were reported to be able to use mitochondrial pathway to induce apoptosis in cancer cell lines (SGC-7901).	Ji & Yu, 2014
20	*Chrysanthemum coronarium*	Crown Daisy	Weak anticancer activities on human cell lines HCT-15, PC-3 and A549 were reported from dihydrochrysanolide derivatives and sesquiterpene lactones isolated from the plant. Essential oils from the plant were reported to be good antiproliferative agents on human colon cancer.	Lee et al., 2003; Zaid et al., 2010b
21	*Momordica dioica*	Small/spiny bitter gourd	Methanolic seed coat extracts of the plant showed antiproliferative activities against lung and breast carcinoma.	Sukumar et al., 2016

TABLE 6.1 *(Continued)*

Sl.N.	Name of Plant	English Name	Demonstrated Anticancer Activities	References
22	*Cichorium intybus*	Chicory	Plant extracts were reported to be able to shield DNA deoxyribose moiety against oxidative damage. Tumor inhibitory activities on Ehrlich ascites carcinoma and leukemia cell lines were reported in mice.	Hazra et al., 2002
23	*Cinnamomum camphora*	Camphor	Solid melanoma in mice skin and human hepatocarcinoma were inhibited by camphorin, a compound isolated from the plant.	Lin et al., 2008
			A monoterpenoid, Subamone, showed anticancer activities against human prostate cancer cell lines (DU-145 and LNCaP) and human lung cancer cell (A549).	
24	*Citrullus colocynthis*	Colocynth	Cucurbitacin glucosides from the plant caused both cell cycle arrest and apoptosis by its pleiotropic activities on cancerous cells. Triterpenoidal glucosides demonstrated potential HepG2 inhibition.	Ayyad et al., 2012; Mukherjee & Patil, 2012; Tannin-Spitz et al., 2007
25	*Cucumis melo*	Cucumber	The plant was reported to have potential to reduce the risk of various forms of chronic diseases including cancers. Triterpenoids of the cucurbitane form demonstrated significant in vitro cytotoxic activities against BEL7402 cells and A549/ATCC proliferation. Reduces risk of cancer and other chronic diseases.	Chen et al., 2009
26	*Eclipta Alba*	Primary liver herb	Hydroalcoholic extract of the plant showed apoptotic, antiproliferative, and anti-invasive activities.	Chaudhary et al., 2011
27	*Matricaria aurea*	Chamomile	Extracts from the plant showed cytotoxic activities on MCF-7, A-549, and PC-3 cell lines. Antigenotoxicity tests were reported to be positive for the plant. The plant contains glucosides of the apigenin group which were reported to be able to inhibit cancer cell growth. Essential oils rich in α-Bisabolol and Bisabolol from the plant demonstrated strong anticancer properties.	Kamatou & Viljoen, 2010; Zaid et al., 2010b; Zu et al., 2010

TABLE 6.1 *(Continued)*

Sl.N.	Name of Plant	English Name	Demonstrated Anticancer Activities	References
28	*Narcissus tazetta*	Bunchflower daffodil	Different parts of the plant demonstrated anticancer activities, with pseudolycorine alkaloids reported to have antileukemic activities. Plant extracts significantly reduced the rate of cancer cell survival on KT1/A3, A3, K562, and HL-60 cell lines. Fractions showed inhibition of solid lymphosarcoma cells and Ehrlich ascites tumors in mice.	Liu et al., 2006; Talib & Mahasneh, 2010; Zaid et al., 2010a, 2010b
29	*Barleria grandiflora*		Juices from the leaves are used to treat mouth ulcers. Extracts were reported to have cytotoxic effects on DLA and A-549 cell lines.	Manglani et al., 2014
30	*Vitis vinifera*	Grapes	Leaf and seed extracts were reported to have cytotoxic activities on MCF-7, A-549, and PC-3 cancer cell lines. Further, seed extracts and stem extracts showed anticancer activities on colon cancers (HT29), breast cancers (MDA-MB-23 and MCF-7), renal cancers (786-0 and Caki-1), HepG2, thyroid cancer (K1).Extracts isolated from the plant seeds and stems demonstrated antitumor activity in human breast cancer cell lines MCF-7 and MDA-MB-23), colon (HT29), renal (786-0and Caki-1),thyroid (K1), oral squamous cell carcinoma, and human fibroblasts.	Zu et al., 2010; Aghbali et al., 2013; Durak et al., 2005; Sahpazidou et al., 2014
31	*Thymus vulgaris*	Thyme	Extract of the plant showed leukemia cell lines growth inhibition potential. They were also reported to have cytotoxic potential against MCF-7, PC-3, and A-549 cancer cell lines.	Zu et al., 2010
32	*Quercus calliprinos*	Palestine Oak	Decoctions from bark and fruits were reported to have anticancer activities, making them useful as cancer treatment.	Saad et al., 2008; Zaid et al., 2010a, 2010b
33	*Triticum aestivum*	Wheat	Wheat bread was reported to have preventive potential against colon tumorigenesis. Wheat bioactive components (isoflaavones, lignans, phenolic acids, and vitamins posses some antioxidant properties and act through mechanisms that inhibit tumor progression. Some research suggests that lignans could be cancer preventive in mice through	Omar et al., 2009; Mathankumar et al., 2015

TABLE 6.1 (*Continued*)

Sl.N.	Name of Plant	English Name	Demonstrated Anticancer Activities	References
			apoptosis. Triticumoside, a phenolic compound from the plant, was reported to have shown apoptotic activities on human lung cancer cell lines. Generally, extracts from the plant were reported to have antiproliferative potential on A549 and HCT 116 cell lines.	
34	*Zingiber officinal*	Ginger	Plant extracts showed cytotoxic potential against A-549, PC-3, and MCF-7 cancer cell lines. Some research has reported the potential of ginger as a supplement in breast neoplasm and liver cancer risk reduction.	Karimi et al., 2015; Zu et al., 2010
35	*Urtica pilulifera*	Stinging Nettle	Some studies have reported cytotoxic activities by the plant extracts. Flavonoid glycosides isolated from the plant showed intracellular killing activities. Root extracts are used to treat benign prostate hyperplasia and aqueous leaf extracts demonstrated the inhibition of adenosine deaminase against prostate cancer.	Zaid et al., 2010a
36	*Viscum cruciatum*	Redberry mistletoe	Hirsutanone, an isolate from the plant fractions showed cytotoxicity against renal, melanoma, and breast cell lines. Mistletoe was reported to have anticancer activity against BJAB cell lines, with IC_{50} of 14.21 mg/ml. Hexane extracts showed cytotoxic effects against larynx HEp2 cell lines.	Assaf et al., 2013; Martin-Cordero et al., 2001; Zaid et al., 2010a, 2010b
37	*Punica granatum*	Pomegranate	Fruit extracts were reported to induce apoptosis and decrease the proliferation of prostate cancer cell lines (DU-145) and also to suppress the invasive potential of PC-3 cell lines. Extracts from the plant were reported to show significant inhibition of the growth of tumors in mice prostate tumor models. They are effective in the inhibition of mouse lung tumorigenesis. In colon cancers, the plant inhibits inflammatory cell signaling. Oils from seeds and peel-offs have been shown to be effective against cell cycle invasion, tumor cell proliferation, and angiogenesis. They also showed antitumor activity on lung cell carcinoma (A549), breast adenocarcinoma cell lines (MCF-7), ovarian cancer cell lines (SKOV3), and prostate adenocarcinoma cell lines (PC-3).	Kohno et al., 2004; Malik & Mukhtar, 2006; Mandal & Bishayee, 2015

TABLE 6.1 *(Continued)*

Sl.N.	Name of Plant	English Name	Demonstrated Anticancer Activities	References
38	*Psoralea corylifolia*		Seed extracts showed anticancer activities against MEC-1 carcinoma and T-cell lymphoma cell lines.	Thapliyal et al., 2018
39	*Solanum nigrum*	Black night shade	Extracts were reported to be hepatoprotective, antiulcerogenic, and cytotoxic properties. Diosgenin is one of the active compounds in the plant.	Patel et al., 2009
40	*Cassia senna*	Senna	Aqueous extracts of the plant circumvent toxicity and mutagenesis induced by hydrogen peroxide in *Escherichia coli* strains and some isolates showed reduced mitogen-activated protein (MAP) kinase activation and proliferation.	Silva et al., 2008
41	*Mesua ferrea*		Hepatoprotective effects were reported for the plant. Some studies reported that α-amyrin or β-amyrin and lupeol mixtures isolated from the plant's dichloromethane extracts showed antitumor activity against KB, MCF-7, and NCI-H187 cancer cell lines.	FRETTM, 2005

6.5 CONCLUSION

The exploration of botanical species targeting the discovery, development, and designing of new or improved drugs against cancers is a venture worth investing in. With the bulk of plants reported to be used traditionally against cancers and preliminary laboratory results that are just staying in the lab, there are many areas that government, health departments, pharmaceutical industries, development partners, PhD students, universities, and research centers can collaboratively work to effectively come up with promising drug leads and/or product from such work. From such botanical information collections, most users use the plants as they are with no regard to the chemical-structural properties that determine the drug-like properties of the molecules. It should be noted that a drug may show bio activity but due to poor molecular properties, the actual active molecules may not be reaching the intended biological targets, in this case, cancers. Robust research in the chemistry of the drug molecules is required from this stage in order to avoid letting patients take herbal formulations that are not at all being bio available due to poor chemical properties.

KEYWORDS

- **cancer therapy**
- **herbal medicine**
- **cell-cycle arrest**
- **chemo preventive**
- **herbal therapy**
- **self-medication**

REFERENCES

Aboul-Enein, A. M., ShanabShalaby, S. M.E. A., Shalaby, E. A., Zahran, M. Z., Lightfoot, D. A., & El-Shemy, H. A. (2014). Cytotoxic and antioxidant properties of active principals isolated from water hyacinth against four cancer cells lines. *BMC Complement Altern Med.,* 14: 397.

Aghbali, A., Hosseini, S. V., & Delazar, A. (2013). Induction of apoptosis by grape seed extract (*Vitis vinifera*) in oral squamous cell carcinoma. *Bosn J Basic Med Sci.,* 13, 186–191.

Ahmed, H. H., Abd-Rabou, A. A., Hassan, A. Z., & Kotob, S. E. (2015). Phytochemical analysis and anti-cancer investigation of *Boswellia serrata* bioactive constituents in vitro. *Asian Pac J Cancer Prev.*, 16, 7179–7188.

Assaf, A. M., Haddadin, R. N., & Aldouri, N. A. (2013). Anti-cancer, anti-inflammatory and anti-microbial activities of plant extracts used against haematological tumours in traditional medicine of Jordan. *J Ethnopharmacol.*, 145, 728–736.

Ayyad, S. N., Abdel-Lateff, A., Alarif, W. M., Patacchioli, F. R., Badria, F. A., & Ezmirly, S. T. (2012). In vitro and in vivo study of *cucurbitacins*-type triterpene glucoside from *Citrullus colocynthis* growing in Saudi Arabia against hepatocellular carcinoma. *Environ Toxicol Pharmacol.*, 33, 245–251.

Brandi, G., Schiavano, G. F., & Zaffaroni, N. (2005). Mechanisms of action and antiprolifera-tive properties of *Brassica oleracea* juice in human breast cancer cell lines. *J Nutr.*, 135, 1503–1509.

Bupesh, G., Manikanandan, E., Thanigaiarul, K., & Magesh, S. (2016). Enhanced antibacterial, anticancer activity from *Terminalia chebula* medicinal plant rapid extract by phytosynthesis of silver nanoparticles core-shell structures. *J Nanomed Nanotechnol.*, 7(1), 1000355.

Butlet, M. S. (2004). The role of natural product chemistry in drug discovery. *J Nat Prod.*, 67, 2141–2153.

Chan, J. M., Wang, F., & Holly, E. A. (2005). Vegetable and fruit intake and pancreatic cancer in a population-based case-control study in the San Francisco bay area. *Cancer Epidemiol Biomarkers Prev.*, 14, 2093–2097.

Chaudhary, H., Dhuna, V., Singh, J., Kamboj, S. S., & Seshadri, S. (2011). Evaluation of hydro-alcoholic extract of *Eclipta alba* for its anticancer potential: An in vitro study. *J Ethnopharmacol.*, 136, 363–7.

Chen, C., Qiang, S., Lou, L., & Zhao, W. (2009). *Cucurbitane*-type triterpenoids from the stems of *Cucumis melo*. *J Nat Prod.*, 72, 824–829.

Colli, J. L., & Amling, C. L. (2009). Chemoprevention of prostate cancer: What can be recommended to patients? Search results. *Curr Urol Rep.*, 10, 165–171.

Durak, I., Cetin, R., Devrim, E., & Erguder, I. B. (2005). Effects of black grape extract on activities of DNA turn-over enzymes in cancerous and non-cancerous human colon tissues. *Life Sci.*, 76, 2995–3000.

Elbatrawy, E. N., Ghonimy, E. A., Alassar, M. M., & Wu, F. S. (2015). Medicinal mushroom extracts possess differential antioxidant activity and cytotoxicity to cancer cells. *Int J Med Mushrooms.*, 17, 471–479.

El-Desouky, S. K., Kim, K. H., Ryu, S. Y., Eweas, A. F., Gamal-Eldeen A. M. & Kim, Y.-K. (2007). A new pyrrole alkaloid isolated from *Arum palaestinum* Boiss and its biological activities. *Article Drug Discovery Arch Pharma Res.*, 30, 927–931.

El-Hallouty, S. M., Fayad, W., Meky, N. H., EL-Menshawi, B. S., Wassel, G. M., Hasabo, A. A. In vitro anticancer activity of some Egyptian plant extracts against different human cancer cell lines. *Int J Pharm Tech Res.* (2015), 8, 267–272.

Foundation of Resuscitate and Encourage Thai Traditional Medicine (FRETTM). Thai Pharmaceutical Book. Bangkok: Pikanate Printing Center Corporation; 2005.

Gao, L. L., Feng, L., & Yao, S. T. (2011). Molecular mechanisms of celery seed extract induced apoptosis via s phase cell cycle arrest in the BGC-823 human stomach cancer cell line. *Asian Pac J Cancer Prev.*, 12, 2601–2606.

Georgiev, V., Weber, J., Eva-Maria, K., Denev, P. N., Bley, T., & Pavlov, A. I. (2010). Antioxidant activity and phenolic content of betalain extracts from intact plants and hairy root cultures of the red beetroot *Beta vulgaris* cv. Detroit dark red. *Plant Foods Hum Nutr.*, 65, 105–111.

Gordanian, B., Behbahani, M., Carapetian, J., & Fazilati, M. (2014). In vitro evaluation of cytotoxic activity of flower, leaf, stem and root extracts of five Artemisia species. *Res Pharm Sci.*, 9, 91–96.

Hazra, B., Sarkar, R., Bhattacharyya, S., & Roy, P. (2002). Tumour inhibitory activity of chicory root extract against Ehrlich ascites carcinoma in mice. *Fitoterapia*, 73, 730–733.

Hui-Qing, Y., Feng, K., Xiao-Ling, W., Young, Y. F. C., Xiao-Yan, H., & Hong-Xiang, L. (2008). Inhibitory effect of acetyl-11-keto-β-boswellic acid on androgen receptor by interference of Sp1 binding activity in prostate cancer cells. *Biochem Pharmacol.*, 75, 2112–2121.

Ji, Y. B., & Yu, L. (2014). N-butanol extract of *Capparis spinosa* L. induces apoptosis primarily through a mitochondrial pathway involving mPTP open, cytochrome C release and caspase activation. *Asian Pac J Cancer Prev.*, 15, 9153–9157.

Kamatou, G. P. P., & Viljoen, A. M. (2010). A review of the application and pharmacological properties of α-bisabolol and α-bisabolol-rich oils. *J Am Oil Chem Soc.*, 87, 1–7.

Karimi, N., Dabidi-Roshan, V., & Fathi-Bayatiyani, Z. (2015). Individually and combined water-based exercise with ginger supplement, on systemic inflammation and metabolic syndrome indices, among the obese women with breast neoplasms. *Iran J Cancer Prev.*, 8, 1–8.

Kim, J. Y., & Kwon, O. (2009). Garlic intake and cancer risk: An analysis using the Food and Drug Administration's evidence-based review system for the scientific evaluation of health claims. *Am J Clin Nutr.*, 89, 257–264.

Kohno, H., Suzuki, R., Yasui, Y., Hosokawa, M., Miyashita, K., & Tanaka, T. (2004). Pomegranate seed oil rich in conjugated linolenic acid suppresses chemically induced colon carcinogenesis in rats. *Cancer Sci.*, 95, 481–486.

Lee, K. D., Park, K. H., Park, K. M., Kim, J. H., Rim, Y. S., & Yang, MS.. (2003). Cytotoxic activity and structural analogues of Guaianolide derivatives from the flower of *Chrysanthemum coronarium* L. *J Appl Biol Chem.*, 46, 29–32.

Lin, R. J., Lo, W. L., Wang, Y. D., & Chen, C. Y. (2008). A novel cytotoxic monoterpenoid from the leaves of *Cinnamomum subavenium*. *Nat Prod Res Former Nat Prod Lett.*, 22, 1055–1059.

Liu, J., Li, Y., Ren, W., Wei-Xin, H. (2006). Apoptosis of HL-60 cells induced by extracts from *Narcissus tazetta* var. chinensis. *Cancer Lett.*, 242, 133–140.

Longo, L., Platini, F., Scardino, A., Alabiso, O., Vasapollo, G., & Tessitore, L. (2008). Therapeutics targets, and development autophagy inhibition enhances anthocyanin-induced apoptosis in hepatocellular carcinoma. *Mol Cancer Ther.*, 7, 2476–2485.

Malik, A., & Mukhtar, H. (2006). Prostate cancer prevention through pomegranate fruit. *Cell Cycle*, 5, 371–373.

Mandal, A., & Bishayee, A. (2015). Mechanism of breast cancer preventive action of pomegranate: Disruption of estrogen receptor and Wnt/β-catenin signaling pathways. *Molecules*, 20, 22315–22328.

Manglani, N., Vaishnava, S., Dhamodaran, P., & Sawarkar, H. (2014). In-vitro and in-vivo anticancer activity of leaf extract of *Barleria grandiflora*. *Int J Pharm Pharm Sci.*, 6(3), 70–72.

Martın-Cordero, C., Lopez-Lazaro, M., Agudo, M. A., Navarro, E., Trujillo, J., & Ayuso, M. J. (2001). A cytotoxic diarylheptanoid from *Viscum cruciatum*. *Phytochemistry*, 58, 567–569.

Mathankumar, M., Tamizhselvi, R., Manickam, V., & Purohit, G. (2015). Assessment of anticarcinogenic potential of *Vitex trifolia* and *Triticum aestivum* Linn by in vitro rat liver microsomal degranulation. *Toxicol. Int.*, 22, 114–118.

Mezni, F., Shili, S., Ben-Ali, N., Larbi-Khouja, M., Khaldi, A., & Maaroufi, A. (2016). Evaluation of *Pistacia lentiscus* seed oil and phenolic compounds for in vitro antiproliferative effects against BHK21 cells. *Pharm Biol.*, 54, 747–751.

Mukherjee, A., & Patil, S. D. (2012). Effects of alkaloid rich extract of *Citrullus colocynthis* fruits on *Artemia salina* and human cancerous (MCF-7 AND HEPG-2) cells. *J Pharm Sci Tech.*, 1, 15–19.

Murganatham, N., Solomon, S., & Senthamilselvi, M. M. (2016). Anticancer activity of *Cucurbita maxima* flowers (Pumpkin) against human liver cancer. *Int J Pharm Sci.*, 6(1), 1356–1359.

Muruganantham, N., Solomon, S., & Sentamnselvi, M. M. (2015). Anticancer activity of *Cassia auriculata* (flowers) against human liver cancer. *Pharmacophore*, 6(1), 19–24.

Nowacki, L., Vigneron, P., & Rotellini, L. (2015). Betanin-enriched red beetroot (*Beta vulgaris* L.) extract induces apoptosis and autophagic cell death in MCF-7 cells. *Phytother Res.*, 29, 1964–1973.

Oberlines, N. H., & Kroll, D. J. (2004). Camptothecins and taxol: Historic achievement in natural products research. *J Nat Prod.*, 67, 129–135.

Omar, R. M., Ismail, H. M., El-Lateef, B. M., Youssif, M. I., Gomaa, N. F., & Sheta, M. (2009). Effect of processing on folic acid fortified Baladi bread and its possible effect on the prevention of colon cancer. *Food Chem Toxicol.*, 47, 1626–1635.

Patel, A., Hafez, E., Elsaid, F., & Amanullah, M. (2014). Anti-cancer action of a new recombinant lectin produced from Acacia species. *Int J Med Sci.*, 5, 1.

Patel, S., Gheewala, N., Suthar, A., & Shah, A. (2009). In-vitro cytotoxicity activity of *Solanum nigrum* extracts against Hela Cell-line and Vero cell-line. *Int J Pharm Pharm Sci.*, 1(1), 38–46.

Saad, B., Azaizeh, H., & Said, O. (2008). Arab herbal medicine. *Bot Med Clin Prac.*, 4, 31.

Sahpazidou, D., Geromichalos, G. D., Stagos D. (2014). Anticarcinogenic activity of polyphenolic extracts from grape stems against breast, colon, renal and thyroid cancer cells. *Toxicol Lett.*, 230, 218–224.

Savio, A. L., da Silva, G. N., & Salvadori, D. M. (2015). Inhibition of bladder cancer cell proliferation by allyl isothiocyanate (mustard essential oil). *Mutat Res.*, 771, 29–35.

Shaikh, A. M., Shrivastava, B., Apte, K. G., Navale, S. D. (2016). Medicinal plants as potential source of anticancer agents: A review. *J Pharmacognosy Phytochem.*, 5(2), 291–295.

Silva, C. R., Monteiro, M. R., Rocha, H. M., (2008). Assessment of antimutagenic and genotoxic potential of senna (*Cassia angustifolia* Vahl.) aqueous extract using in vitro assays. *Toxicol In Vitro.*, 22, 212–218.

Subhadradevi, V., Khairunissa, K., Asokkumar, K., Umamaheswari, M., Sivashanmugam, A., & Jagannath, P. (2011). Induction of apoptosis and cytotoxic activities of *Apium graveolens* Linn. Using in vitro models. *Middle-East J Sci Res.*, 9, 90–94.

Sukumar, G., Divyalaxmi, D., Mithula, S., & Rupachandra, S. (2016). In-vitro anticancer effect of methanolic seed coat extracts of *Momordica dioica* against human carcinoma cell lines. *Int J Chemtech Res.*, 9(2), 284–289.

Tabolacci, C., Cordella, M., & Turcano, L. (2015). Aloe-emodin exerts a potent anticancer and immunomodulatory activity on BRAF-mutated human melanoma cells. *Eur J Pharmacol.*, 762, 283–292.

Takezaki, T., Gao, C. M., Ding, J. H., Liu, T. K., Li, M. S., & Tajima, K. (1999). Comparative study of lifestyles of residents in high and low risk areas for gastric cancer in Jiangsu Province, China; with special reference to allium vegetables. *J Epidemiol.*, 9, 297–305.

Talib, W. H., & Mahasneh, A. M. (2010). Antimicrobial cytotoxicity and phytochemical screening of Jordanian plants used in traditional medicine. *Molecules.*, 15, 1811–1824.

Tannin-Spitz, T., Grossman, S., Dovrat, S., Gottlieb, H. E., & Bergman, M. (2007). Growth inhibitory activity of cucurbitacin glucosides isolated from *Citrullus colocynthis* on human breast cancer cells. *Biochem Pharmacol.*, 73, 56–67.

Thapliyal, A., Khar, R. K., & Chandra, A. (2018). Overview of cancer and medicinal herbs used for cancer therapy. *Asian J Pharm.*, 12(1), S1–S8.

Venkateswaralu, G., Swarupa-Rani, T., Vani, M., & Vineela, P. A. J. (2015). In-Vitro anticancer activity of petroleum ether extracts of *Cynodon dactylon*. *J Pharmacognosy Phytochem.*, 4(1), 164–168.

Yonehara, A., Tanaka, Y., Kulkeaw, K., Era, T., Nakanishi, Y., & Sugiyama, D. (2015). *Aloe vera* extract suppresses proliferation of neuroblastoma cells in vitro. *Anticancer Res.*, 35, 4479–4485.

Zaid, H., Rayan, A., Said, O., & Saad, B. (2010a). Cancer treatment by Greco-Arab and Islamic herbal medicine. *Open Nutraceuticals J.*, 3, 203–212.

Zaid, H., Said, O., & Saad, B. (2010b). Cancer treatment in the Arab-Islamic medicine: Integration of tradition with modern experimental trails. *JAMIA*, 14, 13–40.

Zu, Y., Yu, H., & Liang, L. (2010). Activities of ten essential oils towards *Propionibacterium acnes* and PC-3, A-549 and MCF-7 cancer cells. *Molecules*, 15: 3200–3210.

CHAPTER 7

Medicinal Mushrooms

TEMITOPE A. OYEDEPO and ADETOUN E. MORAKINYO[*]

Department of Biochemistry, Adeleke University, Ede, Nigeria

[*]Corresponding author. E-mail: topeoyedepo@adelekeuniversity.edu.ng

ABSTRACT

Mushrooms have been used in many aspects of the human activity for many years. Some of these mushrooms have been termed "medicinal mushrooms" due to their various morphological, physiological, and ecological characteristics that are also responsible for their diversity. Medicinal mushrooms produce a wide range of bioactive compounds that confer certain pharmacological activities on medicinal mushrooms for the amelioration of chronic diseases associated with oxidative stress. Hence, mushrooms have gained a lot of attention as a functional food and for the development of drugs and nutraceuticals. This chapter discusses chemical composition, nutritional composition, and bioactivities of medicinal mushrooms. The chapter also reviews selected medicinal mushrooms' nutraceuticals with proven pharmacological properties.

7.1 INTRODUCTION

Mushrooms are the fleshy part of a fungus which are usually produced above ground on soil or on a substrate. They are the spore-producing fruiting body of a macrofungi which are produced mainly by the Basidiomycota and Ascomycota group (Ayeka et al., 2017). This means that they are large enough to be seen with the naked eye (Changs and Miles, 1989). Mushrooms abound exceedingly worldwide and are highly diversified with 500,000 to more than 5 million species (Changs and Miles, 1989). In the past, fungi were classified as a plant but currently fungi are classified as an independent group of organisms under the kingdom Mycota. This is basically because they have chitin in their cell walls (Changs and Miles, 1989).

About 160,000 species of mushrooms have been recommended for general use (Hawksworth, 2012). However, only about 3000 of the recommended mushrooms are edible, approximately 700 exhibit pharmacological properties while about 1% are poisonous (Borchers et al., 2008). At first, mushrooms were only used as food but subsequently, their medicinal properties were elucidated (Rathee et al., 2012). Medicinal mushrooms can therefore be defined as fungal organisms that are considered health foods, nutritional supplements and nutraceuticals. They are classified with a taxonomic group called basidiomycetes (Wasser, 2014). Higher basidiomycete mushrooms possess different types of biologically active compounds (e.g., alkaloids, lactones, and triterpenes) which are found in cultured broth, cultured mycelia, and fruit bodies (Chang and Wasser, 2012; De Silva et al., 2013).

Over the past few decades, there has been an increased interest about the role of the human immune system in promoting good health. Along this line, different species of mushrooms have been examined for their immunomodulating activities. Most of these mushrooms are rich sources of polysaccharides (e.g., β-glucans) and polysaccharide– protein complexes with immunostimulating properties. The main purpose of this chapter is to review the current positions about the bioactivities of medicinal mushroom. The chapter also reviews progress made so far in the study of selected medicinal mushroom nutraceuticals with proven pharmacological properties.

7.2   COMPOSITION OF MUSHROOMS

Most mushrooms are composed of approximately 90% water by weight. Carbohydrates and crude proteins form the principal components of the dry matter (DM) in most types of mushrooms. The fruit bodies of mushrooms are very rich in carbohydrates with a concentration ranging between 20% and 70% of the DM. Typically, most plants store their polysaccharide in the form of starch but mushrooms differ in that they store their polysaccharides in the form of glycogen which contributes about 5%–10% of the DM (Kalac, 2009). Mushrooms also contain high content of insoluble fiber which contribute immensely to its nutritional benefit. The cell wall of mushroom is made up of Chitin (a water-insoluble polysaccharide which contributes about 80%–90%), hemicelluloses and pectin (Xiao et al., 2011). The protein content of medicinal mushroom is about 44.93% of their dry weight. Thus, they provide low calories in diet and lack cholesterol (Mhanda et al., 2015). Mushroom proteins contain all the essential amino acids and they are especially rich in lysine and leucine (Yıldız et al., 2017).

The compositions of carbohydrate and many trace elements vary widely among mushroom species (El Enshasy et al., 2013). The mineral contents are mainly potassium, calcium, phosphorus, magnesium, selenium, iron, zinc, and copper (Huang et al., 2010). The typical contents of some of these trace elements are shown in Table 7.1.

TABLE 7.1 Trace Elements in Mushrooms

Trace Element	Content (mg/kg DM)
Ascorbic acid	150–300
Thiamine	1.7–6.3
Riboflavin	2.6–9.0
Pyridoxine	1.4–5.6
Niacin	63.8–83.7
Vitamin B2	12.68
Ergosterol	1.98
Vitamin D2	16.88

In addition, mushrooms are rich in many bioactive compounds, such as terpenoids, steroids, phenols, nucleotides, glycoproteins, and polysaccharides. Most of the biological activities exhibited by mushrooms have been attributed to the presence of bioactive polysaccharides, specifically, the β-glucans which have immune-stimulating properties (Batbayar et al., 2012). Mushrooms that are rich in β-glucans have positive impact on the immune system and confer healthful properties that distinguishes such mushrooms from others (Sanodiya et al., 2009).

The total lipid contents (i.e., crude fat) of mushrooms are low when compared with protein and carbohydrate contents. Lipid contents range mostly from 20 to 30 g/kg DM. According to Barros et al. (2007), the major fatty acids of *Agaricus arvensis*, *Lactarius deliciosus*, *Leucopaxillus giganteus*, *Sarcodon imbricatus*, and *Tricholoma portentosum* are linoleic acid and oleic acid. The other branched chain acids and hydroxyl fatty acids have also been identified in some mushrooms but their concentrations are low (Nedelcheva et al., 2007). Fatty acids analysis of most mushrooms showed that the unsaturated fatty acids were at higher concentrations than saturated fatty acids (Barros et al. 2007). The low total fat content and high proportion of polyunsaturated fatty acids relative to the total fatty acids of mushrooms are considered significant contributors to the medicinal value of mushrooms (Sanodiya et al., 2009).

Mushrooms possess substantial antioxidant activity coupled with strong metal chelating ability owing to their phenolic composition (Islam et al., 2016).

Phenolic acids that have been identified in mushrooms include gallic acid, gentisic acid, *p*-hydroxybenzoic acid, protocatechuic acid, syringic acid, and vanillic acid (Robbins, 2003).

7.3 BIOLOGICALLY ACTIVE COMPOUNDS OF MUSHROOMS

The biological activities of medicinal mushrooms have been attributed to the presence of some bioactive compounds (Wasser and Weis, 1999a). Most of these bioactive compounds function as biological response modifiers (BRMs). BRMs are substances that can stimulate the body's response to infection and disease. Since the body cannot produce these substances in appreciable quantity, it is necessary to obtain more through diet or dietary supplements. Nutraceuticals prepared from mushrooms can therefore serve as the source of this exogenous supply (Wasser and Weis, 1999a).

Biologically active compounds of mushrooms can stimulate the immune system as well as modulate specific cellular responses through interference in particular transduction pathways (Wasser and Weis, 1999b). This has made mushrooms to be a vital force in cancer treatment (Chang and Wasser, 2012; Dotan et al., 2011; Ruimi et al., 2010).

7.3.1 *POLYSACCHARIDES*

Fundamentally, polysaccharides are antitumor, antioxidant, antibacterial, antiviral, and immune-stimulating substances. Mushroom polysaccharides have been attracting the interest of chemists and immunobiologists for many years because they are potent substances with antitumor and immune-modulating properties.

Many polysaccharides that have been isolated from mushrooms have been documented to be BRMs which modify immune responses (Kim et al., 2006). It is worthy to note however that the biological activity of mushroom polysaccharides has a direct relationship with the length, branching of the chain, chain rigidity, and helical conformation of the polysaccharide (Zhang et al., 2007). Immunological alterations which are induced by mushroom polysaccharides include:

1. Inhibition of prostaglandin synthesis
2. Reduction in pro-inflammatory cytokines
3. Activation of immune cells
4. Increased antibody production

5. Increased interferon production
6. Increased immune activity against a range of cancers
7. Inhibition of tumor metastasis (Saman et al., 2016)

7.3.1.1 β-D-GLUCANS

β-D-glucans are a naturally occurring structural constituent of the cell walls of mushrooms, mycelium, yeast, and certain bacteria. About 50% of the mass for the cell wall in a fungus is made up of β-glucans (Rop et al., 2009; Wasser, 2002; Wasser and Weis, 1999a). Research into the active polysaccharides in basidiomycetes has identified them as β-D-glucans (Rop et al., 2009; Vetvicka et al., 2008). β-glucans contain a glucose polymer-chain core which are held together by a linear linkage. The glycosidic linkage can present as (β-1 → 3), (β-1 → 4), (β-1 → 6), or a mixture of these (Volman et al., 2010a). However, the designation of the branching depends on the species of the mushroom (Rop et al., 2009). This is the reason why β-glucans from different sources have different structures. For instance, the structure for the β-glucans of mushrooms which have β-1 → 3 and β-1 → 6 side branches is different from those of bacteria that have β-1 → 4 side branches (Sakagami and Aoki, 1991).

Other differences in the structure of β-glucans across species include

1. the core chain length of β-glucans
2. the types of side chain branching
3. the complexity of side chain branching (Volman et al., 2010b).

These differences in structure have a great influence on the β-glucan's function and mechanism of action. Higher level of structural complexity indicates a very strong immunomodulatory and anticancer activities (Ooi, 2008).

β-glucans in fungi primarily usually have (1 → 6)-linked branches coming off the β(1 → 3) backbone. There are other structures that also contain (1 → 3)-β and (1 → 6)-β linkages. The structural complexity of β-glucans in fungi also varies and is considered a primary determinant of activity (Volman et al., 2010b). This is the main accepted theory of why some basidiomycetes are more active than others and why basidiomycete β-glucans are more immunologically active than cereal β-glucans (Chan et al., 2009). The biological activities of medicinal mushrooms have been attributed to the (1 → 3)-β-glucan or a mixture of (1 → 3) and (1 → 6)- β-glucan. (1 → 3)-β-glucan is reported to have the best effect on immune stimulation (Rop et al., 2009).

The attributes of these β-glucans indicate that they activate or potentiate both innate and adaptive immunity. β-glucans act on a variety of receptors with functions related to immunity, and they can also induce a wide array of immune responses. When β-glucans activate the immune system, the numbers of macrophages, NK cells, subsets of T-cells are increased and their functions are enhanced, which subsequently get rid of infectious agents (Chan et al., 2009).

Furthermore, fungal β-glucans have been shown to reduce total cholesterol as well as LDL cholesterol level in blood. This is in addition to achieving a mild increase in the level of HDL cholesterol while positively affecting the metabolism of fats and sugars (Rop et al., 2009). Another study documented the fact that β-glucans improve resistance against allergies by increasing the numbers of Th1 lymphocytes in the blood (Rop et al., 2009).

While it has been established that β-glucans from mushrooms have anticancer activity, the alpha-glucans does not exhibit this effect (Volman et al., 2010b). Some mushroom preparations that have demonstrated significant effect against human cancers include active hexose correlated compound (AHCC) and Lentinan (from *Lentinula edodes*), Maitake D-fraction (from *Grifola frondosa*), Schizophyllan (SPG); from *Schizophyllum commune*), and polysaccharide-Krestin or polysaccharide of P (PSK/PSP) from *Trametes versicolor* (Guggenheim et al., 2014).

7.3.1.2   PROTEOGLYCANS

The mushroom-derived "polysaccharide peptides," also known as proteogly-cans, are polypeptide chains with polysaccharide β-D-glucan chains attached to them. Two proteoglycans from Trametes (*Coriolus versicolor*) which have been studied extensively are protein-bound PSK and PSP. They are isolated and purified from *C. versicolor* (Maehara et al., 2012). These two glycopro-teins do not have the same measure of biological activities when compared with pure β-glucans (Volman et al., 2010b). Series of clinical studies have established the fact that PSK and PSP have anticancer activity and are well tolerated in human without any significant side effects (Chan et al., 2009).

7.3.1.3   LECTINS

Lectins have been demonstrated to have anticancer activities with both in vitro experiments and in clinical trials (Liu et al., 2010). They function by binding to membrane carbohydrates. Its therapeutic purpose is to bind to the membrane of a mutant cell or its receptors, leading to the death of the cancer cells and therefore, encouraging the reduction of tumor (Cheung et al., 2012;

Wang et al., 1995). Lectins can also ameliorate the unwanted side effects that usually accompany chemotherapy in addition to their ability to eradicate tumor cells by increasing the activities of tumor-infiltrating lymphocytes (Cheung et al., 2012).

7.3.2 TERPENOIDS

Terpenes from mushrooms is one of the potent bioactive compounds that can be obtained from mushrooms. Categories of terpenes from mushrooms include monoterpenoids, diterpenoids, triterpenoids, and sesquiterpenoid (Duru and Çayan, 2015). These terpenes have been associated with various biological activities that include antioxidant, antibacterial, antifungal, anti-infectious, anti-inflammatory, antimalarial, cytotoxic, hypocholesterolemic, hypoglycemic, and hypotensive effects (Li et al., 2011). Among all the microbes, sesquiterpenoids are found only in basidiomycetes, and some of them are similar to plant terpenoid compounds (Duru and Çayan, 2015).

The primary activities of triterpenoids include antioxidant, liver protection, improved atherogenic index, anti-inflammatory and inhibition of histamine release. Triterpenoids also play a complementary role with β-glucans in immunomodulatory activities (Wachtel-Galor et al., 2011). The mechanism of anticancer action by mushroom triterpenoids may include:

1. inhibition of tumor cell growth
2. inhibiting metastasis
3. inhibiting angiogenesis (Petrovi´c et al., 2014; Xiao et al., 2009).

Another important terpenoid that has been isolated from mushrooms are the ganoderic acid derivatives. Ganoderic acids U, V, W, X, and Y have cytotoxicity activities, ganoderic acids C and D have demonstrated anti-histamine releasing activities, ganoderic acid F have exhibited inhibitory activity against angiotensin converting enzyme, while Ganoderic acid A exhibited hepatoprotective activity. In addition, ganoderic acid A and methyl ganoderate A act as Farnesyl Protein Transferase Inhibitors which inhibit the activity of Ras oncoproteins (Shi et al., 2010).

7.3.3 STEROIDS

Ergosterol is present in mushroom membrane, although this varies among species. It is a steroid that can be found in all fungi and is a corollary to cholesterol in humans. Ergosterol has even discovered to have antioxidant and antitumor properties (Dalla-Santa et al., 2012). Ergosterol is a precursor to vitamin D and exposure to UV light converts ergosterol to vitamin D_2. It is an essential marker for fungal presence which has been in use for years by the grain industry to check for fungal contamination (Ng, 2008).

Other steroids that have been isolated from mushrooms include ergosterol peroxide, ergosterol, and trametenolic acid. These mushroom steroids are known to inhibit the production of inflammatory mediators (Taofiq et al., 2016). Two known sterols that were isolated from the fruiting bodies of *Agrocybe aegerita* by Zhang et al. (2003) were discovered to have cyclooxygenase inhibitory and antioxidant activities.

7.3.4 POLYKETIDES

Polyketides are secondary metabolites with structurally complex organic compounds which are produced by most living organisms. Higher fungi are important producers of this metabolite (Volman et al., 2010b). Polyketides are extremely active biologically and this has made them to be a good source of drug discovery. Hence, many pharmaceuticals including antibiotic, anticancer, antifungal, hypolipidemic, and immunosuppressive drugs are derived from polyketides (Hung and Nhi, 2012; Staunton and Weissman, 2001).

7.3.5 STATINS

Over the years, researchers have discovered that certain mushrooms, especially *Pleurotus*, have the ability to lower cholesterol (Bobek et al., 2001). Further work has discovered statins, the secondary metabolites, as the most likely compound responsible for this action (Alam et al., 2011). Lovastatin (mevinolin) is a statin that occurs in most of the basidiomycetes just like many other fungi. It is a powerful inhibitor of HMG-CoA reductase, the main rate-limiting enzyme in cholesterol production. Lovastatin analysis could serve as another marker for species with elevated levels of this compound. This should, however, be merged with the fact that quite high amounts of mushroom powder would be necessary to effect any action (Lo et al., 2012).

7.3.6 NUCLEOSIDES

Nucleosides such as adenosine, guanosine, uridine, cytidine, etc. exist in varying amounts in most of the basidiomycetes (Yuan et al., 2008). It is therefore possible to utilize them as markers. Some researchers have been able to create fingerprints using up to ten of these compounds (Phan et al., 2018).

7.3.7 PHENOLIC COMPOUNDS

Mushrooms have phenolic compounds which are important phytochemicals with numerous bioactive properties. Due to their diversity and multifunctional properties, the polyphenol compounds in mushrooms are notable for providing protection against several chronic diseases such as cardiovascular disease and cancer. They can also prevent the cytotoxicity caused by H_2O_2 and the oxidative damage caused by free radicals. The phenolic compounds are also useful in reducing LDL oxidation, DNA damage, and the incidence of cancer (Hertog et al., 1993; Lee et al., 2007). Many of the phenolic compounds and some of their derivatives that have been isolated from mushroom extracts have also been documented to have anti-inflammatory activity (Taofiq et al., 2016).

7.3.8 OTHER METABOLITES IN MEDICINAL MUSHROOMS

There are many other biologically active metabolites present in mushrooms. These include alkaloids, anthraquinones, flavonoids, saponins, tannins and steroids (Adebayo et al., 2012; Egwim et al., 2011; Ehssan and Saadabi, 2012). The presence of these metabolites makes mushrooms to be very valuable as a food source that can also contribute to drug development.

Alkaloids are found useful in the treatment of many deadly human diseases. Some of the biological activities of alkaloids include:

1. regulation of Na^+ ions and channels;
2. antimicrobial activity;
3. immune-stimulating properties;
4. induction of immunogenic cell death;
5. inhibition of angiogenesis which inhibit the growth of cancerous cells (Harborne and Williams, 2000; Mandal et al., 2008).

The antioxidant properties of medicinal mushrooms can also be attributed to the presence of flavonoids which provide protection against oxidative stress

(El Enshasy and Hatti-Kaul, 2013; Harborne and Williams, 2000). Similarly, tannins in many medicinal mushrooms have strong antioxidant potentials which have a very strong link with their anticarcinogenic and antimutagenic potentials (Castro et al., 2014).

7.4 TRADITIONAL ROLES OF MEDICINAL MUSHROOMS

Plants have been used traditionally as a medical system and as such herbal medicines constitute an important part of traditional and evidence-based medicine worldwide. In the ancient times, mushroom use was connected with mysticism (Griensven, 2009). The first record of mushroom use, which dates back to the Paleolithic period (i.e., about 7000– 9000 years ago), was as a hallucinogenic agent by the Yoruba tribe of Nigeria in Africa (Griensven, 2009; Samorini, 1992). There were other ancient documentations on the use of edible and medicinal mushrooms kept in other countries such as China and Japan. These were kept in the earliest Chinese material medical book, Shennong Bencao Jing, and other Chinese medical books that followed (Zhu, 2009). Mushrooms have been used traditionally for the treatment of various diseases affecting the rural populations of eastern European countries (Wasser, 2014).

However, most of the information on indigenous applications of African mushrooms was not properly documented as they were passed orally from one generation to the other (Akpaja et al., 2003). Mushrooms belonging to the species of *Calvatia, Coriolopsis, Ganoderma, Lentinus, Lenzites, Pleurotus, Pycnoporus, Termitomyces,* and *Trametes* have reported to be in use in Nigerian folk medicine (Akpaja et al., 2003; Ayodele et al., 2009; Ezekiel et al., 2013; Oyetayo, 2011). Similarly, mushrooms of the species *Agaricus, Boletus, Cantharellus, Ganoderma, Geastrum, Macrolepiota, Pleurotus, Termitomyces,* and *Tricholoma* have been reported to be in use in some other African countries such as Tanzania, Western Uganda, and Burkina Faso (Beiersmann et al., 2007; Kamatenesi-Mugisha and Oryem-Origa, 2007; Tibuhwa, 2012).

7.5 MODERN FUNCTIONS OF MEDICINAL MUSHROOMS

Medicinal mushrooms that have been playing an important role in several aspects of human activity are thought to possess over a hundred medicinal functions (Ayodele et al., 2009). A very important use of medicinal mushrooms is the prevention and treatment of immune disorders, especially in patients with immunodeficiency and immunosuppression. They are also used to ameliorate the effect of chemotherapy or radiotherapy in cancer patients undergoing treatments. In addition, mushrooms are used to treat chronic viral infections such as Hepatitis B, C, and D, Herpes, and acquired immuno-deficiency syndrome (AIDS). They are equally used to treat different types of anemia, chronic fatigue syndrome, dementia, chronic gastritis, and gastric ulcers caused by *Helicobacter pylori* (Chang and Wasser, 2012; Dai et al., 2009; Lo and Wasser, 2011; Wasser, 2010). Hence, medicinal mushrooms are used today as:

1. Dietary food
2. dietary supplement
3. drugs which are termed "mycopharmaceuticals"
4. natural bio-control agents for plant protection
5. cosmeceuticals.

Medicinal mushrooms are also found useful in cosmetics industry. Some of the secondary metabolites found in mushrooms (e.g., β-glucans) are now being used by cosmetic companies due to their antioxidative, anti-allergic, antibacterial and anti-inflammatory activities. Other reasons for using medicinal mushrooms in cosmetics industry include:

1. film-forming ability
2. activation of epidermal growth factor
3. stimulation of collagen activity
4. inhibition of autoimmune vitiligo
5. treatment of acne (Chang and Wasser, 2012; Dai et al., 2009; Hyde et al., 2010; Lo and Wasser, 2011).

7.6 THE USE OF MUSHROOMS AS PREBIOTICS

Prebiotics are nondigestible ingredients found in food which stimulate the growth or action of probiotics (i.e., beneficial microorganisms). Probiotics play an important role in the overall health of the gut. They help with the breakdown and digestion of food and the regulation of the immune system, and they also inhibit the growth of pathogens. Prebiotics are obtained from the nondigestible fiber in certain plant-based foods. The key compounds in prebiotics are galactooligosaccharide, oligosaccharides, and inulin. Mushroom is one of the good sources of prebiotics because they are rich in polysaccharides such as chitin, galactans, hemicellulose, mannans, α- and β-glucans, and xylans

(Singdevsachan et al., 2015). These nondigestible mushroom polysaccharides have the potential to enhance the growth of probiotic bacteria in the gut, and this will inhibit the proliferation of pathogens (Bhakta and Kumar, 2013).

The composition of gut microbiota can modify gut barrier and affects the regulation of energy metabolism and adipose tissue proliferation. Several metabolic dysregulations that may lead to inflammation of the brain, liver and intestine have been traced to the gut microbiota (Geurts et al., 2014). With the use of prebiotics, these can be corrected since prebiotics have the ability to depress endogenous pathogens found within the gastrointestinal (GI) tract. This will increase the competency of immune system to resist microorganisms from external sources (De Sousa et al., 2011). The roles which prebiotics play to modulate the human gut microbiota and attenuate several disease conditions are well documented (De Sousa et al., 2011).

More than 380 species of mushrooms have been documented to possess medicinal properties and this as a result of their high content of prebiotics (Geurts et al., 2014). Mushrooms prebiotics are known to improve the antioxidant status as a result of alterations in the composition of gut microbiota. Mushrooms play a vital role in immune response during the treatment of respiratory diseases, atherosclerosis, cancer, and other metabolic diseases (Koyyalamudi et al., 2009a, 2009b; Varshney et al., 2013). Prebiotics from mushrooms also have hypocholesterolemia properties that help reduce lipogenic gene expression (Hmgcr, Fasn, Srebp1c, and Acaca) and genes responsible for reverse cholesterol transport (Abcg5 and Abcg8) significantly, as well as an increase in Low Density Lipoprotein Receptor gene expression in the liver (Meneses et al., 2016). *Ganoderma lucidium* is a mushroom that has been documented to reduce obesity in mice by altering the composition of gut microbiota (Xu and Zhang, 2015).

7.7 PHARMACOLOGICAL EFFECTS OF MEDICINAL MUSHROOMS

7.7.1 ANTIOXIDANT ACTIVITIES

The antioxidant activity of edible mushrooms could be directly applied to daily life because it is associated with natural prevention of oxidative stress which is often as a result of lifestyle habits (Sakano et al., 2009). A number of in vitro and in vivo studies have reported antioxidant potentials of various mushroom species that enable them to neutralize free radicals (Ferreira et al., 2009). Assays involving chromogen compounds (e.g., ABTS and DPPH methods) are the most commonly used methods to measure mushrooms antioxidant activity (Sánchez, 2016). The findings of those studies demonstrated that the antioxidant components are found in fruit bodies, mycelium and culture both of the mushrooms. These often show good activity as a scavenger of DPPH radical and reactive oxygen species (hydroxyl radical, superoxide radical, and hypochlorous acid). They also act as an xanthine oxidase inhibitor which has indications for various therapeutic procedures, including cancer therapy since inhibition of Xanthine Oxidase can inhibit the oxidation of 6-mercaptopurine and potentiate antitumor properties (Pacher et al., 2006; Ribeiro et al., 2007). The metabolites responsible for the antioxidant activities include polysaccharides, phenolic compounds, carotenoids, ergosterol, tocopherols, ascorbic acid and many others (Sánchez, 2016; Zhang et al., 2016a).

7.7.2 *IMMUNOMODULATORY AND ANTITUMOR ACTIVITIES*

Immunomodulators, also known as BRMs, are compounds used to regulate or normalize the immune system (Jong and Yang, 1999). BRMs have been isolated from the fruiting bodies of mushrooms as well as the stalk, spores, and mycelium. They can also be isolated from fermentation broth when cultivated in submerged culture. In some studies, BRMs were applied simultaneously with conventional chemotherapy and radiotherapy during cancer treatment to increase their efficiency. Mushroom BRMs have been classified into four main categories according to their chemical structure:

1. lectins
2. terpenoids
3. polysaccharides
4. fungal immunomodulator proteins (El Enshasy, 2010).

Cancer is one of the leading causes of death all over the world especially in low and middle- income countries where resources available for diagnosis, prevention, and treatment of cancer is limited. For many years, treatment of cancer was limited to chemotherapy in conjunction with radiotherapy. However, development of resistance to chemotherapy and severe side effects linked to the therapy became major pitfalls for these treatments (Brenner and Stevens, 2010). Some of the reported side effects include reduced caloric intake and decreased absorption of nutrients that could endanger the life of cancer patients (Shervington and Lu, 2008). In addition, chemotherapy often results in damage and weakening of patient's natural immunological defenses (Orsine et al., 2012).

All these led to the consideration of using natural products for the prevention and treatment of various chronic diseases. For over three decades, scientific and medical research studies have confirmed the bioactive compounds which can be extracted from mushrooms and are useful in the management and treatment of cancer and other related diseases (Lee et al., 2013). Mushrooms are judged to strengthen the immune system by applying their effects on cellular activities through secondary production of chemical compounds that boost the immune system (El Enshasy, 2010). Mushrooms polysaccharides (mainly α- or β-glucans and glycoproteins) demonstrated immunomodulatory activities through:

1. activation of cytotoxic lymphocyte, that is, natural killer (NK) cells
2. regulation of cytokines production (i.e., IL-10, IL-12p70, and IL-12p40) by dendritic cells
3. increased production of TNF-a, IL-1, IL-6, IL-8, IL- 12p40, and NO, and expression of iNOS by macrophages (Borchers et al., 2008).

Compounds of mushrooms in crude and pure form have effectively demonstrated antitumor and immunomodulatory activities (Krishnamoorthy and Sankaran, 2016; Vetvicka et al., 2008) coupled with anti-angiogenesis properties (Liu et al., 2012). One of the mechanisms by which mushrooms exert protection against cancer is through stimulation of the immune system response. This is because β-glucan polysaccharide activates immune cells that destroy tumor cells (Vetvicka et al., 2008). However, an intact T-cell component is required for the antitumor action of β-glucan polysaccharides since their activity is controlled by a thymus-dependent immune mechanism (Vetvicka et al., 2008).

Drugs prepared from mushroom polysaccharides which were applied as immunotherapeutic agents have shown positive results from in vitro and in vivo studies on cancer treatment (Lee and Kim, 2014; Liu et al., 2012). The anticancer role of mushrooms can be described from the following mechanisms:

1. Cell cycle arrest
2. Induction of cell apoptosis to prevent cancer cell proliferation (Song et al., 2013).

Mushrooms with demonstrated anticancer properties include the following genus: Agaricus, Albatrellus, Antrodia, Calvatia, Clitocybe, Cordyceps, Flammulina, Fomes, Funlia, Ganoderma, Inocybe, Inonotus, Lactarius, Phellinus, Pleurotus, Schizophyllum, Russula, Suillus, Trametes, and Xerocomus.

7.7.3 HYPOLIPIDEMIC PROPERTIES

Adding mushrooms to diet has been documented to significantly decrease the atherogenic lipid profiles and this may provide health benefits for cardiovascular-related complications. (Chen et al., 2012; Jeong et al., 2010; Wani et al., 2010). Dietary fiber is an active polysaccharide, which is part of the edible portion of most mushrooms. Such fibers cannot be decomposed by lytic enzyme or digested in human alimentary tract (Chang and Miles, 1989). This water-soluble dietary fiber can be combined with cholesterol or cholic acid by adsorption, reducing the amount of cholic acid returned to liver and increasing the metabolism of cholesterol and its transformation into cholic acid. The inevitably leads to reduction in cholesterol concentration and absorption of lipids in the small intestine is disturbed. Consequently, the lipid content in blood reduces and the probability of cardiovascular disease is diminished (Anderson et al., 2009). Studies have shown that feeding mice with 1% water-soluble fiber can effectively reduce lipids and triglyceride (TG) in their blood liver (Gallaher et al.,1993; Yeh et al., 2014).

Water-soluble fiber has better ability to regulate hypolipidemic activity when compared with nonwater-soluble fiber because water-soluble dietary fiber is capable of combining with bile salts and produce a hypolipidemic effect (Yeh et al., 2014). The polyphenol compounds in mushrooms which have been documented to reduce the risk of cardiovascular disease and cancer have also been demonstrated to be useful in reducing low density lipoprotein (LDL) oxidation (Chen et al., 2011).

7.7.4 ANTIDIABETIC PROPERTIES OF MEDICINAL MUSHROOMS

Mushrooms are known to have bioactive compounds which help the functions of vital organs such as liver, pancreas, and other endocrine glands (Wani et al., 2010). This consequently leads to healthy metabolic functioning (Chen et al., 2012). Mushroom polysaccharides (especially β-glucans) can restore the functions of pancreatic tissues, thereby increasing insulin output by the functional β-cell and also improving the sensitivity of peripheral cells to circulating insulin. The combination of these effects will reduce blood glucose levels (Xiao et al., 2011). In addition, medicinal mushrooms contain natural compounds and enzymes which act like insulin and help to improve insulin resistance and break down sugar or starch in foods (Kim et al., 2005, 2010a, 2010b).

Research studies on some mushrooms have reported their hypoglycemic and antidiabetic activities in type 2 diabetic patients (Kim et al., 2010a; Li et al., 2011; Lu et al., 2010; Shokrzadeh et al., 2017). The possible mechanisms for antidiabetic effects of mushrooms may include:

1. Increased adiponectin concentrations after taking fruiting body extract (Hsu et al., 2007). Plasma adiponectin concentrations predict insulin sensitivity toward glucose metabolism.
2. Inhibition of glucose absorption
3. Protection of β-cell damage
4. Increase of insulin release
5. Enhancement of antioxidant defense. Oxidative stress is a mainstream effect of the metabolic mechanisms by which alimentation can lead to insulin resistance (Zhai et al., 2011).
6. Attenuation of inflammation. A study by Niwa et al. (2011) suggested that the antidiabetic effects exhibited by mushrooms are through the suppression of oxidative stress and pro-inflammatory cytokine, TNF-α.
7. Modulation of carbohydrate metabolism pathway. Diabetic patients with dysfuntional metabolic control are susceptible to pulmonary complications with micro- and macro vascular disorders (Kaparianos et al., 2008).
8. Regulation of insulin-dependent and insulin-independent signaling pathways. (Lo and Wasser, 2011; Oh et al., 2010).

7.7.5 ANTIMICROBIAL ACTIVITIES

Mushrooms need antibacterial and antifungal compounds which makes them to survive in their natural environment. This is the reason why antimicrobial compounds with strong activities could be isolated from many edible and nonedible mushrooms which could be of benefit for human. Mushroom compounds with antimicrobial activities could be low-molecular weight or high-molecular weight compounds. The low-molecular weight compounds are metabolites such as oxalic acid, quinolines, steroids, terpenes, anthraquinone derivatives and benzoic acid derivatives. The high- molecular weight compounds are mainly peptides and proteins (Alves et al., 2012).

The antimicrobial property of several mushrooms has been documented and the most susceptible Gram-positive bacteria to mushroom inhibitory action are *Bacillus cereus, Bacillus subtilis* and *Staphylococos aureus* (Alves et al., 2012; Ezeronye, et al., 2005; Ofodile et al., 2008; Oyetayo, 2009; Ramesh and

Manohar, 2010; Waithaka et al., 2017). For example, ethanol and methanol extracts of *Pleurotus ostreatus*, a common edible mushroom, were able to inhibit *Bacillus cerreus*, *Escherichia coli*, *Listeria innocua*, and other Gram-positive and Gram-negative bacteria just as it also inhibited fungi (Vamanu et al., 2013). Methanolic extract of *Fistulina hepatica* successfully inhibited *Proteus vulgaris*, *E. coli* (Giri et al., 2012). Similarly, *P. ostreatus* inhibited *S. aureus*, *B. subtilis*, *P. vulgaris* and *E. coli*. Acetone extract of *Ganoderma lucidum* was found to be as effective as gentamycin sulfate against bacteria (Yamac and Bilgili, 2006). Furthermore, the ability of *Lentinula edodes* extracts to improve oral health has been extensively studied because it has a strong bactericidal effect upon *Streptococcus mutans*, the bacteria which is strongly indicated for dental caries and tooth decay (Signoretto et al., 2013).

7.8 MUSHROOM NUTRACEUTICALS

Medicinal mushrooms have become a novel class of products with different names such as dietary supplements, functional foods, nutraceuticals, myco-pharmaceuticals, and functional foods which have found their ways into many homes for health benefits. A mushroom nutraceutical can therefore be defined as refined, or partially refined, extract or dried biomass from the mycelium or the fruiting body of a mushroom which is packaged as a dietary supplement in the form of capsules or tablets with potential therapeutic applications (Chang and Buswell, 1996; Wasser, 2010). Cases of toxicity and overdose are not common with most of these nutraceuticals since they are extracted from edible mushroom species.

Several types of dietary supplements from mushrooms are available in the market today in form of:

1. Naturally grown, dried mushroom from fruit bodies prepared in the form of capsules or tablets
2. Fruit body powders from artificially cultivated mushrooms.
3. Dried and pulverized preparations of the combined substrate, mycelium, and primordial mushroom fruit bodies.
4. Biomass or extracts from mycelium harvested from submerged liquid culture grown in a fermentation tank or bioreactor
5. Mushroom spores and their extracts (Lindequist, 2013).

The important edible mushrooms with outstanding nutraceutical properties include the species of *Agaricus, Auricularia, Cordyceps, Flammulina, Grifola, Hericium, Lactarius, Lentinus (Lentinula), Pisolithus, Pleurotus, Russula,*

Tremella, and many others. Specifically, mushroom preparations like lentinan, SPG, AHCC, Maitake D-fraction, polysaccharide-K, and polysaccharide-P have shown clinically significant efficacy against human cancers (Barros et al., 2008). Some of these are discussed below.

7.8.1 LENTINAN

Lentinan (also known as Shiitake) is a water-soluble polysaccharide that is extracted from the fruiting body of *Lentinula edodes* which has been documented to have antitumor activities and proven to potentiate human immunity (Tanigawa et al., 2016). Its chemical formula is $C_{42}H_{72}O_{36}$ (Figure 7.1) with a molecular weight of 400,000–1,000,000 daltons (Ngai and Ng, 2003). It is a common BRM which has been approved for use as an adjuvant for cancer treatments in several countries (Ngai and Ng, 2003). The bioactive ingredient of lentinan is β-1, 3-D-glucan polymer with β-1, 6 or β-1, 4 branches. The oral bioavailability of Lentinan is reportedly limited so its administration is usually by parental injection even though it is sometimes taken orally as nutritional products.

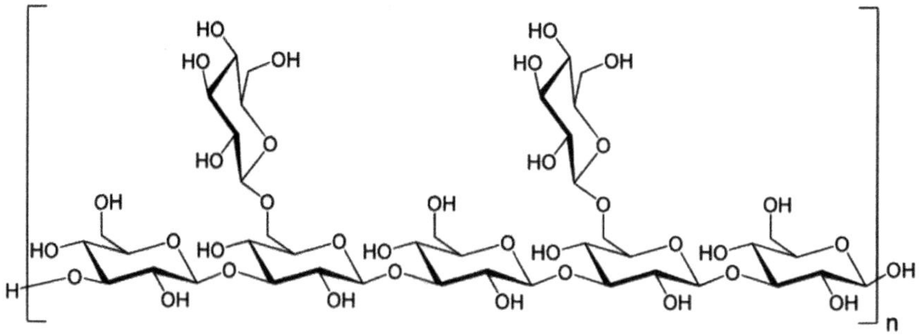

FIGURE 7.1   Lentinan.

Lentinan is a nonspecific BRM so its immunomodulatory effects is beneficial to various kinds of cancers (Wang et al., 2017). Studies carried out by Yoshino et al. (2000) have suggested that Lentinan action may be adjustment of the Th1/Th2 imbalance by inducing an upregulation of T helper 1 (Th1) response and a downregulation of Th2 response. Lentinan can also increase B-cell count and increase the level of Interferon gamma (IFN-γ), a cytokine that is vital for innate and adaptive immunity (Yoshino et al., 2007).

In vitro and in vivo studies have indicated that Lentinan can have direct anticancer and immunomodulatory effects. Anticancer effects of Lentinan can, therefore, be summarized as follows:

1. Lentinan alone can inhibit cancer cell proliferation and induce cell apoptosis (Sun et al., 2015)
2. Using Lentinan in combination with chemotherapy drugs or monoclonal antibodies enhances their ability to destroy cancer cells (Ren et al., 2014).
3. Lentinan can generate immune-modifying effects by activating macrophages and other monocytes and thus induce antibody-dependent, cell-mediated cytotoxicity (Yoshino et al., 2016).

7.8.2 SCHIZOPHYLLAN

SPG (also known as Sonifilan, Sizofiran, or Sizofilan) is produced as extracellular polysaccharide by the basidiomycete *Schizophyllum commune*. It is another β-1 $\rightarrow$ 3, β-1 $\rightarrow$ 6 D-glucan (Figure 7.2) with β-1 $\rightarrow$ 6 branching and the molecular formula is $(C_6H_{10}O_5)$. It has a molecular weight of about 450,000 daltons which makes it too large for effective oral administration and is therefore usually administered by intramuscular injection (Mansour et al., 2012; Ngai and Ng, 2003).

SPG was discovered by Komatsu et al. (1969) who demonstrated that it exhibits host-mediated antitumor activity against Sarcoma 180. SPG can also induce various immunostimulatory effects and activate NK cells, spleen cells, lymphoid cells and bone marrow cells. In addition, it enhances the production of antitumor and immunostimulatory cytokines such as interleukines 1, 2, and 3 (Zhang et al., 2013b). The bioactivity of SPG is founded on the improvement of cell-mediated immune response with stimulation of T-lymphocytes and macrophages in order to bring about T-cell activation in target organs, and also improve cytokine production (Ooi and Liu, 2000). SPG has been clinically tested as adjuvants that can be used alongside chemotherapeutic agents such as mitomycin and 5-fluorouracil.

SPG is used in cosmetics as an anti-aging ingredient because it can provide hydration by underpinning the viability of epidermal cells. It also reduces wrinkles due to stabilization and protection of the extracellular matrix (Muthny and Moravcova, 2013).

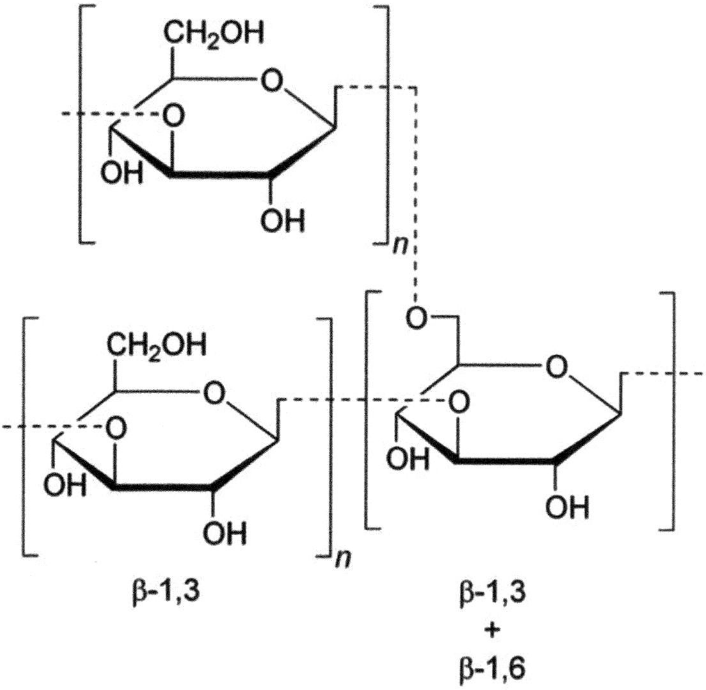

FIGURE 7.2   Schizophyllan.

7.8.3 *ACTIVE HEXOSE CORRELATED COMPOUND*

AHCC is an extract prepared from co-cultured mycelia of several species of Basidiomycete mushrooms through enzyme fermentation (Kidd, 2000; Ritz et al., 2006). AHCC contains polysaccharides, amino acids, and minerals (Figure 7.3), and is available orally. The predominant bioactive components of AHCC are oligosaccharides which constitute about 74% of the total dry weight, out of which 20% are partially acetylated α-1,4-glucans. The glucans in AHCC have low molecular weight of around 5000 daltons (Kidd, 2000). Both of these attributes are peculiar for immunoactive mushroom glucans (Gao et al., 2006; Kamiyama et al., 2013).

AHCC has demonstrated positive effects on immune function in humans and animal models in the following ways:

1. decreased tumor formation (Gao et al., 2006);
2. increased resistance to exogenous pathogens (Aviles et al., 2008);

3. enhanced NK cell activity with increased dendritic cell function (Nogusa et al., 2009; Terakawa et al., 2008);
4. increased T-cell proliferation in addition to altered T-cell activity (Aviles et al., 2008);
5. altered cytokine production (Aviles et al., 2008);
6. suppression of glucocorticoid-induced thymocyte apoptosis (Arora et al., 2014; Hirahara et al., 2012);
7. increased production of nitric oxide by peritoneal cells which is an important mediator in many biological processes, including macro-phage-mediated cellular host defense (Aviles et al., 2008); and
8. antioxidant and anti-inflammatory activities (Daddaoua et al., 2007).

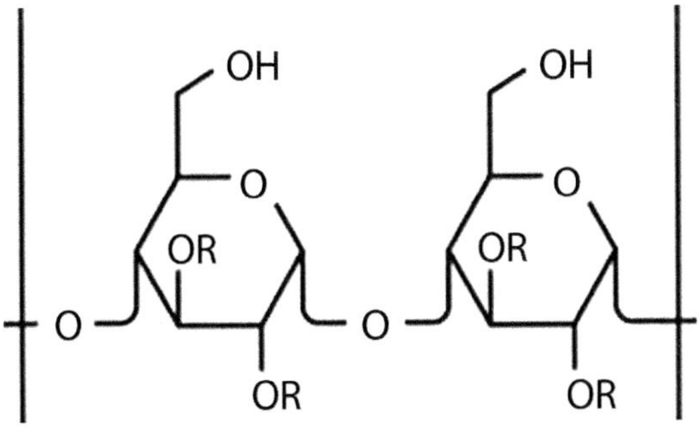

FIGURE 7.3 Active hexose correlated compound (AHCC).

Clinical effects and safety of AHCC have been demonstrated in cancer patients (Kawaguchi, 2009), as well as in healthy subjects (Yin et al., 2010). AHCC is usually administered as an adjuvant in combination with surgery and chemotherapy or radiation (Kawaguchi, 2009). AHCC can stimulate an increase in dendritic cell (DC) number and function. However, it has no significant effect on T-cell proliferation in response to mitogen stimulation (Terakawa et al., 2008). Furthermore, AHCC does not affect basal NK cell activity but enhances NK cell induction when there is an overlying infection (Nogusa et al., 2009). Th effect of AHCC on tumor cells was linked to the increase in CD4+/CD8+T-cell proliferation, CD8+T-cell production of interferon (IFN-γ), and increased numbers of NK and γ-δ T-cells (Gao et al., 2006; Suknikhom et al., 2017).

Some other studies have shown that consumption of raw or inadequately cooked shiitake can cause extensive erythematous rash and an allergic response known as "shiitake dermatitis" (Bachmeyer and Rouff, 2017). Merely handling the raw mushroom can cause contact dermatitis. Ingestion might also cause eosinophilia and GI problems in some patients (Levy et al., 1998).

7.8.4   *MAITAKE GRIFOLA POLYSACCHARIDE OR D-FRACTION*

Maitake Grifola polysaccharide (MGP) or D-fraction is a mixed β-D-glucan compound (Figure 7.4) that is usually prepared from the maitake mushroom (*Grifola frondosa*). It is available orally with established safety reports (Deng et al., 2009). Maitake D-fraction contains a mixture of β-D-glucan material with β-1 → 6 main chains and β-1 → 4 branchings, and the common β-1 → 3 main chains and β-1 → 6 branchings (Alonso et al., 2017).

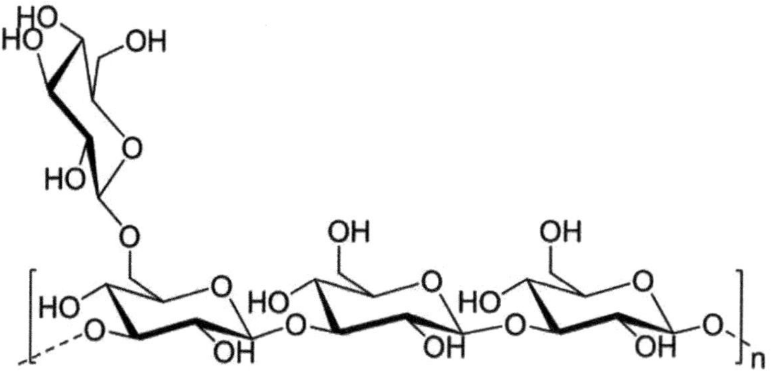

FIGURE 7.4   *Maitake Grifola* polysaccharide.

MGP has demonstrated anticancer activities in vitro and in vivo (Alonso et al., 2017).

It enhances the body's ability to destroy tumor cells by improving the body's specific and nonspecific immunity. MGP can also induce the production of inflammatory markers like cytokines and chemokines. MGP does not have a direct action on tumor cells, but it boosts the body's innate and adaptive immunity to inhibit tumor growth, prevent metastasis, and also protect normal cells from becoming cancerous (Alonso et al., 2017). Furthermore, when MGP was combined with chemotherapy or radiotherapy, immune-competent cell activities increased 1.2–1.4 times and reduced the side effects of chemotherapy or radiotherapy, compared with chemotherapy or radiotherapy alone (He et al., 2017). The crude polysaccharides are sometimes found to be more active in inhibiting tumor than purified fractions

(Mao et al., 2015; Zhang et al., 2016b). The mechanism of MGP action against cancer is similar to that of other fungus polysaccharides. However, the cancer-fighting ability of MGP was superior to polysaccharides from *G. lucidum*, lentinan, and *C. versicolor* (He et al., 2017; Yan et al., 2010).

Another important function of MGP is that it could decrease blood malondialdehyde (MDA) and TG levels, with no significant increase in alanine transaminase (ALT) and aspartate transaminase (AST). This is a clear indication that MGP has the protective effect on alcoholic oxidative injury in mice (Cao et al., 2010; He et al., 2017). The possible process might be that since the oxidation of ethanol was weak, it only caused lipid peroxidation on the membrane and did not reach the degree of leaking of transaminase from the cytoplasm. That is a further indication that MGP could decrease the serum activities of ALT, AST, lactate dehydrogenase, and MDA in liver tissue while increasing the serum activities of catalase in mice with CCl_4-induced hepatic injury, thereby confirming that MGP showed protective effect on mice with acute hepatic injury (Cao et al., 2010).

Many studies have documented that Maitake polysaccharides could lower the blood sugar level. Ma et al. (2014) showed that the MT-α-glucan from maitake had an antidiabetic effect on KKAy mice. The mechanism might be that MT-α-glucan affects enzymes related to glucose metabolism, protects pancreas and pancreatic islet cells, and increases the number of insulin receptors on targeted-cell membrane to improve insulin sensitivity and reduce the role of insulin resistance. It was found that MGP could reduce fasting blood glucose in alloxan-induced diabetic mice and enhance the sugar tolerance in diabetic mice (Cao et al., 2010; Wang et al., 2017). Lei et al. (2010) observed that MT-α-glucan had a hypoglycemic effect, which correlates with its inhibition of α-glucosidase activity.

Furthermore, *M. grifola* polysaccharide can affect hemodynamics, lower blood pressure, and protect the heart (He et al., 2017; Ni et al., 2009). In addition, Zhang et al. (2013a) found that MGP of different doses had antihypertensive activity and protect the heart and kidney from disorders caused by high blood pressure. A study in 2008 documented that MGP showed notable hypolipidemic effects on high-fat diet-induced hypercholesterolemic mice (Mori et al., 2008).

7.8.5 POLYSACCHAROPEPTIDES

T. versicolor is a basidiomycete that is well known and used traditionally in many countries (Hor et al., 2011; Sun et al., 2014). Polysaccharides from

T. versicolor exhibit several biological activities including hepatoprotective (Yeung and Or, 2012), anti-inflammatory (Kamiyama et al., 2013), anti-oxidant (Jhan et al., 2016), antigenotoxic (Knežević et al., 2015), immuno-regulatory (Chang et al., 2017), antitumor (Chay et al., 2017), and antiviral (Donatini, 2014) abilities. Intracellular polysaccharopeptides (PSPs), such as Polysaccharide Krestin (PSK) and PSP are the best known commercial preparations of *T. versicolor*.

PSPs are protein-bound polysaccharides which can be extracted from *T. versicolor* mycelia through batch fermentation. They are nontoxic with prolonged use in the treatment of cancers. The antitumor activity is achieved by suppressing DNA/RNA synthesis and enhancing the immune function (Yeung and Or, 2012). PSP/PSK are also recognized as BRMs and are therefore used as adjuncts to conventional cancer therapy (Chay et al., 2017).

PSK is a protein-bound polysaccharide, which is used as an immune system boosting agent (Maehara et al., 2012; Meng et al., 2016). The commercial name for PSK is Krestin. It contains approximately 62% polysaccharide and 38% protein, although this ratio is not constant. The approximate molecular weight of PSK is 100,000 daltons, and it has oral bioavailability (Maehara et al., 2012). The molecular weight of polysaccharides has a great influence on their biological activities (Liu et al., 2013). The bioactive component of PSK is β-D-glucan that consists of a β-1 $\rightarrow$ 4 main chain and β-1 $\rightarrow$ 3 side chains, with β-1 $\rightarrow$ 6 side chains that bond to a polypeptide moiety through *O*- or *N*-glycosidic bonds. The polypeptide portion is very rich in aspartic, glutamic, and other acidic amino acids. Polysaccharide is absorbed within 24 hours following administration. Subacute and chronic Toxicological studies indicate that PSK is nontoxic with a low oral LD_{50} (Oba et al., 2007). Animal studies indicated that PSK can effectively be taken orally, intravenously, or intraperitoneally (Meng et al., 2016).

PSP and PSK are both approved as an adjuvant for conventional cancer therapy (Chay et al., 2017). Oral administration of PSK and PSP was able to control carcinomas from in vivo studies (Fisher and Yang, 2002; Lin et al., 2008). Intraperitoneal administration of *T. versicolor* PSPs increased the SOD activity in the lymphocytes and thymus of normal mice.

Furthermore, PSP also exhibits analgesic activities (Meng et al., 2016). The best reported immunomodulatory effect of PSP is its induction of predominantly pro-inflammatory cytokines (Saleh et al., 2017). Interperitoneal administration of PSP was able to increase serum tumor necrosis factor (TNF)-α levels in healthy rats (Jedrzejewski et al., 2015). The ability of PSP to induce cytokines associated with TNF-α was also demonstrated through its induction of IL-12, a T helper 1 (Th1)-related cytokine capable of enhancing NK and CD8+T-cell cytotoxic activities and their expression of TNF-α (Saleh et al., 2017). Other physiological effects that has been linked to the use of PSPs include the following:

1. Amelioration of the immunosuppressive effects generated by chemo-therapy, radiotherapy, and blood transfusion in cancer patients (Chan et al., 2009; Meng et al., 2016).
2. Inhibition of tumor-induced immunosuppression (Frttz et al., 2015).
3. Inhibition of cancer proliferation through immune enhancement, increased production of superoxide dismutase (SOD), glutathione peroxidase (Meng et al., 2016).
4. Appetite stimulation and improvement of liver function in cancer patients (Chan et al., 2009).
5. Relaxation of the central nervous system (Frttz et al., 2015).
6. Enhancement of pain threshold (Chan et al., 2009; Meng et al., 2016)
7. Remediation of intestinal disorders as a result of opportunistic micro-bial infections (Chan et al., 2009; Meng et al., 2016).
8. Induction of enzymes that mop up free radicals and mitigate oxida-tive damage (Hsieh et al., 2011).
9. Immunopotentiation by stimulating the production of Interleukin (IL)-6, Interferons (IFNs), immunoglobulin-G, macrophages, and T-lymphocytes (Chan et al., 2009; Meng et al., 2016).

7.9 CONCLUSION

Medicinal mushrooms have been in use as medicine and food nutrient for many years because of their remarkable bioactivities. There are myriads of edible and nonedible mushrooms with established physiological activities. Medicinal mushrooms are therefore generating a lot of attention as potential natural agents for the prevention and treatment of many diseases. Bioactive compounds from a good number of these medicinal mushrooms have been identified, extracted and prepared in the form of nutraceuticals and dietary supplement.

However, there are still many more medicinal mushroom species with the potential to be developed into functional foods or nutraceuticals which are not common, and enough studies are yet to be carried out on such species. Hence, medicinal mushroom research will be enhanced by the establishment of a germplasm bank of well-authenticated wild mushrooms in different countries and continents. This will help to study, conserve, and characterize the genetic diversity of wild mushrooms. Such bio bank shall be a reliable source of mushroom strains for researchers.

KEYWORDS

- **anticancer**
- **bioactive molecules**
- **immunomodulation**
- **β-glucans**
- **medicinal mushrooms**
- **mushroom nutriceuticals**
- **polysaccharides**

REFERENCES

Adebayo, E.A.; Oloke, J.K.; Ayandele, A.A.; Adegunlola, C.O. Phytochemical, antioxidant and antimicrobial assay of mushroom metabolite from *Pleurotus pulmonarius. J. Microbiol. Biotechnol. Res.* 2012, 2(2), 366–374.

Akpaja, E.A.; Isikhuemhen, O.S.; Okhuoya, J.A. Ethnomycology and usage of edible and medicinal mushrooms among the Igbo people of Nigeria. *Int. J. Med. Mushrooms* 2003, 5 (3), 313–319.

Alam, N.; Yoon, K.N.; Lee, T.S.; Lee, U.Y. Hypolipidemic activities of dietary *Pleurotus ostreatus* in hypercholesterolemic rats. *Mycobiology,* 2011, 39(1), 45–51.

Alonso, E.N.; Ferronato, M.J.; Gandini, N.A.; Fermento, M.E.; Obiol, D.J.; Lopez Romero, A.; Arévalo, J.; Villegas, M.E.; Facchinetti, M.M.; Curino, A.C. Antitumoral effects of D-fraction from *Grifola frondosa* (maitake) mushroom in breast cancer. *Nutr. Cancer.* 2017, 69, 29–43.

Alves, M.J.; Ferreira, I.C.; Dias, J.; Teixeira, V.; Martins, A.; Pintado, M. A review on antimicrobial activity of mushroom (Basidiomycetes) extracts and isolated compounds. *Planta Médica.* 2012, 78(16), 1707–1718.

Anderson, J.W.; Baird, P.; Davis, R.H.; Ferreri, S.; Knudtson, M.; Koraym, A.; Waters V.; Williams, C.L. Health benefits of dietary fiber. *Nutr. Rev.* **2009**, 67(4), 188–205

Arora, R.; Yates, C.; Gary, B.D., McClellan, S.; Tan, M.; Xi, Y.; Reed, E.; Piazza, G.A.; Owen, L.B.; Dean-Colomb, W. Panepoxydone targets NF-κB and FOXM1 to inhibit proliferation, induce apoptosis and reverse epithelial to mesenchymal transition in breast cancer. *PLoS One.* 2014, 9, e98370.

Aviles, H.; O'Donnell, P.; Orshal, J.; Fujii, H.; Sun, B.; Sonnenfeld, G. Active hexose correlated compound activates immune function to decrease bacterial load in a murine model of intramuscular infection. *Am. J. Surg.* 2008, 195, 537–545.

Ayeka, P.A.; Bian, Y.; Githaiga, P.M.; Zhao, Y. The immunomodulatory activities of licorice polysaccharides (*Glycyrrhiza uralensis* Fisch.) in CT 26 tumor-bearing mice. BMC Complement. Altern*at. Med.* 2017, 17 (1), 536.

Ayodele, S.M.; Akpaja, E.O. Adamu, Y. In some edible and medicinal mushrooms found in Igala land in Nigeria and their sociocultural and ethnomycological uses. *Proc. 5th Int. Med. Mushroom Conf.*, Nantong, China, 2009, 526–531.

Bachmeyer, C.; Rouff, E. Shiitake dermatitis. *CMAJ.* 2017, 189(11), E439.

Barros, L.; Baptista, P.; Correia, D.M.; Casal, S.; Oliveira, B.; Ferreira, I.C.F.R. Fatty acid and sugar compositions, and nutritional value of five wild edible mushrooms from Northeast Portugal. *Food Chem.* 2007, 105(1), 140–145.

Barros, L., Cruz, T., Baptista, P., Estevinho, L.M. and Ferreira, I.C. Wild and commercial mushrooms as source of nutrients and nutraceuticals. *Food Chem. Toxicol.* 2008, 46(8), 2742–2747.

Batbayar, S.; Lee, D.H.; Kim, H.W. Immunomodulation of fungal β-glucan in host defense signaling by dectin-1. *Biomol. Ther.* 2012, 20, 433–445.

Beiersmann, C.; Sanou, A.; Wladarsch, E.; de Allegri, M.; Kouyate, B.; Muller O. Malaria in rural Burkina Faso: Local illness concepts, patterns of traditional treatment and influence on health-seeking behaviour. *Malar. J.* 2007, 6, 106–114.

Bhakta, M.; Kumar, P. Mushroom polysaccharides as a potential prebiotics. *Int. J. Health Sci. Res.* 2013, 3, 77–84.

Bobek, P.; Nosalova, V.; Cerna, S. Effect of pleuran (β-glucan from *Pleurotus ostreatus*) in diet or drinking fluid on colitis in rats. *Nahrung* 2001, 45, 360–363.

Borchers, A.T.; Stern, J.S.; Hackman, R.M.; Keen, C.L.; Gershwin, M.E. Mushrooms, tumors, and immunity. *Proc. Soc. Exp. Biol. Med.* 1999, 221, 281–293.

Borchers, A.T.; Krishnamurthy, A.; Keen, C.L.; Meyers, F.J.; Gershwin, M.E. The immunobiology of mushrooms. *Exp. Biol. Med. (Maywood).* 2008, 233, 259–276.

Brenner, G.M.; Stevens, C.W. Antineoplastic drugs. In: *Pharmacology.* 3rd ed. Philadelphia, PA: Elseiver, 2010; pp. 493–511.

Cao, X.; Zhu, H.; Wang, C.; Lu, Y.; Zeng, B.; Wang, A. Protective effect of exo- polysaccharides produced by *Grifola frondosa* on carbon tetrachloride induced liver injury in mice. *Nat. Prod. Res. Dev.* 2010, 22(5), 777–780.

Castro, A.J.G.; Castro, L.S.E.W.; Santos, M.S.N.; Faustino, M.G.; Pinheiro, T.S.; Dore, C.M.G.; Baseia, I.G.; Leite, E.L. Anti-inflamatory, anti-angiogenenic and antioxidant activities of polysaccharide-rich extract from fungi *Caripia montagnei*. *Biomed. Prevent. Nutr.* 2014, 4(2), 121–129.

Chan, G.C.F.; Chan, W.K.; Sze, D.M.Y. The effects of β-glucan on human immune and cancer cells. *J. Hematol. Oncol.* 2009, 2, 25.

Chang, R.T.; Buswell, J.A. Mushroom nutriceuticals. *World J. Microbiol. Biotechnol.* 1996, 12 (5), 473–476.

Chang S.T.; Miles, D.G. The nutritional attributes and medicinal value of edible mushroom. In: *Edible Mushrooms and Their Cultivation*, Boca Raton, FL: CRC Press, 1989; pp. 27–40.

Chang, S.T.; Wasser, S.P. The role of culinary-medicinal mushrooms on human welfare with a pyramid model for human health. *Int. J. Med. Mushrooms* 2012, 1, 95–134.

Chang, Y.; Zhang, M.; Jiang, Y.; Liu, Y.; Luo, H.; Hao, C.; Zeng, P.; Zhang, L. Preclinical and clinical studies of *Coriolus versicolor* polysaccharopeptide as an immunotherapeutic in China. *Discov. Med.* 2017, 23, 207–219.

Chay, W.Y.; Tham, C.K.; Toh, H.C.; Lim, H.Y.; Tan, C.K.; Lim, C.; Wang, W.W.; Choo, S.P. Coriolus versicolor (*Yunzhi*) use as therapy in advanced hepatocellular carcinoma patients with poor liver function or who are unfit for standard therapy. *J. Alternat. Complement. Med.* 2017, 23(8), 648–652.

Chen, G.; Luo, Y.C.; Ji, B.P.; Li, B.; Su, W.; Xiao, Z.L.; Zhang, G.Z. Hypocholesterolemic effects of Auricularia auricula ethanol extract in ICR mice fed a cholesterol-enriched diet. J. Food Sci. Technol. (Mysore). 2011, 48, 692–698.

Chen, J.; Mao, D.; Yong, Y.; Li, J.; Wei, H.; Lu, L. Hepatoprotective and hypolipidemic effects of water-soluble polysaccharidic extract of Pleurotus eryngii. *Food Chem.* 2012, 130, 687–694.

Cheung, Y.H.; Sheridan, C.M.; Lo, A.C.Y.; Lai, W.W. Lectin from *Agaricus bisporus* inhibited S phase cell population and Akt phosphorylation in human RPE cells. *Retin. Cell. Biol.* 2012, 53(12), 7469–7475.

Daddaoua, A.; Enrique, M.P.; Lopez-Posadas, R.; Vieites, J.M.; González, M.; Requena, P.; Zarzuelo, A.; Suárez, M.D.; de Medina, F.S.; Martínez-Augustin, O. Active hexose correlated compound acts as a prebiotic and is anti-inflammatory in rats with hapten-induced colitis. *J. Nutr.* 2007, 137, 1222–1228.

Dai, Y.C.; Yang, Z.L.; Ui, B.K.; Yu, C.H.; Zhou, L.W. Species diversity and utilization of medicinal mushrooms and fungi in China (review). *Int. J. Med. Mushrooms* 2009, 11, 287–302.

Dalla-Santa, H.S.; Rubel, R.; Vitola, F.M.D.; Rodriguez-Leon, J.A.; Dalla-Santa, O.R.; Brand, D.; Alvarez, D.C.; Macedo, R.E.F.; Carvalho, J.C.; Soccol, C.R. Growth parameters of *Agaricus brasiliensis* mycelium on wheat grains in solid-state fermentation. *Biotechnol.* 2012, 11, 144–153.

De Silva, D.D.; Rapior, S.; Sudarman, E.; Stadler, M.; Xu, J.; Alias, S.A.; Hyde, K.D. Bioactive metabolites from macrofungi: Ethnopharmacology, biological activities and chemistry. *Fungal Divers.* 2013, 62(1), 1–40.

De Silva, M.C.S.; Naozuka, J.; da Luz, J.M.R.; de Assunção, L.S.; Oliveira, P.V.; Vanetti, M.C.D.; Bazzolli, D.M.S.; Kasuya, M.C.M. Enrichment of *Pleurotus ostreatus* mushrooms with selenium in coffee husks. *Food Chem.* 2012, 131, 558–563.

De Sousa, V.M.C.; Dos Santos, E.F.; Sgarbieri, V.C. The importance of prebiotics in functional foods and clinical practice. *Food Nutr. Sci.* 2011, 2(4), 133–144.

Deng, G.; Lin, H.; Seidman, A.; Fornier M.; D'Andrea, G.; Wesa, K.; Yeung, S.; Cunningham-Rundles, S.; Vickers, A.J.; Cassileth, B. SA phase I/II trial of a polysaccharide extract from *Grifola frondosa* (Maitake mushroom) in breast cancer patients: Immunological effects. *J. Cancer Res. Clin. Oncol.* 2009, 135(9), 1215–1221.

Donatini B. Control of oral human papillomavirus (HPV) by medicinal mushrooms, *Trametes versicolor* and *Ganoderma lucidum*: A preliminary clinical trial. *Int. J. Med. Mushrooms.* 2014,16, 497–498.

Dotan, N.; Wasser, S.P.; Mahajna, J. The culinary-medicinal mushroom *Coprinus comatus* as a natural antiandrogenic modulator. *Integr. Cancer Ther.* 2011, 10, 148–159.

Duru, M.E.; Çayan, G.T. Biologically active terpenoids from mushroom origin: A review. *Records Nat. Prod.* 2015, 9(4), 456.

Egwim, E.C.; Ellen, R.C.; Egwuche, R.U. Proximate composition, phytochemical screening and antioxidant activity of ten selected edible mushrooms. *Am. J. Food Nutr.* 2011, 1(2), 89–94.

Ehssan, H.O.; Saadabi, A.M. Screening of antimicrobial activity of wild mushrooms from Khartoum State of Sudan. *Microbiol. J.* 2012, 2, 64–69.

El Enshasy, H.A.; Hatti-Kaul, R. Mushroom immunomodulators: Unique molecules with unlimited applications. *Trends Biotechnol.* 2013, 31(12), 668–677.

El Enshasy, H.; Elsayed, E.A.; Aziz, R.; Wadaan, M.A. Mushrooms and truffles: Historical biofactories for complementary medicine in Africa and in the Middle East. *Evid Based Complement. Alternat. Med.* 2013, 2013: 620451.

El Enshasy, H. Immunomodulators. In: The Mycota: Industrial Applications. Berlin, Germany: Springer, 2010; Vol 10; pp. 165–194.

Ezekiel, C.N.; Sulyok, M.; Frisvad, J.C.; Somorin, Y.M.; Warth, M.; Houbraken, J., Samson, R.A.; Krska, R.; Odebode, A.C. Fungal and mycotoxin assessment of dried edible mushroom in Nigeria. *Int. J. Food Microbiol.* 2013, 162, 231–236.

Ezeronye, O.U.; Daba, A.S.; Okwujiako, A.I.; Onumajuru, I.C. Antibacterial of crude polysaccharide extracts from sclerotium and fruitbody (sporophore) of *Pleurotus tuber*-regium (Fried) Singer on some clinical isolates. *Int. J. Mol. Med. Adv. Sci.* 2005, 1(3), 202–205.

Ferreira, I.C.F.R.; Barros, L.; Abreu, R.M.V. Antioxidants in wild mushrooms. *Curr. Med. Chem.* 2009, 16, 1543–1560.

Fisher, M.; Yang, L.X. Anticancer effects and mechanisms of polysaccharide-K (PSK): Implications of cancer immunotherapy. *Anticancer Res.* 2002, 22 (3), 1737–1754.

Frttz, H.; Kennedy, D.A.; Ishii, M.; Fergusson, D.; Fernandes, R.V.; Cooley, K.; Seely M.R. Polysaccharide K and Coriolus versicolor extracts for lung cancer: A systematic review. Integr. Cancer Ther. 2015, 14(3), 210–211.

Gallaher, D.D.; Hassel, C.A.; Lee, K.J. Relationships between viscosity of hydroxypropyl methylcellulose and plasma cholesterol in hamsters. *J. Nutr.* 1993, 123(10), 1732–1738.

Gao, Y.; Zhang, D.; Sun, B.; Fujii, H.; Kosuna, K.I.; Yin, Z. Active hexose correlated compound enhances tumor surveillance through regulating both innate and adaptive immune responses. *Cancer Immunol. Immunother.* 2006, 55(10), 1258–1266.

Geosel, A.; Sipos, L.; Stefanovits-Banyai, E.; Kokai, Z.; Gyorfi, J. Antioxidant, polyphenol, and sensory analysis of *Agaricus bisporus* and *Agaricus subrufescens* cultivars. *Acta Alimentaria.* 2011, 40, 33–40.

Geurts, L.; Neyrinck, A.M.; Delzenne, N.M. Gut microbiota controls adipose tissue expansion, gut barrier and glucose metabolism: Novel insights into molecular targets and interventions using prebiotics. *Benef. Microbes.* 2014, 5, 3–17.

Giri, S.; Biswas, G.; Pradhan, P.; Mandal, S.C.; Acharya, K. Antimicrobial activities of basidiocarps of wild edible mushrooms of West Bengal, India. *Int. J. Pharm. Tech. Res.* 2012, 4, 1554–1560.

Griensven, L.V. In mushrooms, must action be taken? *Proceeding of the 5th International Medicinal Mushroom Conference*, Nantong, China, 2009, pp. 407–412.

Guggenheim, A.G.; Wright, K.M.; Zwicky, H.L. Immune modulation from five major mushrooms: application of integrative oncology. *Integrative Med.* 2014, 13(1), 32–44.

Harborne, J.B.; Williams, C.A. Advances in flavonoid research since 1992. *Phytochemistry.* 2000, 55 (6), 481–504.

Hawksworth, D.L. Global species number of fungi: Are tropical studies and molecular approaches contributing to a more robust estimate? *Biodivers. Conserv.* 2012, 21, 2425–2433.

He, X.; Wang, X.; Fang, J.; Chang, Y.; Ning, N.; Guo, H.; Huang, L.; Huang, X.; Zhao, Z. Polysaccharides in *Grifola frondosa* mushroom and their health promoting properties: A review. *Int. J. Biol. Macromol.* 2017, 101, 910–921.

Hertog, M.G.L.; Hollman, P.C.H.; Katan, M.B.; Kromhout, D. Intake of potentially anticarcinogenic flavonoids and their determinants in adults in The Netherlands. *Nutr. Cancer.* 1993, 20 (1), 21–29.

Hor, S.Y.; Ahmad, M.; Farsi, E.; Lim, C.P.; Asmawi, M.Z.; Yam, M.F. Acute and subchronic oral toxicity of *Coriolus versicolor* standardized water extract in Sprague-Dawley rats. *J. Ethnopharmacol.* 2011, 137, 1067–1076.

Hsieh, Y.C.; Rao, Y.K.; Whang-Peng, J.; Huang, C.Y.; Shyue, S.K.; Hsu, S.L.; Tzeng, Y.M. Antcin B and its ester derivative from *Antrodia camphorata* induce apoptosis in hepatocellular carcinoma cells involves enhancing oxidative stress coincident with activation of intrinsic and extrinsic apoptotic pathway. *J. Agric. Food Chem.* 2011, 59, 10943–10954.

Hsu, Y.L.; Kuo, P.L.; Cho, C.Y.; Ni, W.C.; Tzeng, T.F.; Ng, L.T.; Kuo, Y.H.; Lin, C.C. *Antrodia cinnamomea* fruiting bodies extract suppresses the invasive potential of human liver cancer cell line PLC/PRF/5 through inhibition of nuclear factor kappaB pathway. *Food Chem. Toxicol.* 2007, 45, 1249–1257.

Huang, C. H.; Chang, Y.Y.; Liu, C.W.; Kang, W.Y.; Lin, Y.L.; Chang, H.C.; Chen, Y.C. Fruiting body of Niuchangchih (*Antrodia camphorata*) protects livers against chronic alcohol consumption damage. *J. Agric. Food Chem.* 2010, 58, 3859–3866.

Hung, P.V.; Nhi, N.Y. Nutritional composition and antioxidant capacity of several edible mushrooms grown in the Southern Viet. *Int. Food Res. J.* 2012, 19(2), 611–615.

Hyde, K.D., Bahkali, A.H., Moslem, M.A. Fungi: An unusual source for cosmetics. *Fungal Divers.* 2010, 43, 1–9.

Ishii, P.L.; Prado, C.K.; Mauro, M.O.; Carreira, C.M.; Mantovani, M.S.; Ribeiro, L.R.; Dichi, J.B.; Oliveira, R.J. Evaluation of *Agaricus blazei in vivo* for antigenotoxic, anti carcinogenic, phagocytic and immunomodulatory activities. *Regul. Toxicol. Pharmacol.* 2011, 59, 412–422.

Islam, T.; Yu, X.; Xu, B. Phenolic profiles, antioxidant capacities and metal chelating ability of edible mushroom commonly consumed in China. *LWT Food Sci. Technol.* 2016, 72, 423–431.

Jedrzejewski, T.; Piotrowski, J.; Kowalczewska, M.; Wrotek, S.; Kozak, W. Polysaccharide peptide from *Coriolus versicolor* induces interleukin 6-related extension of endotoxin fever in rats. *Int. J. Hyperthermia.* 2015, 31(6), 626–634.

Jeong, S.C.; Jeong, Y.T.; Yang, B.K.; Islam, R.; Koyyalamudi, S.R.; Pang, G.; Cho, K.Y.; Song, C.H. White button mushroom (*Agaricus bisporus*) lowers blood glucose and cholesterol levels in diabetic and hypercholesterolemic rats. *Nutr. Res.* 2010, 30, 49–56.

Jhan, M. H.; Yeh, C.H.; Tsai, C.C.; Kao, C.T.; Chang, C.K.; Hsieh, C. W. Enhancing the antioxidant ability of *Trametes versicolor* polysaccharopeptides by an enzymatic hydrolysis process. *Molecules.* 2016, 21, E1215.

Jong, S.; Yang, X. PSP—a powerful biological response modifier from the mushroom Coriolus versicolor. In: Advanced Research in PSP, Yang, Q., ed., Hong Kong: Hong Kong Association for Health Care Ltd, 1999, pp. 16–28.

Kalac, B.P. Chemical composition and nutritional value of European species of wild growing mushrooms: A review. *Food Chem.* 2009, 113 (1), 9–16.

Kamatenesi-Mugisha, M.; Oryem-Origa, H. Medicinal plants used to induce labour during childbirth in western Uganda. *J. Ethnopharmacol.* 2007, 109 (1), 1–9.

Kamiyama, M.; Horiuchi, M.; Umano, K.; Kondo, K.; Otsuka, Y.; Shibamoto. T. Antioxidant/ anti -inflammatory activities and chemical composition of extracts from the mushroom Trametes versicolor. *Int. J. Food Sci. Nutr.* 2013, 2, 85–91.

Kamiyama, M.; Horiuchi, M,; Umano, K.; Kondo, K.; Otsuka, Y.; Shibamoto, T. Antioxidant/ anti-inflammatory activities and chemical composition of extracts from the mushroom *Trametes versicolor. Int. J. Nutr. Food Sci.* 2013, 2, 85–91.

Kaparianos, A.; Argyropoulou, E.; Sampsonas, F.; Karkoulias, K.; Tsiamita, M. Spiropoulos, K. Pulmonary complications in diabetes mellitus. Chronic Res. *Dis.* 2008, 5, 101–108.

Kawaguchi, Y. Improved survival of patients with gastric cancer or colon cancer when treated with active hexose correlated compound (AHCC): Effect of AHCC on digestive system cancer. *Nat. Med. J.* 2009, 1(1), 1–6.

Kidd, P.M. The use of mushroom glucans and proteoglycans in cancer treatment. *Alternative Med. Rev.* 2000, 5(1), 4–27.

Kim, G.Y.; Lee, J.Y.; Lee, J.O.; Ryu, C.H.; Choi, B.T.; Lee, K.W.; Jeong, Y.H.; Choi, Y.H. Partial characterization and immunostimulatory effect of a novel polysaccharide protein complex extracted from *Phellinus linteus*. *Biosci. Biotechnol. Biochem.* 2006, 70 (5), 1218–1226.

Kim, H.M.; Kang, J.S.; Kim, J.Y.; Park, S.K.; Kim, H.S.; Lee, Y.J.; Yun, J.; Hong, J.T.; Kim, Y.; Han, S.B. Evaluation of antidiabetic activity of polysaccharide isolated from *Phellinus linteus* in non-obese diabetic mouse. *Int. Immunopharmacol.* 2010a, 10, 72–78.

Kim, J. I.; Kang, M.J.; Im, J.; Seo, Y.J.; Lee, Y.M.; Song, J.H.; Lee, J.H.; Kim, M.E. Effect of King Oyster mushroom (*Pleurotus eryngii*) on insulin resistance and dyslipidemia in db/db mice. *Food Sci Biotechnol.* 2010b, 19, 239–242.

Knežević, A.; Živković, L.; Stajić, M.; Vukojević, J.; Milovanović, I.; Spremo-Potparević, B. Antigenotoxic effect of *Trametes spp.* extracts against DNA damage on human peripheral white blood cells. *Sci. World J.* 2015, 2015, 146378.

Komatsu, N.; Okubo, S.; Kikumoto, S.; Kimura, K.; Saito, G.; Sakai, S. Host-mediated antitumor action of *Schizophyllan*, a glucan produced by *Schizophyllum* commune. *GANN Japan. J. Cancer Res.* 1969, 60(2), 137–144.

Koyyalamudi, S.R., Jeong, S.C., Cho, K.Y., Pang, G. Vitamin B12 is the active corrinoid produced in cultivated white button mushrooms (*Agaricus bisporus*). *J. Agric. Food Chem.* 2009a, 57, 6327–6333.

Koyyalamudi, S.R.; Jeong, S.C.; Song, C.H.; Cho, K.Y., Pang, G. Vitamin D2 formation and bioavailability from *Agaricus bisporus* button mushrooms treated with ultraviolet irradiation. *J. Agric. Food Chem.* 2009b, 57, 3351–3355.

Krishnamoorthy, D.; Sankaran, M. Modulatory effect of *Pleurotus ostreatus* on oxidant/ antioxidant status in 7, 12-dimethylbenz (a) anthracene induced mammary carcinoma in experimental rats—a dose-response study. *J. Cancer Res. Ther.* 2016, 12, 1–10.

Lee, A.H.; Pasalich, M.; Su, D.; Tang, L.; Tran, V.D.; Binns, C.W. Mushroom intake and risk of epithelial ovarian cancer in southern Chinese women. *Int. J. Gynecol. Cancer.* 2013, 23, 1400–1405.

Lee, D.H.; Kim, H.W. Innate immunity induced by fungal β-glucans via dectin-1 signaling pathway. *Int. J. Med. Mushrooms.* 2014, 16, 1–16.

Lee, I.K.; Kim, Y.S.; Jang, Y.W.; Jung, J.Y.; Yun, B.S. New antioxidant polyphenols from the medicinal mushroom *Inonotus obliquus*. *Bioorg. Med. Chem. Lett.* 2007, 17, 6678–6681.

Lei, H.; Wang, Y.; Cai, L.; Wu, W. Hypoglycemic activity of polysaccharides from fruit body of *Grifola frondosa* and its effect on α-glucosidase activity. *Food Sci.* 2010, 31(11), 263–267.

Levy, A.M.; Kita, H.; Phillips, S.F.; Schkade, P.A.; Dyer, P.D.; Gleich, G.J.; Dubravec, V.A. Eosinophilia and gastrointestinal symptoms after ingestion of shiitake mushrooms. *J. Allergy Clin. Immunol.* 1998, 101(5), 613–620.

Li, H.J.; Chen, H.Y.; Fan, L.L.; Jiao, Z.H.; Chen, Q.H.; Jiao, Y.C. *In vitro* antioxidant activities and *in vivo* anti-hypoxic activity of the edible mushroom *Agaricus bisporus* (*Lange*) *Sing.* Chaidam. *Molecules.* 2015, 20, 17775–17788.

Li, T.H.; Hou, C.C.; Chang, C.L.T.; Yang, W.C. Anti-hyperglycemic properties of crude extract and triterpenes from *Poria cocos*. *Evid. Based Complement. Alternat. Med.* 2011, 128402, 10.

Lin, F.Y.; Lai, Y.K.; Yu, H.C.; Chen, N.Y.; Chang, C.Y.; Lo, H.C.; Hsu, T.H. Effects of *Lycium barbarum* extract on production and immunomodulatory activity of the extracellular polysaccharopeptides from submerged fermentation culture of *Coriolus versicolor*. *Food Chem.* 2008, 110(2), 446–453.

Lindequist, U. The merit of medicinal mushrooms from a pharmaceutical point of view. *Int. J. Med. Mushrooms* 2013, 15(6), 1–12.

Liu, B.; Bian, H.J.; Bao, J.K. Plant lectins: Potential antineoplastic drugs from bench to clinic. *Cancer Lett.*, 2010, 287(1), 1–12.

Liu, B.; Zhong, M.; Lun, Y.; Wang, X.; Sun, W.; Li, X.; Ning, A.; Cao, J.; Zhang, W.; Lei, L.; Huang, M. A novel apoptosis correlated molecule: Expression and characterization of protein latcripin-1 from *Lentinula edodes* C91-3. *Int. J. Mol. Sci.* 2012, 13, 6246–6265.

Liu, D.; Sheng, J.; Li, Z.; Qi, H.; Sun, Y.; Duan, Y.; Zhang, W. Antioxidant activity of polysaccharide fractions extracted from *Athyrium multidentatum* (Doll.) *Ching. Int. J. Biol. Macromol.* 2013, 56, 1–5.

Lo, H.C.; Wasser, S.P. Medicinal mushrooms for glycemic control in diabetes mellitus: History, current status, future perspectives, and unsolved problems (review). *Int. J. Med. Mushrooms,* 2011, 13(5),401–426.

Lo, Y.; Lin, S.; Ulziijargal, E.; Chen, S.; Chien, R.; Tzou, Y.; Mau J. Comparative study of contents of several bioactive components in fruiting bodies and mycelia of culinary medicinal mushrooms. *Int. J. Med. Mushrooms.* 2012, 14(4), 357–363.

Lu, X.M.; Chen, H.X.; Dong, P.; Fu, L.L.; Zhang, X. Phytochemical characteristics and hypoglycaemic activity of fraction from mushroom *Inonotus obliquus*. *J. Sci. Food Agric.* 2010, 90(2), 276–280.

Ma, G.; Yang, W.; Mariga, A.M.; Fang, Y.; Ma, N.; Pei, F.; Hu, Q. Purification, characterization and antitumor activity of polysaccharides from *Pleurotus eryngii* residue. *Carbohydr. Polym.* 2014, 114, 297–305.

Maehara, Y.; Tsujitani, S.; Saeki, H. Biological mechanism and clinical effect of protein-bound polysaccharide K (KRESTIN): Review of development and future perspectives. *Surg. Today.* 2012, 42, 8–28.

Mandal, S.; Nayak, A.; Banerjee, S.K.; Banerji, J.; Banerji, A. A new carbazole alkaloid from *Murraya koenigii* Spreng (Rutaceae). *Nat. Prod. Commun.* 2008, 3 (10), 1679–1682.

Mansour, A.; Daba, A.; Baddour, N.; El-Saadani, M.; Aleem, E. Schizophyllan inhibits the development of mammary and hepatic carcinomas induced by 7,12 dimethylbenz (α) anthracene and decreases cell proliferation: Comparison with tamoxifen. *J. Cancer Res. Clin. Oncol.* 2012, 138, 9, 1579–1596.

Mao, G.; Ren, Y.; Feng, W.; Li, Q.; Wu, H.; Jin, D.; Zhao, T.; Xu, C.; Yang, L.; Wu, X. Antitumor and immunomodulatory activity of a water-soluble polysaccharide from *Grifola frondosa*. Carbohyd. *Polym.* 2015, 134, 406–412.

Meneses, M.E.; Carrera, M.D.; Torres, N. Hypocholesterolemic properties and prebiotic effects of Mexican *Ganoderma lucidum* in C57BL/6 mice. *PLoS One.* 2016, 11: e0159631.

Meng, X.; Liang, H.; Luo, L. Antitumor polysaccharides from mushrooms: A review on the structural characteristics, antitumor mechanisms and immunomodulating activities. *Carbohydr. Res.* 2016, 424, 30–41.

Mhanda, F.N.; Kadhila-Muandingi, N.P.; Ueitele, I.S.E. Minerals and trace elements in domesticated Namibian *Ganoderma* species. *Afr. J. Biotechnol.* 2015, 14, 3216–3218.

Mori, K.; Kobayashi, C.; Tomita, T.; Inatomi, S.; Ikeda, M. Antiatherosclerotic effect of the edible mushrooms *Pleurotus eryngiii* (Eringi), *Grifola frondosa* (Maitake) and *Hypsizygus marmoreus* (Bunashimeji) in apolipoprotein deficient mice. *Nutr. Res.* 2008, 28, 335–342.

Muthny, T.; Moravcova, M. Skin aging in the context of sun damage and immune response alterations. *SOFW J.* 2013, 4, 2–8.

Nedelcheva, D.; Antonova, D.; Tsvetkova S.; Marekov, I.; Momchilova, S.; Nikolova-Damyanova, B.; Gyosheva, M. TLC and GC-MS probes into the fatty acid composition of some *Lycoperdaceae* mushrooms. *J. Liq. Chromatogr. Relat. Technol.* 2007, 30 (18), 2717–2727.

Ng, H.E. Estimation of fungal growth using the ergosterol assay: A rapid tool in assessing the microbiological status of grains and feeds. *Lett. Appl. Microbiol.* 2008, 46(1), 113–118.

Ngai, P.H.; Ng, T.B. Lentin, a novel and potent antifungal protein from shiitake mushroom with inhibitory effect on activity of human immunodefiency Virus-1 reverse transcriptase and proliferation of Leukemia cells. *Life Sci.* 2003, 73, 3363–3374.

Ni, Y.; Zhang, L.; Yang, J.; Cui, D.; Shao, Y.; Gao, SR. Effect of *Grifola frondosa* polysaccharide on cardiac hemodynamicsin hypertensive rats. *J. Anhui. Agri. Sci.* 2009, 37(25), 12037–12038.

Niwa, A.; Tajiri, T.; Higashino, H. *Ipomoea batatas* and *Agarics blazei* ameliorate diabetic disorders with therapeutic antioxidant potential in streptozotocin-induced diabetic rats. *J. Clin. Biochem. Nutr.* 2011, 48(3), 194–202.

Nogusa, S.; Gerbino, J.; Ritz, B. Low-dose supplementation with active hexose correlated compound improves the immune response to acute influenza infection in C57BL/6 mice. *Nutr. Res.* 2009, 29, 139–143.

Oba, K.; Teramukai, S.; Kobayashi, M., Matsui, T., Kodera Y, Sakamoto J. Efficacy of adjuvant immunochemotherapy with polysaccharide K for patients with curative resections of gastric cancer. *Cancer Immunol. Immunother.* 2007, 56 (6), 905–911.

Ofodile, L.N.; Simmons, S.J.; Grayer, R.J.; Uma, N.U. Antimicrobial activity of two species of the genus *Trametes* Fr. (Aphyllophoro mycetideae) from Nigeria. J. Med. Mushrooms. 2008, 10(3), 265–268.

Oh, T.W.; Kim, Y.A.; Jang, W.J.; Byeon, J.I.; Ryu, C.H.; Kim, J.O.; Ha, Y.L. Semi- purified fractions from the submerged-culture broth of *Agaricus blazei* Murill reduce blood glucose levels in streptozotocininduced diabetic rats. *J. Agric. Food Chem.* 2010, 58(7), 4113–4119.

Ooi, V.E.C. Antitumor and immunomodulatory activities of mushroom polysaccharides. In: *Mushrooms as Functional Food*, New York, NY: John Wiley & Sons, 2008; pp. 147–198.

Ooi, V.E.C.; Liu, F. Immunomodulation and anti-cancer activity of polysaccharide-protein complexes. *Curr. Med. Chem.* 2000, 7 (7), 715–729.

Orsine J.V.C.; Novaes, M.R.C.G.; Asquieri, E.R. Nutritional value of *Agaricus sylvaticus*; mushroom grown in Brazil. *Nutr. Hosp.* 2012, 27 (2), 449–455.

Oyetayo, F.L. Responses of plasma lipids to edible mushroom diets in albino rats, *Af. J. Biotechnol.* 2006, 5(13), 263–1266.

Oyetayo, V.O. Medicinal uses of mushrooms in Nigeria: Towards full and sustainable exploitation. *Afr. J. Tradit. Complement. Altern. Med.* 2011, 8 (3), 267–274.

Oyetayo, V.O. Free radical scavenging and antimicrobial properties of extracts of wild mushrooms. *Brazilian J. Microbiol.* 2009, 40, 380–386.

Pacher, P.; Nivorozhkin, A.; Szabó, C. Therapeutic effects of xanthine oxidase inhibitors: renaissance half a century after the discovery of allopurinol. *Pharmacol. Rev.* 2006, 58(1), 87–114.

Petrovi´c, J.; Stojkovi´c, D.; Reis, F.S.; Barros, L.; Glamoclija, J.; Ciric, A.; Ferreira, I.C.F.R.; Sokovic, M. Study on chemical, bioactive and food preserving properties of *Laetiporus sulphureus* (Bull.: Fr.) *Murr.* Food Funct. 2014, 5, 1441–1451.

Phan, C.W.; Wang, J.K.; Cheah, S.C.; Naidu, M.; David, P.; Sabaratnam, V. A review on the nucleic acid constituents in mushrooms: Nucleobases, nucleosides and nucleotides. *Crit. Rev. Biotechnol.* 2018, 38(5), 762–777.

Ramesh, C.; Manohar, G.P. Antimicrobial properties, antioxidant activity and bioactive compounds from six wild edible mushrooms of Western Ghats of Karnataka, India. *Pharmacogn. Res.* 2010, 2, 107–112.

Rathee, S.; Rathee, D.; Rathee, D.; Kumar, V.; Rathee, P. Mushrooms as therapeutic agents. *Rev. Bras. Farmacogn.* 2012, 22, 457–474.

Ren, M.; Ye, L.; Hao, X.; Ren, Z.; Ren, S.; Xu, K.; Li, J. Polysaccharides from *Tricholoma matsutake* and *Lentinus edodes* enhance 5-fluorouracil-mediated H22 cell growth inhibition. *J. Trad. Chin. Med.* 2014, 34(3), 309–316.

Ribeiro, B.; Valentão, P.; Baptista, P.; Seabra, R.M.; Andrade, P.B. Phenolic compounds, organic acids profiles and antioxidative properties of beefsteak fungus (*Fistulina hepatica*). *Food Chem. Toxicol.* 2007, 45(10), 1805–1813.

Ritz, B.W.; Nogusa, S.; Ackerman, E.A.; Gardner, E.M. Supplementation with active hexose correlated compound increases the innate immune response of young mice to primary influenza infection. *J. Nutr.* 2006,136, 2868–2873.

Robbins, R.J. Phenolic acids in foods: An overview of analytical methodology. *J. Agric. Food Chem.* 2003, 51, 2866–2887.

Rop, O.; Vlcek, J.; Jurikova, T. Beta-glucans in higher fungi and their health effects. *Nutr. Rev.* 2009, 67 (11), 624–631.

Ruimi, N.; Petrova, R.D.; Agbaria, R.; Sussan, S.; Wasser, S.P.; Reznick, A.Z.; Mahajna, J. Inhibition of TNF α-induced iNOS expression in HSV-tk transduced 9L glioblastoma cell lines by *Marasmius oreades* substances through NF-κB- and MAPK dependent mechanisms. *Mol. Biol. Rep.* 2010, 37, 3801–3812.

Sakagami, H.; Aoki, T. Induction of immunopotentiation activity by a protein-bound polysaccharide, PSK (review). *Anticancer Res.* 1991, 11, 993–1000.

Sakano, N.; Wang, D.H.; Takahashi, N., Wang, B.; Sauriasari, R.; Kanbara, S.; Sato, Y.; Takigawa, T.; Takaki, J.; Ogino, K. Oxidative stress biomarkers and lifestyles in Japanese healthy people. *J. Clin. Biochem. Nutr.* 2009, 44, 185–195.

Saleh, M. H.; Rashedi, I.; Keating, A. (2017). Immunomodulatory properties of *coriolus versicolor*: The role of polysaccharopeptide. *Front. Immunol.*, 8, 1087.

Saman, P.; Chaiongkarn, A.; Moonmangmee, S.; Sukcharoen, J.; Kuancha, C.; Fungsin, B. Evaluation of prebiotic property in edible mushrooms. *Biol. Chem. Res.* 2016, 3, 75–85.

Samorini, G. The oldest representations of hallucinogenic mushrooms in the world (Sahara Desert, 9000–7000 B.P.). *Integration.* 1992, 2 (3), 69–78.

Sánchez C. Reactive oxygen species and antioxidant properties from mushrooms. *Synth. Syst. Biotechnol.* 2016, 2(1), 13–22.

Sanodiya, B.S.; Thakur, G.S.; Baghel, R.K.; Prasad, G.B.; Bisen, P.S. *Ganoderma lucidum*: A potent pharmacological macrofungus. *Curr. Pharm. Biotechnol.* 2009, 10(8), 717–742.

Shervington, A.; Lu, C. Expression of multidrug resistance genes in normal and cancer stem cells. *Cancer Invest.* 2008, 26, 535–542.

Shi, L.; Ren, A.; Mu, D.; Zhao, M. Current progress in the study on biosynthesis and regulation of ganoderic acids. *Appl. Microbiol. Biotechnol.* 2010, 88(6),1243–1251.

Shokrzadeh, M.; Azdo, S.; Ahmadi, M.A.; Habibi, E. Anti-diabetic effect of methanol extract of *Trametes versicolor* on male mice. *J. Mazandaran Univ. Med. Sci.,* 2017, 26, 165–175.

Signoretto, C.; Marchi, A.; Bertoncelli, A.; Burlacchini, G.; Milli, A.; Tessarolo, F.; Caola, I.; Papetti, A.; Pruzzo, C.; Zaura, E.; Lingström, P. Effects of mushroom and chicory extracts on the shape, physiology and proteome of the cariogenic bacterium Streptococcus mutans. *BMC Complement. Alternat. Med.* 2013, 13(1), 117.

Singdevsachan, S.K.; Mishra, P.A.J.; Baliyarsingh, B.; Tayung, K.; Thatoi, H. Mushroom polysaccharides as potential prebiotics with their antitumor and immunomodulating properties: A review. *Bioact. Carbohydr. Diet Fibre.* 2015, 7, 1–14.

Song, F.Q.; Liu, Y.; Kong, X.S.; Chang, W.; Song, G. Progress on understanding the anticancer mechanisms of medicinal mushroom: *Inonotus obliquus. Asian Pac. J. Cancer Prev.*2013, 14(3), 1571–1578.

Staunton, J.; Weissman, K.J. Polyketide biosynthesis: a millennium review. *Nat. Product Rep.* 2001, 18(4), 380-416.

Suknikhom, W.; Lertkhachonsuk, R.; Manchana, T. The effects of active hexose correlated compound (AHCC) on levels of CD4+ and CD8+ in patients with epithelial ovarian cancer or peritoneal cancer receiving platinum based chemotherapy. *Asian Pac. J. Cancer Prev.* 2017, 18(3), 633–638.

Sun, M.; Zhao, W.; Xie, Q.; Zhan, Y.; Wu, B. Lentinan reduces tumor progression by enhancing gemcitabine chemotherapy in urothelial bladder cancer. *Surg. Oncol.* 2015, 24, 28–34.

Sun, X.; Sun, Y.; Zhang, Q.; Zhang, H.; Yang, B.; Wang, Z.; Zhu, W.; Li, B.; Wang, Q.; Kuang, H. Screening and comparison of antioxidant activities of polysaccharides from *Coriolus versicolor. Int. J. Biol. Macromol.* 2014, 69, 12–19.

Tanigawa, K.; Itoh, Y.; Kobayashi, Y. Improvement of QOL and immunological function with *Lentinula edodes* mycelia in patients undergoing cancer immunotherapy: An open pilot study. *Altern. Ther. Health Med.* 2016, 22(4), 36–42.

Taofiq, O.; Martins, A.; Barreiro, M.F.; Ferreira, I.C. Anti-inflammatory potential of mushroom extracts and isolated metabolites. *Trends Food Sci. Technol.* 2016, 50, 193–210.

Terakawa, N.; Matsui, Y.; Satoi, S.; Yanagimoto, H.; Takahashi, K.; Yamamoto, T.; Yamao, J.; Takai, S.; Kwon, A.H.; Kamiyama, Y. Immunological effect of active hexose correlated compound (AHCC) in healthy volunteers. *Nutr. Cancer.* 2008, 60(5), 643–651.

Tibuhwa, D. D. Folk taxonomy and use of mushrooms in communities around Ngorongoro and Serengeti National Park, Tanzania. *J. Ethnobiol. Ethnomed.* 2012, 8, 36–45.

Vamanu, E. Studies on the antioxidant and antimicrobial activities of *Pleurotus ostreatus* PSI101109 mycelium. *Pak. J. Bot.* 2013, 45(1), 311–317.

Varshney, J.; Ooi, J.H.; Jayarao, B.M. White button mushrooms increase microbial diversity and accelerate the resolution of *Citrobacter rodentium* infection in mice. *J. Nutr.* 2013, 143, 526–532.

Vetvicka, V.; Vashishta, A.; Saraswat-Ohri, S.; Vetvickova, J. Immunological effects of yeast- and mushroom-derived β-glucans. *J. Med. Food.* 2008, 11 (4), 615–622.

Volman, J.J.; Mensink, R.P.; Ramakers J.D.; de Winther, M.P.; Carlsen, H.; Blomhoff, R.; Buurman, W.A.; Plat, J. Dietary (1→3), (1→4)-β-d-glucans from oat activate nuclear factor-kappa B in intestinal leukocytes and enterocytes from mice. Nutr. *Res.* 2010a 30 (1), 40–48.

Volman, J.J.; Mensink, R.P.; van Griensven, L.J.; Plat, J. Effects of alpha-glucans from *Agaricus bisporus* on *ex vivo* cytokine production by LPS and PHA-stimulated PBMCs; a placebo-controlled study in slightly hypercholesterolemic subjects. *Eur. J. Clin. Nutr.* 2010b, 64, 720–726.

Wachtel-Galor, S.; Yuen, J.; Buswell, J.A.; Benzie, I.F. *Ganoderma lucidum* (Lingzhi or Reishi): A medicinal mushroom. In: Herbal Medicine: Biomolecular and Clinical Aspects, 2nd ed. Boca Raton, FL: CRC Press, 2011.

Waithaka, P.N.; Gathuru, E.M.; Githaiga, B.M.; et al. Antimicrobial activity of mushroom (*Agaricus bisporus*) and Fungal (*Trametes gibbosa*) extracts from mushrooms and fungi of Egerton Main Campus, Njoro Kenya. *J. Biomed. Sci.* 2017, 6, 3.

Wang, H.; Cai, Y.; Zheng, Y.; Bai, Q.; Xie, D.; Yu, J. Efficacy of biological response modifier lentinan with chemotherapy for advanced cancer: A meta-analysis. *Cancer Med.*, 2017, 6 (10), 2222–2233.

Wang, H.X.; Ng, T.B.; Liu, W.K.; Ooi, V.E.C.; Chang, S.T. Isolation and characterization of two distinct lectins with antiproliferative activity from the cultured mycelium of the edible mushroom *Tricholoma mongolicum*. *Int. J. Pept. Protein Res.* 1995, 46 (6), 508–513.

Wang, J.; Wang, C.; Li, S.; Li, W.; Yuan, G.; Pan, Y.; Chen, H. Anti-diabetic effects of *Inonotus obliqus* polysaccharides in streptozotocin-induced type-2-diabetic mice and potential mechanism via P13K-Akt signal pathway. *Biomed. Pharmacother.* 2017, 95, 1669.

Wani, B.A.; Bodha, R.H.; Wani, A.H. Nutritional and medicinal importance of mushrooms. *J Med. Plants Res.* 2010, 4, 2598–2604.

Wasser, S.P. Medicinal mushroom science: History, current status, future trends, and unsolved problems. *Int. J. Med. Mushrooms.* 2010,12, 1–16.

Wasser, S.P. Medicinal mushrooms as a source of antitumour and immunostimulating polysaccharides. *Appl. Microbiol. Biotech.* 2002, 60, 258–274.

Wasser, S.P. Medicinal mushroom science: Current perspectives, advances, evidences, and challenges. *Biomed. J.* 2014, 37 (6), 345–356.

Wasser S.P.; Weis, A.L. Medicinal properties of substances occurring in higher Basidiomycetes mushrooms: Current perspectives. *Int. J. Med. Mushrooms.* 1999a, 1, 31–62.

Wasser, S.P.; Weis, A.L. Therapeutic effects of substances occurring in higher basidiomycetes mushrooms a modern perspective. *Crit. Rev. Immunol.* 1999b, 19 (1), 65–96.

Xiao, C.; Wu, Q.P.; Tan, J.B.; Cai, W.; Yang, X.B.; Zhang, J.M. Inhibitory effects on alpha-glucosidase and hypoglycemic effects of the crude polysaccharides isolated from 11 edible fungi. *J. Med. Plants Res.* 2011, 5, 6963–6967.

Xiao, J.H.; Xiao, D.M.; Sun, Z.H.; Xiong, Q.; Liang, Z.Q.; Zhong, J.J. Chemical compositions and antimicrobial property of three edible and medicinal *Cordyceps* species. *J. Food Agric. Environ.* 2009, 7(3&4), 91–100.

Xu, X.; Zhang, X. *Lentinula edodes*-derived polysaccharide alters the spatial structure of gut microbiota in mice. *PLoS One.* 2015, 10: e0115037.

Yamac, M.; Bilgili, F. Antimicrobial activities of fruit bodies and/or mycelial cultures of some mushroom isolates. *Pharm. Biol.* 2006, 44, 660–667.

Yamac, M.; Kanbak, G.; Zeytinoglu, M.; Senturk, H.; Bayramoglu, G.; Dokumacioglu, A.; Van Griensven, L.J.L.D. Pancreas protective effect of button mushroom Agaricus bisporus (J.E. Lange) Imbach (Agarico mycetidae) extract on rats with streptozotocin induced diabetes. Int. J. Med. Mushroom. 2010, 124, 379–389.

Yan, Y.; Yang, F.; Yao, W.; Guo, J. The study on the immunoregulation and anti-radiating damage of maitake polysaccharide in mice. *Chin. J. Radiol. Health.* 2010, 19(1), 6–8.

Yeh, M.; Ko, W.; Lin, L. Hypolipidemic and antioxidant activity of Enoki mushrooms (*Flammulina velutipes*). *Biomed. Res. Int.* 2014, 352385, 1–6

Yeung, J.H.K.; Or, P.M.Y. Polysaccharide peptides from *Coriolus versicolor* competitively inhibit model cytochrome p450 enzyme probe substrates metabolism in human liver microsomes. *Phytomedicine* 2012, 19, 457–463.

Yin, Z.; Fujii, H.; Walshe, T. Determining the frequency of CD4+ and CD8+ T cells producing IFN-γ and TNF- in healthy elderly people using flow cytometry before and after active hexose correlated compound (AHCC) intake. *Human Immunol.* 2010, 71(12), 1187–1190.

Yıldız, S.; Yılmaz, A.; Can, Z.; Tabbouche, S.A.; Kılıç, A.O.; Sesli, E. Some bioactive properties of wild and commercial mushroom species. *J. Food Health Sci.* 2017, 3(4), 161–169.

Yoshino, S.; Nishikawa, K.; Morita, S.; Takahashi, T.; Sakata, K.; Nagao, J.; Nemoto, H.; Murakami, N.; Matsuda, T.; Hasegawa, H.; Shimizu, R. Randomised phase III study of S-1 alone versus S-1 plus lentinan for unresectable or recurrent gastric cancer (JFMC36-0701). *Eur. J. Cancer.* 2016, 65, 164–171.

Yoshino, S.; Tabata, T.; Hazama, S.; Iizuka, N.; Yamamoto, K.; Hirayama M.; Tangoku, A.; Oka, M. Immunoregulatory effects of the antitumor polysaccharide lentinan on Th1/Th2 balance in patients with digestive cancers. *Anticancer Res.* 2000, 20(6C), 4707–4711.

Yoshino, S.; Watanabe, H.; Imano, M.; Yagi, M.; Suga, T.; Nakazawa, S. Evaluation of efficacy for immune adjuvant (Lentinan (β-l, 3-glucan))-containing supplementary food in patients with unresectable and recurrent gastric cancer. Biotherapy. 2007, 21, 265–273.

Yuan, J.P.; Zhao, S.Y.; Wang, J.H.; Kuang, H.C.; Liu, X. Distribution of nucleosides and nucleobases in edible fungi. J. Agric. Food Chem. 2008, 56(3), 809–815.

Zhai, L.; Ballinger, S.W.; Messina, J.L. Role of reactive oxygen species in injury- induced insulin resistance. Mol. Endocrinol. 2011, 25, 492–502.

Zhang, J.; Meng, G.; Zhai, G.; Yang, Y.; Zhao, H.; Jia, L. Extraction, characterization and antioxidant activity of polysaccharides of spent mushroom compost of Ganoderma lucidum. Int. J. Biol. Macromol. 2016a, 82, 432–439.

Zhang, L.; Yang, J.; Ni, Y.; Li, C.; Shao, Y. Effect of Grifola frondosa polysaccharide on blood pressure and the ultrastructure of heart and kidney of hypertensive rats. Chin. J. Gerontol. 2013a, 33(19), 4764–4766.

Zhang, M.; Cui, S.W.; Cheung, P.C.K.; Wang, Q. Antitumor polysaccharides from mushrooms: A review on their isolation process, structural characteristics and antitumor activity. Trends Food Sci. Technol. 2007, 18 (1), 4–19.

Zhang, Y.; Kong, H.; Fang, Y.; Nishinari, K.; Phillips, G.O. Schizophyllan: A review on its structure, properties, bioactivities and recent developments. Bioact. Carbohydr. Dietary Fibre. 2013b, 1, 53–71.

Zhang, Y.; Meng, M.; Wang, M.; Wang, C. Isolation, purification and in vitro antitumor activity of polysaccharides GFD-1 from *Grifola frondosa*. Food Res. Dev. 2016b, 37 (15), 1–5.

Zhang, Y.; Mills, G.L.; Nair, M.G. Cyclooxygenase inhibitory and antioxidant compounds from the fruiting body of an edible mushroom, *Agrocybe aegerita*. Phytomedicine. 2003, 10 (5), 386–390.

Zhu, P. In: The present status and prospects of medicinal fungal research and development in China, *Proceedings of the 5th International Medicinal Mushroom Conference*, Nantong, China, 2009, pp. 26–33.

CHAPTER 8

Herbal Product Development and Characteristics

MIRIAN PATEIRO[1], RUBÉN DOMÍNGUEZ[1], PREDRAG PUTNIK[2],
DANIJELA BURSAĆ KOVAČEVIĆ[2], FRANCISCO J. BARBA[3],
PAULO S. E. MUNEKATA[1], ELENA MOVILLA FIERRO[4], and
JOSÉ M. LORENZO[1*]

[1]*Centro Tecnológico de la Carne de Galicia, Rúa Galicia No. 4,
Parque Tecnológico de Galicia, San Cibrao das Viñas,
32900 Ourense, Spain*

[2]*Faculty of Food Technology and Biotechnology, University of Zagreb,
Pierottijeva 6, 10000 Zagreb, Croatia*

[3]*Nutrition and Food Science Area, Preventive Medicine and Public Health,
Food Science, Toxicology and Forensic Medicine Department,
Faculty of Pharmacy, Universitat de València, Avda. Vicent Andrés
Estellés, s/n, 46100 Burjassot, València, Spain*

[4]*Complejo Hospitalario Universitario de Ourense, Ourense, Spain*

Corresponding author. E-mail: jmlorenzo@ceteca.net

ABSTRACT

The arising consumer's concern for health together with the unhealthy lifestyle have brought back to the use of herbs, commonly used as food and in traditional medicine. There are several properties associated with these products such as anti-inflammatory, antioxidant, antimicrobial, anti-carcinogenic, antihypertensive, gluco-regulatory, or antithrombotic effects, which would have potential as alternative therapies for the common health problems. These properties are especially attributed to the presence of bioactive compounds (e.g., phenolic compounds) isolated from the green leaves of these plants. Although there are healthy eating models with increased consumption of herbs, as the case with Mediterranean diet, this would not be enough to obtain the desired effects. Therefore, there is a need to incorporate them as dietary supplements to achieve the aforementioned beneficial effects on health using nonfood matrix, for example, pills, capsules, powder, essential oils, extracts, etc. Moreover, these herbal products are useful tools for food production as they tend to extend the shelf life of foods due to their antioxidant and antimicrobial properties.

8.1 GLOBAL PERSPECTIVE OF HERBAL NUTRACEUTICALS

From ancient times, certain foods have been recognized as beneficial for health (Poojary et al., 2017; Putnik et al., 2019). This belief is still valid, thanks to the scientific researches carried out in this field, showing that health and nutrition go hand in hand (Lorenzo et al., 2018a). This is mainly due the existence of certain bioactive components of foods (Montesano et al., 2018). Thus, foods containing beneficial compounds can help human health based on two fundamental aspects: health promotion (prevention of diseases) and improvement of the physical functionality. These functions together with their natural origins and the absence of undesirable secondary effects make them potential substitutes for synthetic medications. Both reasons justify the increased use of nutraceuticals (Barba et al., 2014; Granato et al., 2017).

A nutraceutical can be defined as a dietary supplement, embedded in a non-nutritional matrix of a concentrated bioactive natural substance (e.g., pills, capsules, powder, etc.). This way these compounds that are usually present in foods are taken in higher doses, as compared to the usual dieting (Luengo, 2007). Hence, they will presumably have greater favorable effects on health than those of a normal food intake could have. In addition, a bioactive component and nutraceuticals can be administered in concentrated forms or added to natural foods, in order to increase the functional properties, in the sense in which we have defined them (Granato et al., 2017).

Herbs have been used as nutraceuticals with different names for a long time, due to their natural origins and their active biological properties, with known health benefits and defined preventive and/or therapeutic capacity. These products were usually isolated and purified by nondenaturing methods, such as nonfood matrix. They contributed to the beneficial effects for human health, for instance, by improving of one or more physiological functions, preventive and/or curative actions, and enhancements of life quality (Barba et al., 2014; Chauhan et al., 2013).

Lamiaceae herbs are one of the most common herbal families (Figure 8.1) used for nutraceuticals, because of their biological and medical applications

(Naghibi et al., 2005). According to their use, this family can be divided into three categories: aromatic (the most commonly used), medicinal, and ornamental (Viuda-Martos et al., 2016).

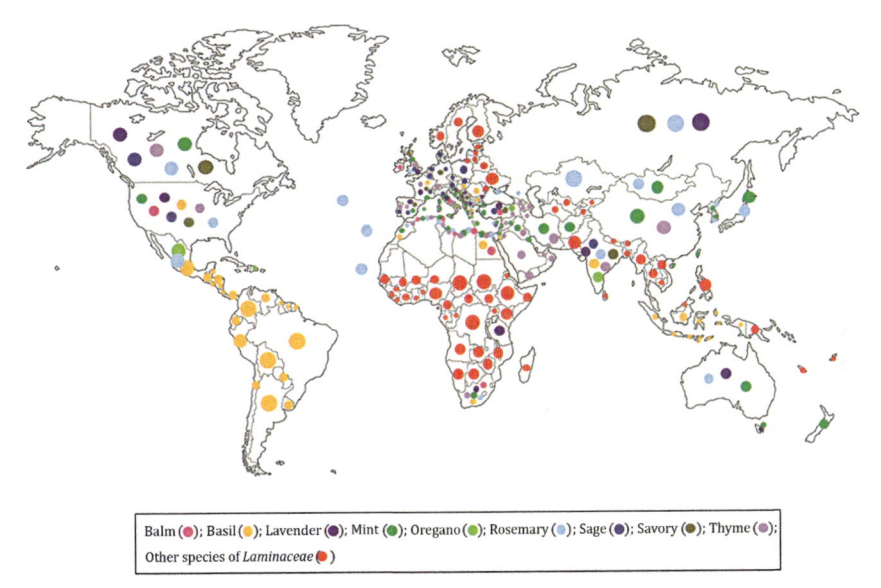

Balm (●); Basil (●); Lavender (●); Mint (●); Oregano (●); Rosemary (●); Sage (●); Savory (●); Thyme (●); Other species of *Laminaceae* (●)

FIGURE 8.1 Countries where *Laminaceae* plants are cultivated.

Lemon balm, basil, lavender, mint, oregano, rosemary, sage, savory, and thyme are among the members of this family having functional benefits, which are mainly attributed to the presence of bioactive compounds (BACs) (Giacometti et al., 2018; Vinceković et al., 2017). BACs are responsible for anti-inflammatory, antioxidant, antimicrobial, anticarcinogenic, antihypertensive, gluco-regulatory, or antithrombotic effects (Lorenzo et al., 2018b). Moreover, their phytochemical composition is also related with their nutraceutical effects (Frezza et al., 2018), while phenolic compounds (flavonoids and polyphenolic compounds), vitamins, and terpenoids (Mustafa et al., 2010; Zivkovic et al., 2009), are mainly responsible for the beneficial effects on health (Dudeja and Gupta, 2016).

It is important to take into account that the use of natural products does not necessarily imply they are safe. For instance, pharmacokinetics and safety profiles have to be clinically evaluated for their efficacy, safety, toxicity, and mechanism of action, and the right dosage must account the age of potential consumers (Žuntar et al., 2018). Moreover, as there is no specific legislation on nutraceuticals, their use must be regulated, so they have to pass quality controls and authorization procedures to be classified as Generally Recognized as Safe. Thus, also allowing their use in food products as safe additives (Jackson and Paliyath, 2011).

Relevant European directives that regulate plant food supplements are: Directive 2002/46/EC and General Food Law (Regulation EC 178/2002) that define responsibility of food business operators for production and food safety. Furthermore, all herbal products prior commercialization need authorization and fulfillment of Herbal Directive, 2004/24/EC that was established to expedite commercialization of traditional herbal products (Žuntar et al., 2018). European Commission of Concerted Action on Functional Bromatology (Functional Food Science in Europe, FUFOSE) was created in the European Union with the aim of developing and establishing a scientific approach for testing developed food products. This was established in order to identify beneficial effects on physiological functions and improvements of health and well-being of an individual, and/or to reduce the risks of developing diseases (FUFOSE, 2018). Moreover, it becomes necessary to establish the guidelines about nutrition and health claims made on food, which are regulated through the Regulation 1924/2006 of the European Parliament.

8.2   BIOACTIVE COMPOUNDS

The main compounds isolated from the *Lamiaceae* plants are polyphenolic compounds (phenolic acids, flavonoids, stilbenes, and lignans) and terpenes (Figure 8.2). In most cases, they do not contain toxic compounds in their chemical composition, such as toxic alkaloids, cyanogenic glucosides, grayanotoxins, cucurbitacins, cardiac glycosides, or phorbol esters (Wyk and Wink, 2004).

These compounds are nitrogen-free secondary metabolites produced as defense mechanisms of plants against adverse conditions (Mazid et al., 2011). The formation of these BACs allows the use of the plants from this family for developing nutraceuticals. In some cases, their beneficial effect is the result of the combination of several BACs from several structural groups, so that each of these compounds contributes to the beneficial activities associated with the plant (Carović-Stanko et al., 2016). These synergistic interactions allow the prevention and/or treatment of a wide range of diseases. The presence of the compound(s) in the *Lamiaceae* member will define the nutraceutical function. Moreover, the chemotype of each plant will depend on the part of the plant selected, the method of extraction, and the harvest season (Lorenzo et al., 2018b).

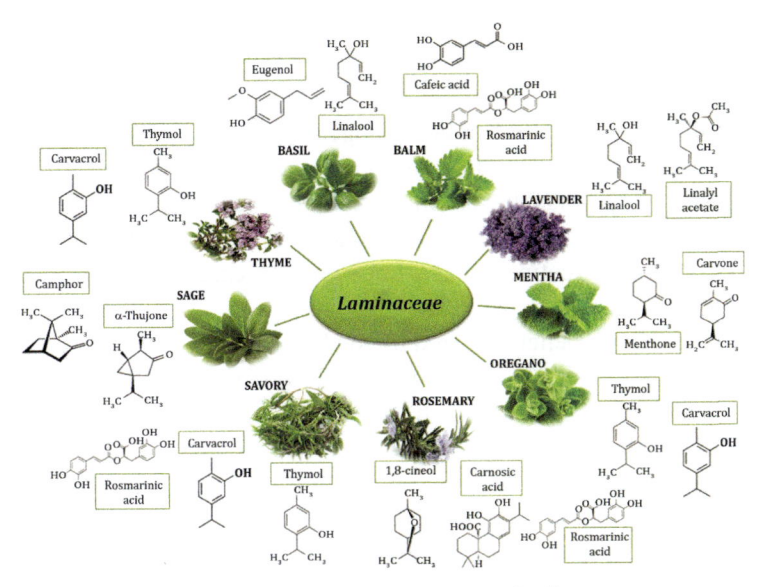

FIGURE 8.2 Main BACs found in plants of *Laminaceae* family.

8.2.1 TERPENES

Concerning terpenes, they are among the main components present in essential oils of the plants of the *Lamiaceae* family. These compounds can have antimicrobial, antitumor, and antioxidant properties (Alves-Silva et al., 2013; Nikolić et al., 2014). Camphor, camphene, carvacol, thymol, α-pinene, ρ-cymene, 1,8-cineol, limonene, γ-terpinene, and terpinen-4-ol are among the most common terpenes found this family.

8.2.2 POLYPHENOLIC COMPOUNDS

Phenolic acids and flavonoids are important members within this group BACs. Phenolic acids are widely present in *Lamiaceae* plants (Pedersen, 2000). As many other BACs, they are secondary plant metabolites with anticarcinogenic, anti-inflammatory, antimicrobial, antioxidant, and cardioprotective effects (Barros et al., 2010; Berdowska et al., 2013; Stagos et al., 2012). Caffeic, ρ-coumaric, ferulic, rosmarinic, and sinapic acids are among the most common compounds isolated from *Lamiaceae* herbs.

Similar functional properties are associated with flavonoids, like anticarcinogenic, anti-inflammatory, antimicrobial, antioxidant, and cardioprotective effects (González-Gallego et al., 2014; Kaleem and Ahmad, 2018; Rodríguez-Meizoso et al., 2006). Carnosol, eriodictyol, kaempferol, and quercetin were quantified in *Lamiaceae* family.

Regarding the potential anticarcinogenic properties, Berdowska et al. (2013) found that polyphenolic compounds are responsible for this activity. Between the compounds that are involved in this reaction are phenolic acids, as rosmarinic acid, carnosol, caffeic acid, ferulic acid, and flavonoids as kaempferol and quercetin. Finally, terpenoids as carvacrol are also responsible of this effect.

The other health benefit offered by the herbs of this family is the prevention of arteriosclerosis (Yamamoto et al., 2005), as one of the most known diseases caused by cholesterol. It causes an accumulation in the arterial walls, reducing their elasticity and diameter. This results with circulatory and cardiac disorders, which are commonly increasing in the population. Several studies have attributed *Ocimum basilicum*, *Rosemary officinalis*, *Thymus vulgaris* or *Salvia miltiorrhiza*, with antithrombotic effects (Harnafi et al., 2013; Ji and Gong, 2008; Yamamoto et al., 2005). Carvacrol and thymol, present in them, are between the compounds responsible for this effect (Viuda-Martos et al., 2016).

8.3   MODE OF ACTION

The nutraceuticals extracted from these plants can be used as fresh leaf extracts (water or alcohol extracts, distillate, or essential oil), decoctions, brews, herbal teas, infusions, dry powders, and others. Additionally, they can even be used as plasters and compresses. Their active components can act either directly or as buffering agents (Dudeja and Gupta, 2016). In former way, they act directly on health, while in the later they relieve the symptoms of the disease.

Chemical composition of the nutraceuticals from *Lamiaceae* family determines their characteristics, and therefore their mode of action. Even though there are several BACs from several structural groups, their composition makes it very difficult to establish a single mechanism of action (Wink, 2015). However, to explain general mode of action, it is usual to associate physiological effects with the main present components. In this case, phenolic compounds, which represent around 85% of their total composition, are the main BACs involved (Bakkali et al., 2008). Therefore, noncovalent modification of proteins, interaction with biomembranes and free-radical scavengers, represent the typical modes of action (Figure 8.3).

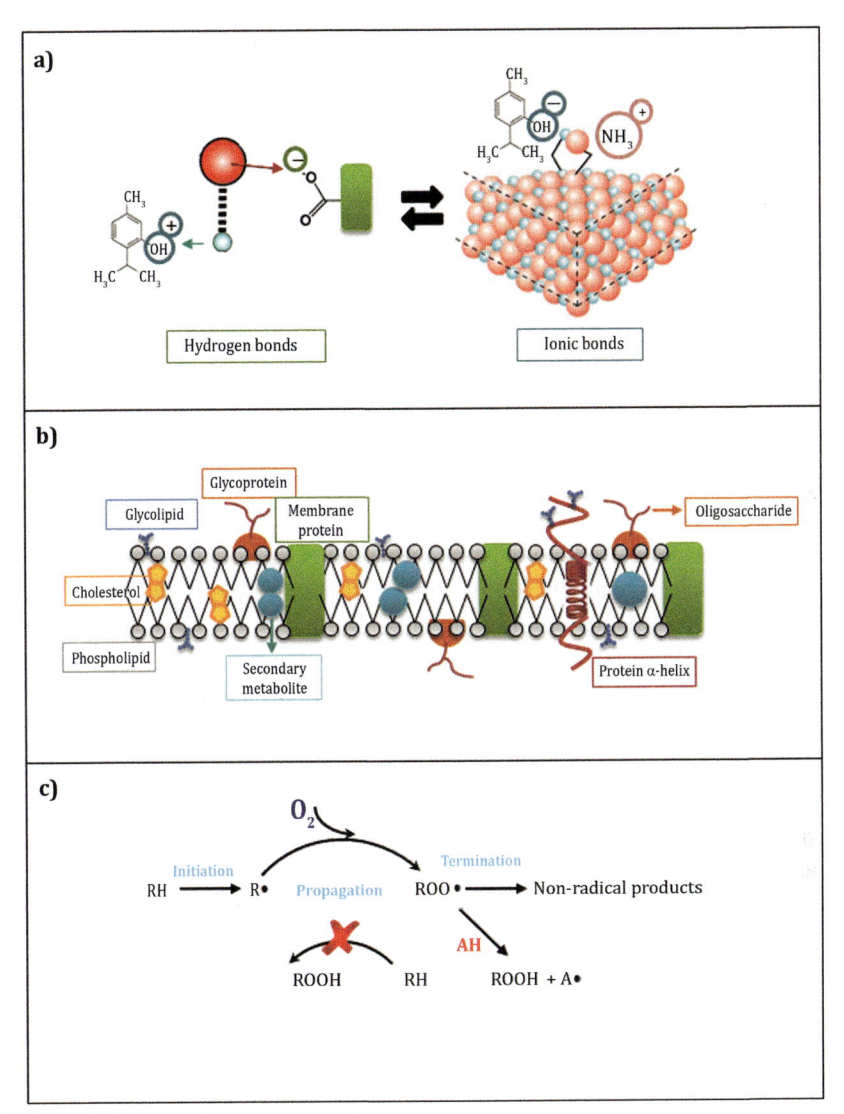

FIGURE 8.3 Mechanism of action of BACs isolate from *Laminaceae* family. Noncovalent modification of proteins (a), interactions of secondary metabolites with biomembranes (b), antioxidant properties: free-radical scavengers (c).

8.3.1 NONCOVALENT MODIFICATION OF PROTEINS

As already mentioned, phenolic compounds are main BACs of *Lamiaceae* family that contain hydroxyl groups capable to form several hydrogen bonds with peptides and proteins through binding with electronegative atoms of oxygen and nitrogen (Wyk and Wink, 2004, 2015). Moreover, a partial dissociation of the phenolic OH-groups from the phenolic compounds can occur (Wink, 2012). This results with negatively charged phenolate ions that allow the formation of ionic bonding with the amino groups from proteins (positively charged). These both types of bonding can modify the structural and functional aspects of proteins and affect physiology (Figure 8.3a) (Wink, 2015). Furthermore, the presence of sugar molecules in phenolic compounds makes additional support between phenolics and proteins by hydrogen bonding.

8.3.2 INTERACTIONS OF SECONDARY METABOLITES WITH BIOMEMBRANES

The biomembranes are cellular structures that surround living organisms. They are formed by a double lipid layer, which confers with one of its most important functions, as regulation of the entry and exit of macromolecules, whether or not this is intracellular or extracellular delivery. In addition, they are also characterized by possessing capacity for selective permeability. Biomembranes also contain three types of proteins, such as ion channels, receptors, and transporters which allow the exchange of substances with other cells or tissues. The presence of polyphenols can modify activity of these proteins, what could change the interactions with phospholipids (Figure 8.3b) (Wink, 2015).

8.3.3 ANTIOXIDANT PROPERTIES: FREE RADICAL SCAVENGERS

Free radicals are reactive oxygen species (ROS) that bind to important components of the cell structures (proteins, lipids, and nucleic acids); thus forming cascade of ROS events that will lead to a structural damages and result with illnesses (Wyk and Wink, 2015). The BACs isolated from *Laminaceae*, as phenolics and terpenoids with conjugated double bonds, are able to inhibit the formation of ROS and other oxygen radicals (Wyk and Wink, 2015; Wyk and Wink, 2004), by giving them a protective character against prooxidant compounds. Therefore, BACs isolated from leaves of this family can be used as antioxidants. Their action as free-radical scavenger would protect the cells against the action of ROS (Figure 8.3c).

Besides their antioxidant properties, the polyphenols present in these plants could have a "pleiotropic effect." This implies that polyphenols included in the diet as supplements could have simultaneously more than one benefit (Kishimoto et al., 2013). For instance, to have effects as multiple benefits on intestine, liver and vascular tissue are attributed to them.

8.3.4 SYNERGISTIC ACTIVITY

The synergistic activity, with abovementioned physiological antioxidant properties, comes from the *Laminaceae* essential oils that are able to extend the shelf life of foods (Pateiro et al., 2018). This effect is higher for mixtures of BACs than what could be derived from the single essential oil (Hyldgaard et al., 2012). The mentioned synergism is important, since it would allow the use of lower doses of essential oils for getting higher antioxidant activities, which has tendency to reduce the adverse physiological side effects (Bag and Chattopadhyay, 2015).

8.4 NUTRACEUTICALS DEVELOPMENT

Development of any new product is costly process with uncertain outcome on the market. For the most of nutraceuticals, first step in the production is process of extraction of BACs from plant matrices. As mentioned above, BACs are usually isolated and purified by nondenaturing methods, and deposited in nonfood matrices. They contribute to the beneficial effects for human health and/or act as food preservatives (Lorenzo et al., 2018c).

The selection of the extraction is essential for obtaining a product of high quality, since these conditions could affect its chemical composition (Tavakolpour et al., 2017). Moreover, the method depends on the final use of the product (Li et al., 2014). Although conventional extraction methods (steam distillation, hydrodistillation, solvent extraction) are still used (Bursać Kovačević et al., 2018; Gavahian et al., 2012; Hashemi et al., 2017; Putnik et al., 2018;), the trend is to replace them with newer alternatives (supercritical fluid extraction, microwave-assisted extraction, and ultrasound-assisted extraction) (Domínguez et al., 2018; Gavahian et al., 2012; Mašković et al., 2017). Current and preferable approaches for extraction of BACs from plants are environment friendly methods, labeled as "green" extractions (Putnik et al., 2017a) that have tendency to avoid negative influences on nutritive value of products (Granato et al., 2018).

The first steps of extraction involve cleaning, drying, and grounding of the plants into fine powders, while sometimes obtained material is immediately ready for the use in infusions. In other cases, these dried plant materials are subjected to solvent extraction process to obtain an extract or an essential oil. Ethanol, methanol, hexane, and water are among the most commonly used solvents (Fidelis et al., 2018).

Other novel technology that is able to enhance the stability and bioavailability (Gabrić et al., 2018; Koubaa et al., 2017) is encapsulation (Gharibzahedi et al., 2019). While protecting the bioactivity of the BACs from degradation in plant material by ambient conditions (Barba et al., 2017; Putnik et al., 2017b), this method can improve organoleptic properties (color, flavors) by masking unfavorable features of foods (Gharibzahedi et al., 2019). Spray drying, fluid-bed, or emulsion techniques are mostly used technologies for the encapsulation of nutraceuticals (Augustin and Sanguansri, 2012; Huang et al., 2010). These encapsulated nutraceuticals can be used as dietary supplements or added in foods to fortify and develop functional foods (Fangueiro et al., 2016). Microencapsulation and nanoencapsulation techniques can be incorporated in food industry to fortify foods with nutraceuticals (Gómez et al., 2018).

8.5 HERBAL HEALTH BENEFITS

8.5.1 LEMON BALM

Lemon balm is the common name of the perennial herbs from *Melissa officinalis* L., cultivated mainly in Europa and Asia (Figure 8.1). Previous studies confirmed its medicinal properties among which stand antidepressant, antispasmodic, antihistaminic, and antiviral benefits (Table 8.1). Besides, it can be used in anxiety and in nervous disorders (Khare, 2008).

This plant contains about 40 secondary metabolites that are isolated from this herb. The secondary plant metabolites responsible for the medicinal activity are the volatile compounds present in its essential oils and phenolic compounds (phenolic acids, flavonoids, and tannins) (Shakeri et al., 2016). Rosmarinic, caffeic, and ferulic acids are among the phenolic acids responsible of its antiviral properties (Figure 8.3). Moreover, antibacterial, anti-inflammatory, and antithrombotic effects are also attributed to rosmarinic acid (Englberger et al., 1988; Iauk et al., 2003). In addition, it should be noted that the important contents of the flavonoid luteolin-3'-*O*-glucuronide are found in lemon balm (Barros et al., 2013; Heitz et al., 2000).

TABLE 8.1 Herbal Health Benefits and Main Bioactive Compounds (BACs) Associated With the Activity of *Lamiaceae* Family

Herb	Specie	Activity	Edible Part	BACs	Main Effects	Regulation of Use	Reference
Lemon balm	*M. officinalis L.*	AChE-inhibitory Analgesic Antidepressant Antianxiety Antidepressant Antiepileptic Anti-inflammatory Antimicrobial Antinociceptive Antioxidant Antispasmodic Antiviral Cognitive functions Cardiovascular functions Cardiovascular Cytotoxic Neuroprotective Relaxant	Aerial part Flowers Leaves Roots Stems	Caffeic acid Ferulic acid Gallic acid Luteolin-3'-O-glucuronide Rosmarinic acid Rutin	– Reduce anxiety symptoms – Reduce insomnia symptoms – Effect on dementia, epilepsy and paralysis – Memory enhancing effect – Hypoglycemic effect – Hypolipidemic effect – Reduce levels of LDL/VLDL and triacylglycerol in blood – Inhibit cell proliferation – Effect on monoamine oxidase – Natural preservative food	FDA	Carović-Stanko et al. (2016) Naghibi et al. (2005) Shakeri et al. (2016)
Basil	*O. gratissimum*	Analgesic Anti-inflammatory Anti-lipidemic Antimicrobial Antinociceptive Antioxidant Antiulcer Relaxant	Flowers Leaves	Thymol ρ-Cymene Eugenol γ-Terpinene 1,8-Cineole	– Reduce blood pressure – Inhibits enzymes mediating inflammation	-	Naghibi et al. (2005) Pandey et al. (2014)
	O. micranthum	Analgesic Anti-inflammatory Antimicrobial Antinociceptive Antioxidant Antiulcer	Flowers Leaves	Eugenol β-Caryophyllene 1,8-Cineole β-Elemene Bicyclogermacrene	– Effect anti-inflammatory and antinociceptive on smooth muscle cells	—	de Pinho et al. (2012) Pandey et al. (2014)
	O. sanctum	Analgesic Anticancer Antidiabetic Anti-inflammatory Antimicrobial Antioxidant Antistress	Flowers Leaves	Eugenol Linalool Methyl eugenol Methyl chavicol β-Caryophyllene	– Chemopreventive in cancer – Reduce glucose and uric acid levels in blood – Peripheral vasodilatory – Treatment of gastric ulcer	—	Carović-Stanko et al. (2016) Pandey et al. (2014) Prakash and Gupta (2000) Singh and Chaudhuri (2018)

TABLE 8.1 *(Continued)*

Herb	Specie	Activity	Edible Part	BACs	Main Effects	Regulation of Use	Reference
Lavender	*L. angustifolia*	Analgesic Anti-inflammatory Antimicrobial Antioxidant Cytotoxic Neuroprotective	Flowering tops Leaves Petals	Linalool Linalyl acetate Camphor Borneol Terpinen-4-ol α-Terpineol 1,8-Cineole	– Treatment of chronic pain (rheumatoid arthritis) – Local anesthetic activity – Use of prophylactic in surface infections – Improve sleep quality	FDA	Carović-Stanko et al. (2016) Danh et al. (2013) Khoury et al. (2016) Naghibi et al. (2005)
	Lavandula latifolia	Antibacterial Expectorant Neurological Stimulant	Flowering tops	Linalool 1,8-Cineole Camphor Linalyl acetate Borneol	– Treatment of chronic pain – Use of prophylactic in surface infections	FDA	Cavanagh and Wilkinson (2002) Erland and Mahmoud (2016)
Mint	*M. piperita*	Analgesic Antimicrobial Anti-inflammatory Antioxidant Sedative	Leaves	Carvone Menthol Menthone	– Natural option in pain therapy – Activity against human pathogenic Gram-positive and Gram-negative bacteria	FDA	Carović-Stanko et al. (2016) Naghibi et al. (2005)
	M. spicata	Analgesic Antimicrobial Anti-inflammatory Cytotoxic Digestive disorder Nervous disorders	Flowering tops	Carvone Limonene Linalool Linalyl acetate Menthone β-Caryophyllene	– Inhibit cell proliferation	FDA	Khoury et al. (2016) Naghibi et al. (2005) Prakash et al. (2016)
Oregano	*O. majorana*	AChE-inhibitory Antimicrobial Antioxidant Cytotoxic Expectorant Sedative Stimulant	Leaves	Terpinen-4-ol Thymol γ-Terpinene p-Cymene 4-Thujanol	– Prevention neurodegenerative diseases – Activity against Gram-positive and Gram-negative bacteria	FDA	Carović-Stanko et al. (2016) García-Beltrán and Esteban (2016) Ortega-Ramirez et al. (2016)
	O. vulgare	Anticancer Antihyperglycemic Antimicrobial Anti-inflammatory Antioxidant Antiviral Cytotoxic Inhibition of thrombin	Flowering tops Leaves	Carvacrol Cymene Thymol α-Pinene	– Activity against Gram-positive and Gram-negative bacteria	FDA	Carović-Stanko et al. (2016) García-Beltrán and Esteban (2016) Naghibi et al. (2005)

TABLE 8.1 (*Continued*)

Herb	Specie	Activity	Edible Part	BACs	Main Effects	Regulation of Use	Reference
Rosemary	*R. officinalis*	AChE-inhibitory Analgesic Anti-inflammatory Antimicrobial Antinociceptive Antioxidant Antitumor Antiulcerogenic Hepatoprotective Hyperglycemic Relaxant	Aerial part Leaves	Borneol Camphor 1,8-Cineole α-Pinene α-Terpineol Limonene Rosmarinic acid Carnosic acid	– Reduce muscular, rheumatic pains – Respiratory diseases (asthma) – Prevention of CNS diseases (Alzheimer, depression and Parkinson) – Activity against Gram-positive and Gram-negative bacteria – Natural preservative food.	EFSA FDA	Carović-Stanko et al. (2016) Khoury et al. (2016) Naghibi et al. (2005)
Sage	*S. lavandulifolia*	AChE-inhibitory Anti-inflammatory Antimicrobial Antimutagenic Antioxidant CNS depressant Oestrogenic	Aerial parts Leaves	Camphor 1,8-Cineole Linalool α-Pinene Rosmarinic acid	– Reduce anxiety symptoms – Sedative effects – Improve cognitive function (Alzheimer) – Inhibit cell proliferation – Inhibit growth of *Listeria monocytogenes*	FDA	Hamidpour et al. (2014) Porres-Martinez et al. (2013)
	S. officinalis	AChE-inhibitory Analgesic Anti-inflammatory Antimicrobial Antimutagenic Antinociceptive Antioxidant Antiseptic Metabolic Cognitive functions Effect on glucose and lipids	Leaves Flowers	Borneol Camphor 1,8-Cineole β-Pinene α-Thujone β-Thujone Rosmarinic acid	– Improve memory and cognitive function (dementia, Alzheimer) – Prevention of cardiovascular diseases – Reduce levels of cholesterol, LDL and triglycerides in blood – Alternative to antibiotics – Natural preservative food	FDA	Carović-Stanko et al. (2016) Ghorbani and Esmaeilizadeh (2017) Hamidpour et al. (2014) Naghibi et al. (2005)
Savory	*S. hortensis*	Analgesic Anti-inflammatory Antimicrobial Antinociceptive Antioxidant Antispasmodic Cytotoxic Antigenotoxic	Leaves	Borneol Carvacrol p-Cymene Thymol γ-Terpinene Rosmarinic acid Caffeic acid Apigenin Rutin	– Reduce anxiety symptoms – Reduce muscular pains – Inhibit proliferation of tumor cell – Hepatoprotective	FDA	Carović-Stanko et al. (2016) Mašković et al. (2017) Naghibi et al. (2005) Tepe and Cilkiz (2016)

TABLE 8.1 *(Continued)*

Herb	Specie	Activity	Edible Part	BACs	Main Effects	Regulation of Use	Reference
	S. montana	AChE-inhibitory BChE-inhibitory Antimicrobial Antioxidant Antitumor Antiviral Cytotoxic Antigenotoxic	Flowered aerial part Leaves	Borneol Carvacrol Methyl carvacrol Methyl thymol Thymol γ-Terpinene ρ-Cymene Rosmarinic acid Caffeic acid	– Control of CNS diseases (Alzheimer) – Inhibit growth HIV-1, HSV-1	FDA	Carović-Stanko et al. (2016) Tepe and Cilkiz (2016)
Thyme	*T. pulegioides*	Anti-inflammatory Antimicrobial Antitumor Cognitive functions Respiratory pathologies	Leaves	Carvacrol Linalool Thymol ρ-Cymene γ-Terpinene Apigenin	– Natural substitute of synthetic anti-inflammatories – Activity against respiratory Gram-positive and Gram-negative bacteria – Natural preservative food	-	Ložienė et al. (2007) De Martino et al. (2009) Senatore (1996)
	T. vulgaris	AChE-inhibitory Anti-inflammatory Antimicrobial Antioxidant Antitumor Cognitive functions Skin disorders Spasmolytic	Flowered aerial part	Carvacrol Linalool Thymol γ-Terpinene ρ-Cymene Apigenin	– Natural substitute of synthetic anti-inflammatories – Hypoglycemic effect – Hypolipidemic effect – Control of diabetes and its complicacies – Inhibit growth of HSV-1 and HSV-2 – Natural preservative food	FDA	Alu'datt et al. (2018) Carović-Stanko et al. (2016) Lorenzo et al. (2018b) Naghibi et al. (2005)

The processing, quality, purity, and effectiveness of the products obtained from *M. officinalis* L. are regulated by the FDA in the USA (FDA, 2018). This plant is commonly administered as extracts in capsules and coated pills (Kennedy et al., 2003, 2004) for mood improvements and cognitive performance.

The lemon balm extract is obtained by the extraction of dried leaves using water at 60 °C and vacuum evaporation at 40 °C, until obtaining the final extract (Scholey et al., 2014). Cyracos® and Bluenesse® are two of the most known commercial formulations of lemon balm extracted for the treatment of stress and sleep disorders, while having positive effects on mental focus and concentration. These effects are observed in treatments consisting of supplementations with 300 mg twice a day for at least 15 days. Recently, innovative formulations are being studied with new food ingredients (Fruit Up®) which contain carbohydrates in its composition for improvements of cognitive function (Scholey et al., 2014).

8.5.2 BASIL

The genus *Ocimum* is recognized in traditional medicine for its various properties, being considered as innovative drugs (Uritu et al., 2018). Between the species of this family having medicinal benefits we find *Ocimum sanctum*, *Ocimum gratissimum*, or *Ocimum micranthum*, specially related to antimicrobial, anti-inflammatory, and immunomodulatory activities (Table 8.1) (Mahajan et al., 2013).

O. sanctum also known as *Ocimum tenuiflorum* is a plant widely used in the traditional medicine of India (Ayurvedic medicine), which is associated with anticancer, antimicrobial, and analgesic properties, especially in neurobiology areas (Pattanayak et al., 2010). The analgesic effect is also associated with the use of *O. gratissimum*. This plant, originally from South America and Africa, has a proved effect in nociceptive, neuropathic, and inflammatory pains (Okiemy-Andissa et al., 2004). Finally, several studies carried out with *O. micranthum* confirmed its anti-inflammatory effects (Pinho et al., 2012).

The compounds responsible for the aforementioned properties are mainly (−)-linalool (30%–40%), eugenol (8%–30%), and methyl chavicol (15%–27%), following the importance by (+)-delta-cadinene, 3-carene, α-humulene, citral, and (−)-trans-caryophyllene (Bhattacharyya and Bishayee, 2013; Li and Chang, 2016). Its administration as dietary supplement is often prescribed in the form of pill or capsule, with a recommended dosage from 300 to 2000 mg per day for general preventative purposes.

Moreover, a fresh leaf extract, a decoction with hot water, or essential oil form (distilled from leaves and flowers of the plant) can be used to obtained beneficial effects (Mondal et al., 2011).

The effectiveness and safety of the products obtained from *O. sanctum* are not regulated by the FDA (FDA, 2018).

8.5.3 *LAVENDER*

Lavandula genus also belongs to *Laminaceae* with over 40 species, mainly existing as hybrids. This genus, originally from Europe, has been widely spreading and nowadays is commonly cultivated in North America and Australia (Figure 8.1) (Erland and Mahmoud, 2016). Among the species from this genus, it is worth emphasizing that *Lavandula angustifolia* and *Lavandula latifolia* are the most frequently used over the last decades, due to their therapeutic properties (Table 8.1) (Pistelli et al., 2017). These plants are often confused because different names are used. *L. angustifolia* is also known as *Lavandula officinalis* or English lavender, French lavender, and true lavender. *Lavandula latifolia* is also called as *Lavandula spica*, spike lavender or spike. The clinical properties of these plants depend on the specie, thereby *L. angustifolia* has sedative, relaxant, and hypotensor properties, while *Lavandula latifolia* can be used as stimulant and expectorant (Buckle, 2015). Another species with medicinal properties in this category is *Lavandula stoechas*. This species is very common in Mediterranean area, and often used for its expectorant and digestive effects, while also being applied topically to disinfect wounds.

As mentioned earlier, *Lavandula* genus has different medicinal properties, which could be related to their chemotype (Figure 8.3). In other words, their nutraceutical effect is related to their BACs profile. This can be seen in the case of *L. angustifolia* having high contents of linalool (10%–50%), linalyl acetate (12%–54%), and low camphor content (0.21%). By contrast, *Lavandula latifolia* has high contents of linalool (26%–44%), 1,8-cineole (25%–36%) and camphor (5%–15%), and low contents of linalyl acetate (0–1.5%) (Clarke, 2009; Erland and Mahmoud, 2016).

In this way, the neurological effects attributed to monoterpenes linalool and linalyl acetate, and antibacterial properties (disrupt bacterial membranes), related with the presence of oxygenated terpenes, have promoted an emerging use of these compounds over the last years. Consequently, these plants can be used as alternative antibiotics against bacteria such as *Staphylococcus aureus* and *Enterococcus faecalis* related to hospital infections (Cavanagh and Wilkinson, 2002). The quality of *L. angustifolia* is regulated by the FDA (FDA, 2018).

8.5.4 MINT

Since ancient times, *Mentha* genus has been cultivated around the world for its properties (Figure 8.1), especially for its medicinal and nutraceutical interests. Hence, it has become one of the most important commercial herbs cultivated in Europe and West of Asia (Brahmi et al., 2017; Prakash et al., 2016).

Mentha piperita (Peppermint), *Mentha spicata* (Spearmint), *Mentha gracilis*, *Mentha arvensis* (Japaneese mint), *Mentha longifolia* (Horse mint), and *Mentha suaveolens* are commonly used (Prakash et al., 2016). *M. piperita* and *M. spicata* are regulated by the FDA from the point of view of their processing until the final quality (FDA, 2018). These plants are being used for its anti-inflammatory, analgesic (methanolic extracts), antiviral (*Herpes simplex* viruses), antibacterial (Gram-positive cocci and rods and Gram-negative rods), stimulant, and anticatharrhal properties (Table 8.1) (Uritu et al., 2018; Venditti et al., 2017). The aforementioned medicinal properties have been tested both in vivo and in vitro, with menthol being mainly responsible for these activities (Brahmi et al., 2017). Regarding its analgesic activity, the effects are similar to those showed by morphine what would allow its use as a new option in pain therapy, with less side effects (Uritu et al., 2018).

The common applications are traditional medicines, dietary supplements, or tea infusions of the dried leaves or essential oils (oil or capsulated). The compounds responsible for nutraceutical and medicine properties are mainly terpenoids as menthol and menthone, followed by 1,8-cineole, carvone, linalyl acetate, and linalool (Brahmi et al., 2017; Prakash et al., 2016). In addition to these compounds are polyphenolics as the caffeic phenolic acid with its derivatives, chlorogenic, and rosmarinic acid, which represent a 60%–80% of total phenolic compounds. The composition depends on many factors as variety and methods of extraction (Lorenzo et al., 2018b).

8.5.5 OREGANO

Oregano is yet another herb of the *Laminaceae* family, cultivated in the Mediterranean area since ancient times (Figure 8.1) (Mahmoud et al., 2004). There are more than 20 species included in the genus *Origanum*, being *Origanum vulgare* L. and *Origanum majorana* L. the most relevant for their medicinal activity (Uritu et al., 2018). For this purpose, dried leaves and flowering tops of the aforementioned species are used (Table 8.1). Among the main beneficial properties of this herb include pasmodic, antimicrobial, digestive, and expectorant activities (Teixeira et al., 2013). Regarding its antimicrobial activity, *O. majorana* L. is able to inhibit the formation of microbial metabolites, and has properties related to cardiovascular disorders, such as vasodilatatory effects, and it is capable to prevent atrial fibrillation (Floria et al., 2015; Marino et al., 2001).

The main phytochemicals accountable of their therapeutic uses are borneol, carvacrol, linalool, α-pinene, β-pinene, α-terpinene, and γ-terpinene (García-Beltrán and Esteban, 2016). The implication of one or other will depend on the target disease. Thymol and carvacrol are the most important compounds isolated from this genus (Figure 8.3). There are several studies that associate thymol with antimicrobial, anti-inflammatory, antioxidant, respiratory, and digestive disorders, relaxing the soft tissue of the throat and stomach. Carvacrol is also associated with the prevention of neurodegenerative pains, antihepatotoxic, antimutagenic, analgesic, antitumor, and antispasmodic (Friedman, 2014; Talavera-Alemán et al., 2016). Other components with nutraceutical properties are: rosmarinic acid responsible of anti-inflammatory and antiviral activity (Shen et al., 2010), 4-terpineol associated with antibacterial and antitumor effects, and thymoquinone with antioxidant, anti-inflammatory and analgesic properties. The last one is a molecule which could be used in the future as an alternative against tumors (Begnini et al., 2014).

Controversial information is found about the compounds responsible for preventive action of *Origanum* genus against neurodegenerative diseases. Some authors affirm that thymol and carvacrol are the main active components (Loizzo et al., 2009), while others attributed main activity to terpenoids (4-terpinenol γ-terpinene, α-terpinene, p-cymene and 1,8-cineol) (Mossa and Nawwar, 2011). Recent studies combined both ideas, suggesting that these effects might be due to mixture of compounds present in the extracts or essential oils from this genus (Loizzo et al., 2009). Somewhat similar is documented for antibacterial activity, mentioning a synergistic effects of terpenoids such as carvacrol, linalool, thymol, γ-terpinene, p-cymene, and 4-terpineol (Burt, 2004).

8.5.6 ROSEMARY

The leaves of the genus *Rosmarinus*, commonly known as rosemary, are mainly cultivated around Mediterranean since the distant past (Figure 8.1).

These plants are used in the traditional medicine and by the pharmaceutical industries due to their analgesic, anti-inflammatory, anticarcinogenic, antirheumatic, spasmolytic, antihepatotoxic, atherosclerotic, carminative, and choleretic uses (Fahim, et al., 1999; Haloui et al., 2000; Mahmoud et al., 2004; Yamamoto et al., 2005). These properties are attributed to its bioactive components (Table 8.1), while the presence of one or the other BAC will depend on the technique used for their extraction. Between these techniques, maceration, hydrodistillation, distillation, or supercritical fluid extraction should be highlighted. Obtained extracts or essential oils have different properties linked to their chemical composition. Within essential oils compounds that stand out are: 1,8-cineole, α-pinene, limonene, camphor, and less presence of camphene, borneol, bornyl acetate, and α-terpineol (Bozin et al., 2006; Hernández et al., 2016). On the contrary, extracts of these plants contain phenolic acids as rosmarinic, caffeic, ursolic, betulinic, and carnosic acids, and terpenes as camphor and carnosol (WHO, 2018).

Among the main compounds isolated from this genus are phenolic acid, for example, rosmarinic acid, with antimicrobial, anti-inflammatory, and antioxidant activities (Naghibi et al., 2005) (Figure 8.3). Several studies suggested that rosmarinic acid may have neuroprotective effects which would be due to their ability to inhibit acetylcholinesterase (AChE) (El Omri et al., 2012; Ozarowski et al., 2013).

The other very important compound for biological properties is carnosic acid with antitumor and antioxidant activities. The antioxidant activity related to preventive medicine could be attributed to free-radical scavenger activity for neutralizing ROS (Andrade et al., 2018). The products obtained from *R. officinalis* L. are regulated by the FDA (FDA, 2018).

8.5.7 SAGE

Aromatic varieties of the genus *Salvia* are used since ancient times for traditional medicine. This genus was cultivated in the Mediterranean, Europe, and North America (Figure 8.1), and includes about 900 species, making it the largest genus of the *Lamiaceae* family (Ghorbani and Esmaeilizadeh, 2017). However, not of all them have beneficial properties for human diet (Altindal and Altindal, 2016).

Sage as medicinal plant has antimicrobial, antioxidative, antimalarial, anti-inflammatory, antidepressant, antidiabetic, cardiovascular, and antitumor effects (Shakeri et al., 2016; Uritu et al., 2018). *Salvia glutinosa* L., *Salvia lavandulifolia* L., *Salvia officinalis* L., and *Salvia verticillata* L., are important species included in this genus. *S. lavandulifolia* L. and *S. officinalis* L. were the most common members of these genera, and the last one is the most valuable, due to the large amount of detected BAC (Pop et al., 2014). The use of *S. officinalis* L. is regulated by the FDA (FDA, 2018).

The nutraceutical effect of the plants from this genus is related to a wide range of BACs included in their chemotype (Table 8.1), containing more than 100 compounds (Fu et al., 2013). Between these BACs, the most abundant compounds are essential oils (Figure 8.3), terpenoids, and flavonoids (Szentmihályi et al., 2014), being α-thujone (10%–60%), β-thujone (4%–36%), camphor (5%–20%), 1,8-cineole (1%–15%), and rosmarinic acid (up to 3.3%). The last one is responsible for the anticancer, anti-inflammatory, antimicrobial, and the antioxidant properties, as well as its effect on the prevention of cerebrovascular and cardiovascular disorders (Ghorbani and Esmaeilizadeh, 2017). Thujone, camphor, and 1,8-cineole are associated with antimicrobial and antimutagenic effect (Delamare et al., 2007; Vujošsević and Blagojević, 2004). Finally, there are minor compounds such as carnosol and carnosic acid that are related to antimicrobial, anti-inflammatory, and antioxidant properties (Cuvelier et al., 1996; Horiuchi et al., 2007; Rodrigues et al., 2012).

Nowadays, the leaves of this plant can be used as alternative therapies against diseases, such important today, such as cancer and Alzheimer. This cognitive effect could be related with *S. officinalis* L., which is able to inhibit the action of the enzyme AChE (Ghorbani and Esmaeilizadeh, 2017). Its use as dietary supplementation is by tablets, essential oils droplets, and infusions (Altindal and Altindal, 2016). The recommended dosage for capsules is 1–2 per day, which contains an extract with 4%–6% essential oils and 2.5% rosmarinic acid.

Regarding the safety of its use, there are no studies related to its toxicity. Only adverse effects could appear in an overdose cases, and they are related with one of the compounds isolated in their composition. This component is thujone that has convulsing effect associated with epileptic seizures (Ghorbani and Esmaeilizadeh, 2017; Halicioglu et al., 2011). This is the reason why it is more appropriate to use *S. lavandulifolia* L., due to the absence of this compound in its composition.

8.5.8 SAVORY

The *Satureja* genus includes numerous species, as *Satureja hortensis* L. (Summer savory), while *Satureja montana* L. (Winter savory) is the most commonly used (Table 8.1). The aerial part of the plant *S. hortensis* L. is known as the traditional medicine because of its antimicrobial, antioxidant, anti-inflammatory, antinociceptive, and analgesic activities, as well as for positive effects on immune system (Jafari et al., 2016). For medicinal purposes, this plant can be used as infusion, or as dried leaves in the form of a pill, whereas dosages depend on the desired effect. Therefore, for the treatment of high blood pressure, the recommended dosage is 250 mg for 60 days in a tablet form (dried leaves), while for preventive effects it is usually used 2–3 cups a day with two teaspoons of the herb (Bojor, 2003; Hamidpour et al., 2014). The processing, quality, purity, safety, and effectiveness of the products and essential oils obtained from *S. hortensis* L. are regulated by the FDA (FDA, 2018).

The main compounds found in its chemical composition that are responsible for medicinal use are terpenoids, phenols, and flavonoids (Fierascu et al., 2018). The presence of one or the other will depend on the way in which it is produced. In this way, thymol, γ-terpinene, carvacrol, and ρ-cymene are the main compounds isolated from essential oil of entire herb (Mihajilov-Krstev et al., 2009), while rosmarinic acid, caffeic acid, naringenin, isoferulic acid, and apigenin are obtained from the methanolic extracts of the aerial parts of the plant (Fierascu et al., 2018). As it can be seen, the chemical composition depends on the method used for the extraction (Fierascu et al., 2018).

Regarding *S. montana* L., several studies showed antimicrobial, antioxidant, antispasmodic, and antidiarrhoeal effects. The BACs that could be responsible for these properties are the same as those found in *S. hortensis* L. Moreover, other compounds can also be isolated as methyl-carvacrol, methyl-thymol, and borneol, which depend on developmental stages of the plant (Tepe and Cilkiz, 2016). It is important to highlight its beneficial effect for Alzheimer's control, since it is able to inhibit the action of two key enzymes related with the progress of the disease (Silva et al., 2009). These enzymes are AChE and butyrylcholinesterase (BChE).

8.5.9 THYME

The use of thyme in traditional medicine is broad. The leaves of the *Thymus* genus were used for antimicrobial and anti-inflammatory properties, as well as the possible use for treatments of dementia or oncological pathologies, through apigenin (Ali et al., 2017; Uritu et al., 2018). Leaves of the two commonly used thymes from this family that are attributed with health benefits are *T. vulgaris* and *Thymus pulegioides* (Table 8.1).

T. vulgaris is the most commonly used for pain, pyrexia, and inflammation, due to its anti-inflammatory properties (Qadir et al., 2016). There have even been studies comparing their effects with those of ibuprofen (Salmalian et al., 2014). In addition, its inflammatory power is useful for the treatment of skin disorders (Alabdullatif et al., 2017; Basch et al., 2004), such as scabies, herpes, wounds, alopecia, dental plaque, and ringworm. The processing, quality, and purity of the products obtained from *T. vulgaris* L. are regulated by the FDA (FDA, 2018).

T. pulegioides is among the most used plants in traditional medicine, and is the second most important plant for respiratory pathologies (Vitalini et al., 2013). In addition, anti-inflammatory (Stalińska et al., 2005) and antifungal properties (Pinto et al., 2006) are also attributed to this plant (Lorenzo et al., 2018b). Thymol, carvacrol, ρ-cymene, γ-terpinene, and linalool are predominant BACs found in essential oils extracted from the plants of this genus, which could be responsible for their nutraceutical effects. In the same way, thymol and carvacrol are also the main isolated compounds. Furthermore their use as extracts, infusion, or capsules are offered as dietary supplements to enhance immune heath. The recommendations are 1–2 capsules three times a day with each capsule containing 450 mg of *T. vulgaris*.

8.6 FOOD PRESERVATION: ANTIOXIDANT ACTIVITY

Modern consumers are increasingly aware of nutritional principles that will foster their well-being (Putnik and Bursać Kovačević, 2017), with favoring use of natural products for food processing (Putnik et al., 2016). Functional properties of foods can be increased by enrichment, fortification, and alteration (Čukelj et al., 2016). To that end, herbal nutraceuticals can be added to food in a concentrated form, as extracts or essential oils. These manufactured products with *Laminaceae* herbs are well accepted by the consumers because of declared use of natural ingredients that are able to reduce food spoilage. However, the use of natural ingredients will avoid the use of synthetic antioxidants and result with clean label products (Embuscado, 2015; Granato et al., 2017). Essential oils and extracts from the *Laminaceae* plants contain strong antioxidants which allow their use as food preservatives that prevent lipid oxidation and microbial growth (Alu'datt et al., 2018; Shah et al., 2014; Zheng and Wang, 2001). Moreover, they decrease the formation of undesirable compounds that could have detrimental effects on health (Embuscado, 2015).

The antioxidant power of these compounds will depend on several factors that are related specific to: chemical composition, extraction method, food matrix, storage temperature, pH, etc. That is the reason why the *Laminaceae* extracts or essential oils are added in greater doses to foods for the same effect as identified in vitro batches (Alu'datt et al., 2018). Carvacrol, eugenol, and thymol are the most effective phenolic compounds isolated in the aforementioned extracts and essential oils (Bakkali et al., 2008).

The mode of action of the BACs present in the extracts and in essential oils is the same as the synthetic antioxidants, commonly used in foods, such as BHA, BHT, and TBHQ. They act as free-radical scavengers, blocking free radicals by donating a hydrogen atom (Embuscado, 2015; Tongnuanchan and Bejakul, 2014). As with clinical properties, phenolic compounds are the antioxidants and effective free-radical scavengers, able to impede initiation and cascade of lipid oxidation (Hyldgaard et al., 2012). They are primary antioxidants that act in three steps against lipid oxidation: initiation, propagation, and termination. Their capacity to donate an electron to the free radical prevents the oxidation of other compounds (Yanishlieva-Maslarova, 2001). They react with free (lipid) radicals leading to nonradical species and the inactivation of peroxyl radicals, therefore inhibiting the cascade reactions leading to termination (Jayasena and Jo, 2014). Similar mechanism of action is characteristic for protein oxidation where phenolic compounds avoid protein carbonylation by joining with the proteins (Siebert et al., 1996).

Regarding the effect of *Laminaceae* as antimicrobial agents in foods, their activity is related to alteration of microbial cellular metabolism with the modification of enzymatic activity, either by protein denaturation or by changing the permeability of microbial cells (Aminzare et al., 2016; Marino et al., 2001) leading to cell death.

Within this family, oregano, rosemary, sage, and thyme are the main sources of extracts and essential oils with antioxidant capacity. *Rosmarinus* genus is characterized by its strong antioxidant power, while being evaluated and included in the list of Food Safety Authority (EFSA) for its use as a natural food preservative (EFSA, 2008). Several studies documented that its leaves can have close to 2.5% of oil (Uritu et al., 2018). Terpenoids such as eucalyptol, α-pinene, limonene, and camphor, with phenolic acids (e.g., rosmarinic acid), isolated in the essential oil are responsible for its antioxidant activity (Almela et al., 2006; Bozin et al., 2007; Hernández et al., 2016), whereas its essential oils are commonly used to improve the shelf life of foods (Pateiro et al., 2018). In fact, several studies evidenced their use against microbial food spoilage originated from Gram-negative (*Pseudomonas*) and Gram-positive (*Lactobacillus*) bacteria (Ouattara et al., 1997).

The *Thyme* genus is highlighted for their great antioxidant power. As mentioned previously, *T. pulegioides* and *T. vulgaris* are mostly used. *T. pulegioides* has an important antioxidant activity (Schaffer et al., 2004), while related radical scavenging is bind to main components present in essential oils. Among them, linalool stands out, as well as thymol, carvacrol, γ-terpinene, p-cymene, geraniol, and geranial (Ložienė et al., 2007). A similar composition and antioxidant activity was observed in the essential oil of *T. vulgaris*. Isolated thymol, carvacrol, ρ-cymene, γ-terpinene, and linalool are also found among more than 60 compounds from their chemical profile (Lorenzo et al., 2018b).

Origanum genus (*O. vulgare L.*) has a great antioxidant ability that allows its use as natural antioxidant for extending the shelf life of foods, for improving their organoleptic characteristics (Krishnan et al., 2014), and for reducing the spoilage and microbe growth. Carvacrol and thymol are the main compounds responsible of these effects, representing 78% of the compounds found in essential oils (O'Grady and Kerry, 2009). Several studies revealed that using both its extracts and its essential oils significantly reduced oxidation in the samples treated with oregano products (Fasseas et al., 2007; Fernandes et al., 2017).

The essential oils extracted from the leaves of *Salvia* genus, especially from *S. officinalis* L., have a great antioxidant activity, but still lesser than those found in other essential oils of the same family (Estevez et al., 2007). However, its effect on lipid oxidation, lipid-derived volatiles, and degradation of PUFA is higher than synthetic antioxidants like BHT, which makes it a good alternative against synthetic antioxidants. Carnosic acid and carnosol are the phenolic compounds responsible of its activity as free-radical scavengers (Estevez et al., 2007).

8.7 CONCLUSIONS

The extensive benefits from herbs of the *Laminaceae* family, both on health and food preservation, foster consideration of their use as nutraceuticals and industrial food additives. Whether or not they are introduced as dietary supplements to dieting, current trends that foster healthy lifestyle will drive demand for herbal products (e.g., nutraceuticals) of the future food markets. In addition to the nutraceutical use, their incorporation in foods would allow design of functional foods that have significant market potential. The BACs isolated from the plants of this family would contribute to the health benefits and will tend to extend the shelf-life of the products due to their antioxidant and antimicrobial effects. As a result, these food products would be classified as clean label, due to the replacement of synthetic additives by natural herbal products.

ACKNOWLEDGMENTS

Paulo E. Munekata acknowledges postdoctoral fellowship support from the Ministry of Economy and Competitiveness (MINECO, Spain) "Juan de la Cierva" program (FJCI-2016-29486).

KEYWORDS

- **herbal**
- *Laminaceae*
- **nutraceutical**
- **health**
- **antioxidants**

REFERENCES

Alabdullatif, M.; Boujezza, I.; Mekni, M.; Taha, M.; Kumaran, D.; Yi, Q-L.; Landoulsi, A.; Ramirez-Arcos, S. Enhancing blood donor skin disinfection using natural oils. *Transfusion.* **2017**, 57(12), 2920–2927.

Ali, F.; Rahul, Naz, F.; Jyoti, S., Siddique, Y.H. Health functionality of apigenin: A review. *Int. J. Food Prop.* **2017**, 20(6), 1197–1238.

Almela, L.; Sánchez-Muñoz, B.; Fernández-López, A.; Roca, M.J.; Rabe, V. Liquid chromatograpic–mass spectrometric analysis of phenolics and free radical scavenging activity of rosemary extract from different raw material. *J. Chromatogr. A.* **2006**, 1120(1-2), 221–229.

Altindal, D.; Altindal, N. Sage (Salvia officinalis) Oils. In: Essential Oils in Food Preservation, Flavor and Safety; Preedy, V. R., Ed.; Elsevier: London, **2016**; pp 715–721.

Alu'datt, M.H.; Rababah, T.; Alhamad, M.N.; Gammoh, S.; Al-Mahasneh, M.A.; Tranchant, C.C.; Rawshdeh, M. Pharmaceutical, Nutraceutical and Therapeutic Properties of Selected Wild Medicinal Plants: Thyme, Spearmint, and Rosemary. In: Therapeutic, Probiotic, and Unconventional Foods; Grumezescu, A. M., Holban, A. M., Eds.; Academic Press, Elsevier: London, **2018**; pp 275–290.

Alu'datt, M.H.; Rababah, T.; Johargy, A.; Gammoh, S.; Ereifej, K.; Alhamad, M.N.; ... Rawshdeh, M. Extraction, optimisation and characterisation of phenolics from *Thymus vulgaris* L.: Phenolic content and profiles in relation to antioxidant, antidiabetic and antihypertensive properties. *Int. J. Food Sci. Tech.* **2016**, 51(3), 720–730.

Alves-Silva, J.M.; dos Santos, S.M.D.; Pintado, M.E.; Pérez-Álvarez, J.A.; Fernández-López, J.; Viuda-Martos, M. Chemical composition and in vitro antimicrobial, antifungal and antioxidant properties of essential oils obtained from some herbs widely used in Portugal. *Food Control*. **2013**, 32(2), 371–378.

Aminzare, M.; Hashemi, M.; Azar, H.H.; Hejazi, J. The use of herbal extracts and essential oils as a potential antimicrobial in meat and meat products: A review. *J. Hum. Environ. Health Promot.* **2016**, 1(2), 63–74.

Andrade, J.M.; Faustino, C.; Garcia, C.; Ladeiras, D.; Reis, C.P.; Rijo, P. *Rosmarinus officinalis* L.: An update review of its phytochemistry and biological activity. *Future Sci. OA.* **2018**, 4(4), FSO283.

Augustin, M.A., Sanguansri, L. Challenges in developing delivery systems for food additives, nutraceuticals and dietary supplements. In: Encapsulation technologies and delivery systems for food ingredients and nutraceuticals; Garti, N., McClements, D. J., Eds.; Woodhead Publishing: Cambridge, UK, **2012**; pp 19–48.

Bag, A.; Chattopadhyay, R.R. Evaluation of synergistic antibacterial and antioxidant efficacy of essential oils of spices and herbs in combination. *PLoS One* **2015**, 10(7), e0131321.

Bakkali, F.; Averbeck, S.; Averbeck, D.; Idaomar, M. Biological effects of essential oils—a review. *Food Chem. Toxicol.* **2008**, 46(2), 446–475.

Barba, F.J.; Esteve, M.J.; Frígola, A.; Bioactive components from leaf vegetable products. *Stud. Nat. Prod. Chem.* **2014**, 41, 321–346.

Barba, F.J.; Putnik, P.; Bursać Kovačević, D.; Poojary, M.M.; Roohinejad, S.; Lorenzo, J.M.; Koubaa, M. Impact of conventional and non-conventional processing on prickly pear (*Opuntia* spp.) and their derived products: From preservation of beverages to valorization of by-products. *Trends Food Sci. Technol.* **2017**, 67, 260–270.

Barros, L.; Dueñas, M.; Dias, M.I.; Sousa, M.J.; Santos-Buelga, C.; Ferreira, I.C.. Phenolic profiles of cultivated, in vitro cultured and commercial samples of *Melissa officinalis* L. infusions. *Food Chem*. 2013, 136(1), 1–8.

Barros, L.; Heleno, S.A.; Carvalho, A.M.; Ferreira, I.C. Lamiaceae often used in Portuguese folk medicine as a source of powerful antioxidants: Vitamins and phenolics. *LWT—Food Sci. Technol.* **2010**, 43(3), 544–550.

Basch, E.; Ulbricht, C.; Hammerness, P.; Bevins, A.; Sollars, D. Thyme (*Thymus vulgaris* L.), thymol. *J. Herb. Pharmacother.* **2004**, 4(1), 49–67.

Begnini, K.R.; Nedel, F.; Lund, R.G.; Carvalho, P.H.D.A.; Rodrigues, M.R.A.; Beira, F.T.A; Del-Pino, F.A.B. Composition and antiproliferative effect of essential oil of *Origanum vulgare* against tumor cell lines. *J. Med. Food*. **2014**, 17, 1129–1133.

Berdowska, I.; Zieliński, B.; Fecka, I.; Kulbacka, J.; Saczko, J.; Gamian, A. Cytotoxic impact of phenolics from Lamiaceae species on human breast cancer cells. *Food Chem.* **2013**, 141(2), 1313–1321.

Bhattacharyya, P.; Bishayee, A. *Ocimum sanctum Linn.* (Tulsi): An ethnomedicinal plant for the prevention and treatment of cancer. *Anti-Cancer Drugs.* **2013**, 24(7), 659–666.

Bojor, O. Guide of medicinal and aromatic plants from A to Z. In: Romanian: ghidul plantelor medicinale şi aromatice de la a la Z; Fiat Lux: Bucharest, Romania, **2003**; pp 94–95.

Bozin, B.; Mimica-Dukic, N.; Samojlik, I.; Jovin, E. Antimicrobial and antioxidant properties of rosemary and sage (*Rosmarinus officinalis L.* and *Salvia officinalis L.*, Lamiaceae) essential oils. *J. Agric. Food Chem.* **2007**, 55(19), 7879–7885.

Bozin, B.; Mimica-Dukic, N.; Simin, N.; Anackov, G. Characterization of the volatile composition of essential oils of some Lamiaceae spices and the antimicrobial and antioxidant activities of the entire oils. *J. Agric. Food Chem.* **2006**, 54, 1822–1828.

Brahmi, F.; Khodir, M.; Mohamed, C.; Pierre, D. Chemical composition and biological activities of Mentha species. In: Aromatic and Medicinal Plants-Back to Nature; El-Shemy, H., Ed.; InTechOpen Limited: London, UK, **2017**; pp 47–79.

Buckle, J. Basic plant taxonomy, basic essential oil chemistry, extraction, biosynthesis, and analysis. In: Clinical Aromatherapy. Essential Oils in Healthcare; Buckle, J., Ed.; Elsevier: London, UK, **2015**; pp 37–72.

Bursać Kovačević, D.; Maras, M.; Barba, F.J.; Granato, D.; Roohinejad, S.; Mallikarjunan, K.; Montesano, D.; Lorenzo, J.M.; Putnik, P. Innovative technologies for the recovery of phytochemicals from Stevia rebaudiana Bertoni leaves: A review. *Food Chem.* **2018**, 268, 513–521.

Burt S. Essential oils: their antibacterial properties and potential applications in foods—a review. *Int. J. Food Microbiol.* **2004**, 94, 223–253.

Carović-Stanko, K.; Petek, M.; Martina, G.; Pintar, J.; Bedeković, D.; Ćustić, M.H.; Šatović, Z. Medicinal plants of the family Lamiaceaeas functional foods—a review. *Czech J. Food Sci.* **2016**, 34(5), 377–390.

Cavanagh, H.M.A.; Wilkinson, J.M. Biological activities of lavender essential oil. *Phytother. Res.*, **2002**, 16, 301–308.

Chauhan, B.; Kumar, G.; Kalam, N.; Ansari, S.H. Current concepts and prospects of herbal nutraceutical: A review. *J. Adv. Pharm. Technol. Res.* **2013**, 4(1), 4–8.

Clarke, S. *Essential Chemistry for Aromatherapy,* 2nd ed.; Churchill Livingstone, Elsevier Health Sciences: London, UK, **2009**.

Čukelj, N.; Putnik, P.; Novotni, D.; Ajredini, S.; Voučko, B.; Duška, Ć. Market potential of lignans and omega-3 functional cookies. *Brit. Food J.* **2016**, 118(10), 2420–2433.

Cuvelier, M.E.; Richard, H.; Berset, C. Antioxidative activity and phenolic composition of pilot-plant and commercial extracts of sage and rosemary. *J. Am. Oil Chemists Soc.* **1996**, 73, 645–652.

Danh, L.T.; Triet, N.D.A.; Zhao, J.; Mammucari, R.; Foster, N. Comparison of chemical composition, antioxidant and antimicrobial activity of lavender (*Lavandula angustifolia* L.) essential oils extracted by supercritical CO_2, hexane and hydrodistillation. Food Bioprocess Tech. **2013**, 6(12), 3481–3489.

De Martino, L.; Bruno, M.; Formisano, C.; De Feo, V.; Napolitano, F.; Rosselli, S.; Senatore, F. Chemical composition and antimicrobial activity of the essential oils from two species of Thymus growing wild in southern Italy. *Molecules* **2009**, 14(11), 4614–4624.

de Pinho, J.P.M.; Silva, A.S.B.; Pinheiro, B.G.; Sombra, I.; de Carvalho Bayma, J.; Lahlou, S.; da Cunha Sousa, P.J.; Magalhães, P.G.C. Antinociceptive and antispasmodic effects of the essential oil of *Ocimum micranthum*: Potential anti-inflammatory properties. *Planta Med.* **2012**, 78(7), 681–685.

Delamare, A.P.L.; Moschen-Pistorello, I.T.; Artico, L.; Atti-Serafini, L.; Echeverrigaray, S. Antibacterial activity of the essential oils of *Salvia offcinalis L.* and *Salvia triloba L.* cultivated in South Brazil. *Food Chem.* **2007**, 100, 603–608.

Dini, I. Spices and Herbs as Therapeutic Foods. In: Food Quality: Balancing Health and Disease; Holban, A.M., Grumezescu, A.M., Eds.; Academic Press, Elsevier: London, UK, **2018**; Vol. 13; pp 433–469.

Directive EC 2004/24/EC of the European Parliament and of the Council of 31 March 2004 amending, as regards traditional herbal medicinal products, Directive 2001/83/EC on the Community code relating to medicinal products for human use. *Off. J. Eur. Union* 2004, L136, 85–90.

Directive EC 2002/46/EC of the European Parliament and of the Council of 10 June 2002 on the approximation of the laws of the Member States relating to food supplements. *Off. J. Eur. Union* 2002, L183, 51–57.

Ditu, L. M.; Grigore, M. E.; Camen-Comanescu, P.; Holban, A. M. Introduction in Nutraceutical and Medicinal Foods. In: Therapeutic, Probiotic, and Unconventional Foods; Grumezescu, A. M., Holban, A. M., Eds.; Academic Press, Elsevier: London, **2018**; pp 1–12.

do Prado, D.Z.; Capoville, B.L.; Delgado, C.H.; Heliodoro, J.C.; Pivetta, M.R.; Pereira, M.S.; … Fleuri, L.F. (2018). Nutraceutical food: composition, biosynthesis, therapeutic properties, and applications. In: Alternative and Replacement Foods; Holban, A. M., Grumezescu, A. M., Eds.; Academic Press, Elsevier: London, UK, 2018; Vol. 17; pp 95–140.

Domínguez, R.; Barba, F.J.; Gomez, B.; Putnik, P.; Bursać Kovačević, D.; Pateiro, M.; Santos, E.M.; Lorenzo, J.M. Active packaging films with natural antioxidants to be used in meat industry: A review. *Food Res. Int.* **2018**, 113, 93–101.

Dudeja, P.; Gupta, R. K. Nutraceuticals. In: Food Safety in the 21st Century; Gupta, R. K., Dudeja, P., Minhas, A. S., Eds.; Elsevier: London, UK, **2016**; pp 491–496.

EFSA. Scientific Opinion of the panel on food additives, flavourings, processing aids and materials in contact with food on a request from the commission on the use of rosemary extracts as a food additive. *EFSA J.* **2008**, 721, 1–29.

El Omri, A.E.L.; Han, J.; Abdrabbah M.B.; Isoda, H. Down regulation effect of Rosmarinus officinalis polyphenols on cellular stress proteins in rat pheochromocytoma PC12 cells. *Cytotechnology.* **2012**, 64(3), 231–240.

Embuscado, M.E. Spices and herbs: Natural sources of antioxidants—a mini review. J. *Funct. Foods.* **2015**, 18, 811–819.

Englberger, W.; Hadding, U.; Etschenberg, E.; Graf, E.; Leyck, S.; Winkelmann, J.; Parnham, M.J. Rosmarinic acid: a new inhibitor of complement C3-convertase with anti-inflammatory activity. *Int. J. Immunopharmacol.* **1988**, 10(6), 729–737.

Erland, L.A.; Mahmoud, S.S. Lavender (Lavandula angustifolia) oils. In: Essential Oils in Food Preservation, Flavor and Safety; Preedy, V. R., Ed.; Elsevier: London, UK, **2016**; pp. 501–508.

Estevez, M.; Ramirez, R.; Ventanas, S.; Cava, R. Sage and rosemary essential oils versus BHT for the inhibition of lipid oxidative reactions in liver pate. *LWT—Food Sci. Technol.* **2007**, 40, 5865.

Fahim, F.A.; Esmat, A.Y.; Fadel, H.M.; Hassan, K.F. Allied studies on the effect of *Rosmarinus officinalis L.* on experimental hepatotoxicity and mutagenesis. *Int. J. Food Sci. Nutr.* **1999**, 50(6), pp 413–427.

Fangueiro, J. F.; Souto, E. B.; Silva, A. M. Encapsulation of nutraceuticals in novel delivery systems. In: Nutraceuticals. Nanotechnology in the Agri-Food Industry; Grumezescu, A. M., Ed.; Elsevier: London, **2016**; pp 303–342.

Fasseas, M.K.; Mountzouris, K.C.; Tarantilis, P.A.; Polissiou, M.; Zervas, G. Antioxidant activity in meat treated with oregano and sage essential oils. *Food Chem.* **2007**, 106, 1188–1194.

FDA. CFR-Code of Federal Regulations Title 21. https://www.accessdata.fda.gov/scripts/cdrh/cfdocs/cfcfr/CFRSearch.cfm?fr=182.10 (accessed October 31, 2018).

Fernandes, R.P.P.; Trindade, M.A.; Tonin, F.G.; Pugine, S.M.P.; Lima, C.G.; Lorenzo, J.M.; de Melo, M.P. Evaluation of oxidative stability of lamb burger with *Origanum vulgare* extract. *Food Chem.* **2017**, 233, 101–109.

Fidelis, M.; Santos, J.S.; Escher, G.B.; Vieira do Carmo, M.; Azevedo, L.; Cristina da Silva, M.; Putnik, P.; Granato, D. In vitro antioxidant and antihypertensive compounds from camu-camu (Myrciaria dubia McVaugh, Myrtaceae) seed coat: A multivariate structure-activity study. *Food Chem. Toxicol.* **2018**, 120, 479–490.

Fierascu, I.; Dinu-Pirvu, C.; Fierascu, R.; Velescu, B.; Anuta, V.; Ortan, A.; Jinga, V. Phytochemical profile and biological activities of *Satureja hortensis L.*: A review of the last decade. *Molecules* **2018**, 23(10), 2458.

Floria, M.; Drug, V.L. Atrial fibrillation and gastroesophageal reflux disease: From the cardiologist perspective. *World J. Gastroentero.* **2015**, 21(10), 3154–3156.

Frezza, C.; Venditti, A.; Bianco, A.; Serafini, M. The nutraceutical importance of Lamiaceae species and its correlation with phytochemistry. Book of Abstracts, 3rd Edition of International Conference on Agriculture and Food Chemistry, Rome, Italy, July 23–24, **2018**; *J. Food Nutrit. Popul Health*, 2.

Friedman, M. Chemistry and multibeneficial bioactivities of carvacrol (4-isopropyl-2-methylphenol), a component of essential oils produced by aromatic plants and spices. *J. Agric. Food Chem.* **2014**, 62, 7652–7670.

Fu, Z.H.; Wang, H.; Hu, X.; Sun, Z.; Han, C. The pharmacological properties of Salvia essential oils. *J. Appl. Pharm. Sci.* **2013**, 3(7), pp 122–127.

Gabrić, D.; Barba, F.J.; Roohinejad, S.; Gharibzahedi, S.M.T.; Radojčin, M.; Putnik, P.; Bursać Kovačević, D. Pulsed electric fields as an alternative to thermal processing for preservation of nutritive and physicochemical properties of beverages: A review. *J. Food Process Eng.* **2018**, 41(1), 1–14.

García-Beltrán, J.M.; Esteban, M.A. Properties and applications of plants of *Origanum* sp. genus. *SM. J. Biol.* **2016**, 2(1), 1006.

Gavahian, M.; Farahnaky, A.; Farhoosh, R.; Javidnia, K.; Shahidi, F. Extraction of essential oils from *Mentha piperita* using advanced techniques: Microwave versus ohmic assisted hydrodistillation. *Food Bioprod. Process.* **2015**, 94, 50–58.

Gavahian, M.; Farahnaky, A.; Javidnia, K.; Majzoobi, M. Comparison of ohmic assisted hydrodistillation with traditional hydrodistillation for the extraction of essential oils from *Thymus vulgaris* L. *Innov. Food Sci. Emerg. Technol.* **2012**, 14, 85–91.

Gharibzahedi, S.M.T.; Hernández-Ortega, C.; Welti-Chanes, J.; Putnik, P.; Barba, F.J.; Mallikarjunan, K.; Escobedo-Avellaneda, Z.; Roohinejad, S. High pressure processing of food-grade emulsion systems: Antimicrobial activity, and effect on the physicochemical properties. *Food Hydrocolloid.* **2019**, 87, 307–320.

Ghorbani, A.; Esmaeilizadeh, M. Pharmacological properties of *Salvia officinalis* and its components. *J. Tradit. Complement. Med.* **2017**, 7(4), 433–440.

Giacometti, J.; Bursać Kovačević, D.; Putnik, P.; Gabrić, D.; Bilušić, T.; Krešić, G.; Stulić, V.; Barba, F.J.; Chemat, F.; Barbosa-Cánovas, G.; Režek Jambrak, A. Extraction of bioactive compounds and essential oils from Mediterranean herbs by conventional and green innovative techniques: A review. *Food Res. Int.* **2018**, 113, 245–262.

Gómez, B.; Barba, F.J.; Domínguez, R.; Putnik, P.; Kovačević, D.B.; Pateiro, M.; Toldrá, F.; Lorenzo, J.M. Microencapsulation of antioxidant compounds through innovative technologies and its specific application in meat processing. *Trends Food Sci. Technol.* **2018**, 82, 135–147.

González-Gallego, J.; García-Mediavilla, M.V.; Sánchez-Campos, S.; Tuñón, M.J. Anti-inflammatory and immunomodulatory properties of dietary flavonoids. In: Polyphenols in Human Health and Disease; Watson, R.R., Preedy, V.R., Zibadi, S., Eds; Academic Press: London, UK, **2014**; pp 435–452.

Granato, D.; Nunes, D.S.; Barba, F.J. An integrated strategy between food chemistry, biology, nutrition, pharmacology, and statistics in the development of functional foods: A proposal. *Trends Food Sci. Technol.* **2017**, 62, 13–22.

Granato, D.; Putnik, P.; Kovačević, D.B.; Santos, J.S.; Calado, V.; Rocha, R.S.; Gomes da Cruz, A.; Jarvis, B.; Rodionova, O.Y.; Pomerantsev, A. Trends in chemometrics: Food authentication, microbiology, and effects of processing. *Compr. Rev. Food Sci. Food Saf.* **2018**, 17(3), 663–677.

Halicioglu, O.; Astarcioglu, G.; Yaprak, I.; Aydinlioglu, H. Toxicity of *Salvia officinalis* in a newborn and a child: An alarming report. *Pediatr. Neurol.* **2011**, 45(4), 259–260.

Haloui, M.; Louedec, L.; Michel, J.B.; Lyoussi, B. Experimental diuretic effects of *Rosmarinus officinalis* and *Centaurium erythraea*. *J. Ethnopharmacol.* **2000**, 71(3), pp 465–472.

Hamidpour, R.; Hamidpour, S.; Hamidpour, M.; Shahlari, M.; Sohraby, M. Summer Savory: From the selection of traditional applications to the novel effect in relief, prevention, and treatment of a number of serious illnesses such as diabetes, cardiovascular disease, Alzheimer's disease, and cancer. *J Tradit. Complement. Med.* **2014**, 4, 140–144.

Harnafi, H.; Ramchoun, M.; Tits, M.; Wauters, J.N.; Frederich, M.; Angenot, L.; Aziz, M.; Alem, C.; Amrani, S. Phenolic acid-rich extract of sweet basil restores cholesterol and triglycerides metabolism in high fat diet-fed mice: A comparison with fenofibrate. *Biomed. Prev. Nutr.* **2013**, 3(4), 393–397.

Hashemi, S.M.B.; Nikmaram, N.; Esteghlal, S.; Khaneghah, A.M.; Niakousari, M.; Barba, F. J.; Roohinejad, S.; Koubaa, M. Efficiency of ohmic assisted hydrodistillation for the extraction of essential oil from oregano (*Origanum vulgare* subsp. *viride*) spices. *Innov. Food Sci. Emerg. Technol.* **2017**, 41, 172–178.

Heitz, A.; Carnat, A.; Fraisse, D.; Carnat, A.P.; Lamaison, J.L. Luteolin 3'-glucuronide, the major flavonoid from *Melissa officinalis subsp. officinalis*. *Fitoterapia* **2000**, 71(2), 201–202.

Hernández, M. D., Sotomayor, J. A., Hernández, A., Jordán, M. J. (2016). Rosemary (Rosmarinus officinalis L.) oils. In: Essential oils in Food Preservation, Flavor and Safety; Preedy, V. R., Ed.; Elsevier: London, 2016; pp 677–688.

Horiuchi, K.; Shiota, S., Kuroda, T.; Hatano, T.; Yoshida, T.; Tsuchiya, T. Potentiation of antimicrobial activity of aminoglycosides by carnosol from Salvia officinalis. *Biol. Pharm. Bull.* **2007**, 30, 287–290.

Huang, Q.; Yu, H.; Ru, Q. Bioavailability and delivery of nutraceuticals using nanotechnology. *J. Food Sci.* **2010**, 75(1), R50–R57.

Hyldgaard, M.; Mygind, T.; Meyer, R.L. Essential oils in food preservation: Mode of action, synergies, and interactions with food matrix components. *Front. Microbiol.* **2012**, 3(12), 1–24.

Iauk, L.; Lo Bue, A.M.; Milazzo, I.; Rapisarda, A.; Blandino, G. Antibacterial activity of medicinal plant extracts against periodontopathic bacteria. *Phytother. Res.* **2003**, 17(6), 599–604.

Jackson, C. C.; Paliyath, G. Functional foods and nutraceuticals. In: Functional Foods, Nutraceuticals, and Degenerative Disease Prevention; Paliyath, G., Bakovic, M., Shetty, K., Eds.; Wiley-Blackwell. John Wiley & Sons, Inc.: Chichester, UK, **2011**.

Jafari, F.; Ghavidel, F.; Zarshenas, M.M. (2016). A critical overview on the pharmacological and clinical aspects of popular Satureja species. *J. Acupunct. Meridian Stud.* 2016, 9(3), 118–127.

Jayasena, D.D.; Jo, C. Potential application of essential oils as natural antioxidants in meat and meat products: A review. *Food Rev. Int.* **2014**, 30, 71–90.

Ji, W.; Gong, B.Q. Hypolipidemic activity and mechanism of purified herbal extract of Salvia miltiorrhiza in hyperlipidemic rats. *J. Ethnopharmacol.* **2008**, 119(2), 291–298.

Kaleem, M.; Ahmad, A. (2018). Flavonoids as nutraceuticals. In: Therapeutic, Probiotic, and Unconventional Foods; Grumezescu, A., Holban, A. M., Eds.; Academic Press, Elsevier: London, United Kingdom, 2018; pp 137–155.

Kennedy, D.; Wake, G.; Savelev, S.; Tildesley, N.; Perry, E.; Wesnes, K.; Scholey, A. Modulation of mood and cognitive performance following acute administration of single doses of *Melissa officinalis* (lemon balm) with human CNS nicotinic and muscarinic receptor-binding properties. *Neuropsychopharmacology* **2003**, 28, 1871–1881.

Kennedy, D.O.; Little, W.; Scholey, A.B. Attenuation of laboratory-induced stress in humans after acute administration of *Melissa officinalis* (lemon Balm). *Psychosom. Med.* **2004**, 66, 607–613.

Khare, C.P. *Indian Medicinal Plants: An Illustrated Dictionary.* Springer Science & Business Media: Berlin, **2008**.

Khoury, M.; Stien, D.; Eparvier, V.; Ouaini, N.; El Beyrouthy, M. Report on the medicinal use of eleven Lamiaceae species in Lebanon and rationalization of their antimicrobial potential by examination of the chemical composition and antimicrobial activity of their essential oils. *Evid. Based Complement. Alternat. Med.* **2016**, ID 2547169.

Kishimoto, Y.; Tani, M.; Kondo, K. Pleiotropic preventive effects of dietary polyphenols in cardiovascular diseases. *Eur. J. Clin. Nutr.* **2013**, 67(5), 532–535.

Koubaa, M.; Barba, F.J.; Bursać Kovačević, D.; Putnik, P.; Santos, M.S.; Queirós, R.P.; Moreira, S.A.; Inácio, R.S.; Fidalgo, L.G.; Saraiva, J. A. Pulsed electric field processing of different fruit juices. In: Fruit Juices; Rajauria, G., Tiwari, B., Eds.; Academic Press: London, **2017**; pp 437–450.

Krishnan, K.R.; Babuskin, S.; Babu, P.A.S.; Fayidh, M.A.; Sabina, K.; Archana, G.; Sivarajan, M.; Sukumar, M. Bio protection and preservation of raw beef meat using pungent aromatic plant substances. *J. Sci. Food Agric.* **2014**, 94(12), 2456–2463.

Li, Q. X.; Chang, C. L. Basil (*Ocimum basilicum* L.) oils. In: Essential Oils in Food Preservation, Flavor and Safety; Preedy, V. R., Ed.; Elsevier: London, **2016**; pp 231–238.

Li, Y.; Fabiano-Tixier, A. S.; Chemat, F. Essential oils: From conventional to green extraction. In: Essential Oils as Reagents in Green Chemistry; Li, Y., Fabiano-Tixier, A. S., Chemat, F., Eds.; Springer International Publishing: Switzerland, **2014**; pp 9–20.

Loizzo, M.R.; Menichini, F.; Conforti, F.; Tundis, R.; Bonesi, M.; Saab, A.M.; Statti, G.A.; de Cindio, B.; Houghton, P.J.; Menichini, F.; Frega, N. G. Chemical analysis, antioxidant, antiinflammatory and anticholinesterase activities of *Origanum ehrenbergii* Boiss and *Origanum syriacum* L. essential oils. *Food Chem.* **2009**, 117(1), 174–180.

Lorenzo, J.M.; Khaneghah, A.M.; Gavahian, M.; Marszałek, K.; Eş, I.; Munekata, P.E.; Ferreira, I.C.F.R.; Barba, F.J. Understanding the potential benefits of thyme and their derived products on food industry and health: From extraction of high-added value compounds to the evaluation of bioaccessibility, bioavailability, anti-inflammatory, and antimicrobial activities. *Crit. Rev. Food Sci. Nutr.* **2018b**, 17, 1–17.

Lorenzo, J.M.; Munekata, P.; Putnik, P.; Bursać Kovačević, D.; Muchenje, V.; Barba, F.J. Sources, chemistry and biological potential of ellagitannins and ellagic acid derivatives. In: Studies in Natural Product Chemistry; Atta-ur-Rahman, Ed.; Elsevier: London, 2018a; Vol. 60.

Lorenzo, J.M.; Pateiro, M.; Domínguez, R.; Barba, F.J.; Putnik, P.; Kovačević, D.B.; Shpigelman, A.; Granato, A.; Franco, D. Berries extracts as natural antioxidants in meat products: A review. *Food Res. Int.* **2018c**, 106, 1095–1104.

Ložienė, K.; Venskutonis, P.R.; Šipailienė, A.; Labokas, J. Radical scavenging and antibacterial properties of the extracts from different *Thymus pulegioides L.* chemotypes. *Food Chem.* **2007**, 103(2), 546–559.

Luengo, E. *Alimentos Funcionales y Nutracéuticos*. Sociedad Española de Cardiología, 2007.

Mahajan, N.; Rawal, S.; Verma, M.; Poddar, M.; Alok, S. A phytopharmacological overview on Ocimum species with special emphasis on Ocimum sanctum. *Biomed. Prev. Nutr.* **2013**, 3(2), 185–192.

Mahmoud, B.S.; Yamazaki, K.; Miyashita, K.; Il-Shik, S.; Dong-Suk, C.; Suzuki, T. Bacterial microflora of carp (*Cyprinus carpio*) and its shelf-life extension by essential oil compounds. *Food Microbiol.* **2004**, 21 (6), 657–666.

Marino, M.; Bersani, C.; Comi, G. Impedance measurements to study the antimicrobial activity of essential oils from *Lamiaceae* and *Compositae*. *Int. J. Food Microbiol.* **2001**, 67(3), 187–195.

Mašković, P.; Veličković, V.; Mitić, M.; Đurović, S.; Zeković, Z.; Radojković, M.; Cvetanović, A.; Švarc-Gajić, J.; Vujić, J. Summer savory extracts prepared by novel extraction methods resulted in enhanced biological activity. *Ind. Crop. Prod.* **2017**, 109, 875–881.

Mazid M.; Khan T.A.; Mohammad F. Role of secondary metabolites in defense mechanisms of plants. *Biol. Med.* **2011**, 3(2), 232–249.

Mihajilov-Krstev, T.; Radnović, D.; Kitić, D.; Zlatković, B.; Ristić, M.; Branković, S. Chemical composition and antimicrobial activity of *Satureja hortensis L.* essential oil. *Open Life Sci.* **2009**, 4, 411–416.

Mondal, S.; Varma, S.; Bamola V.D.; Naik, S.N., Mirdha, B.R., Padhi, M.M., … Mahapatra, S.C. Double-blinded randomized controlled trial for immunomodulatory effects of Tulsi (*Ocimum sanctum Linn.*) leaf extract on healthy volunteers. *J. Ethnopharmacol.* **2011**, 136, 452–456.

Montesano, D.; Rocchetti, G.; Putnik, P.; Lucini, L. Bioactive profile of pumpkin: an overview on terpenoids and their health-promoting properties. *Curr. Opin. Food Sci.* **2018**, 22, 81–87.

Mossa, A.T.H.; Nawwar, G.A.M. Free radical scavenging and antiacetylcholinesterase activities of *Origanum majorana* L. essential oil. *Hum. Exp. Toxicol.* **2011**, 30(10), 1501–1513.

Mustafa, R.A.; Hamid, A.A.; Mohamed, S.; Abubakar, F.A. Total phenolic compounds, flavonoids, and radical scavenging activity of 21 selected tropical plants. *J. Food Sci.* **2010**, 75(1), 28–35.

Naghibi, F.; Mosaddegh, M.; Motamed, M.M.; Ghorbani, A. Labiatae family in folk medicine in Iran: From ethnobotany to pharmacology. *Iran. J. Pharm. Res.* **2005**, 2, 63–79.

Nikolić, M.; Glamočlija, J.; Ferreira, I.C.F.R.; Calhelha, R.C.; Fernandes, Â.; Marković, T.; Marković, D.; Giweli, A.; Soković, M. Chemical composition, antimicrobial, antioxidant and antitumor activity of *Thymus serpyllum L.*, *Thymus algeriensis* Boiss. and Reut and *Thymus vulgaris L.* essential oils. *Ind. Crop. Prod.* **2014**, 52, 183–190.

O'Grady, M. N.; Kerry, J. P. Using antioxidants and nutraceuticals as dietary supplements to improve the quality and shelf-life of fresh meat. In: Improving the Sensory and Nutritional Quality of Fresh Meat; Kerry, J. P., Ledward, D., Eds.; CRC Press: Cambridge, England, **2009**; pp 372–377.

Okiemy-Andissa, N.; Miguel, M.L.; Etou, A.W.; Ouamba, J.M.; Gbeassor, M.; Abena, A.A. Analgesic effect of aqueous and hydroalcoholic extracts of three congolese medicinal plants: Hyptis suavolens, Nauclea latifolia and Ocimum gratissimum. *Pak. J. Biol. Sci.* **2004**, 7(9), 1613–1615.

Ortega-Ramirez, L.A.; Rodriguez-Garcia, I.; Silva-Espinoza, B.A.; Ayala-Zavala, J.F. Oregano (*Origanum* spp.) Oils. In: Essential Oils in Food Preservation, Flavor and Safety; Preedy, V. R., Ed.; Elsevier: London, **2016**; pp 625–630.

Ouattara, B.; Simard, R.E.; Holley, R.A.; Piette, G.J.P.; Bégin, A. Antibacterial activity of selected fatty acids and essential oils against six meat spoilage organisms. *Int. J. Food Microbiol.* **1997**, 37(2–3), 155–162.

Ozarowski, M.; Mikolajczak, P.L.; Bogacz, A.; Gryszczynska, A.; Kujawska, M.; Jodynis-Liebert, J.; … Bartkowiak-Wieczorek, J. *Rosmarinus officinalis L.* leaf extract improves memory impairment and affects acetylcholinesterase and butyrylcholinesterase activities in rat brain. *Fitoterapia.* **2013**, 91, 261–271.

Pandey, A.K.; Singh, P.; Tripathi, N.N. Chemistry and bioactivities of essential oils of some *Ocimum* species: an overview. *Asian Pac. J. Trop. Biomed.* **2014**, 4(9), 682–694.

Pateiro, M.; Barba, F.J.; Domínguez, R.; Sant'Ana, A.S.; Khaneghah, A.M.; Gavahian, M.; Gómez, B.; Lorenzo, J.M. Essential oils as natural additives to prevent oxidation reactions in meat and meat products: A review. *Food Res. Int.* **2018**, 113, 156–166.

Pattanayak, P.; Behera, P.; Das, D.; Panda, S.K. *Ocimum sanctum* Linn. A reservoir plant for therapeutic applications: An overview. *Pharmacogn Rev.* **2010**, 4(7), 95–105.

Pedersen, J.A. Distribution and taxonomic implications of some phenolics in the family Lamiaceae determined by ESR spectroscopy. *Biochem. Syst. Ecol.* **2000**, 28(3), 229–253.

Pinto, E.; Pina-Vaz, C.; Salgueiro, L.; Gonçalves, M.J.; Costa-de-Oliveira, S.; Cavaleiro, C.; Palmeira, A.; Rodrigues, A.; Martinez-de-Oliveira, J. Antifungal activity of the essential oil of *Thymus pulegioides* on Candida, Aspergillus and Dermatophyte species. *J. Med. Microbiol.* **2006**, 55(10), 1367–1373.

Pistelli, L.; Najar, B.; Giovanelli, S.; Lorenzini, L.; Tavarini, S.; Angelini, L.G. Agronomic and phytochemical evaluation of lavandin and lavender cultivars cultivated in the Tyrrhenian area of Tuscany (Italy). *Ind. Crop. Prod.* **2017**, 109, 37–44.

Poojary, M.M.; Putnik, P.; Kovačević, D.B.; Barba, F.J.; Lorenzo, J.M.; Dias, D.A., Shpigelman, A. Stability and extraction of bioactive sulfur compounds from *Allium* genus processed by traditional and innovative technologies. *J. Food Compos. Anal.* **2017**, 61, 28–39.

Pop, A.V.; Tofană, M.; Socaci, S.A.; Nagy, M.; Fărcaș, A.; Borș, M.D.; Salanță, L.; Feier, D.; Vârva, L. Comparative study regarding the chemical composition of essential oils of some *Salvia* species. *Hop Med. Plants* **2014**, 22(1/2), 79–91.

Porres-Martínez, M.; Carretero-Accame, M.E.; Gómez-Serranillos, M.P. Pharmacological activity of *Salvia lavandulifolia* and chemical components of its essential oil. A review. *Lazaroa* **2013**, 34, 237–254.

Prakash, J.; Gupta, S.K. Chemopreventive activity of *Ocimum sanctum* seed oil. *J. Ethnopharmacol.* **2000**, 72(1–2), 29–34.

Prakash, O.; Chandra, M.; Pant, A.K.; Rawat, D.S. Mint (*Mentha spicata* L.) Oils. In: Essential Oils in Food Preservation, Flavor and Safety; Preedy, V. R., Ed.; Elsevier: London, **2016**; pp 561–572.

Putnik, P.; Barba, F.J.; Lorenzo, J.M.; Gabrić, D.; Shpigelman, A.; Cravotto, G.; Bursać Kovačević, D. An integrated approach to mandarin processing: food safety and nutritional quality, consumer preference and nutrient bioaccessibility. *Compr. Rev. Food Sci. Food Saf.* **2017b**, 16(6), 1345–1358.

Putnik, P.; Bursać Kovačevć, D. Fresh-cut apples spoilage and predictive microbial growth under modified atmosphere packaging. In: Food Safety and Protection; Rai, R.; Aswathanarayan, J.B., Eds.; CRC Press: Boca Raton, FL, **2017**; p 728.

Putnik, P.; Bursać Kovačević, D.; Herceg, K.; Levaj, B. Influence of respiration on predictive microbial growth of *Aerobic mesophilic* bacteria and *Enterobacteriaceae* in fresh-cut apples packaged under modified atmosphere. *J. Food Safety* **2016**, 37(1), e12284.

Putnik, P.; Bursać Kovačević, D.; Režek Jambrak, A.; Barba, F.J.; Cravotto, G.; Binello, A.; Lorenzo, J.M.; Shpigelman, A. Innovative "green" and novel strategies for the extraction of bioactive added value compounds from citrus wastes—A review. *Molecules* **2017a**, 22(5), 680.

Putnik, P.; Gabric, D.; Roohinejad, S.; Barba, F.J.; Granato, D.; Mallikarjunan, K.; Lorenzo, J.M.; Bursać Kovačević, D. An overview of organosulfur compounds from *Allium* spp.: From processing and preservation to evaluation of their bioavailability, antimicrobial, and anti-inflammatory properties. *Food Chem.* **2019**, 276, 680–691.

Putnik, P.; Lorenzo, J.M.; Barba, F.J.; Roohinejad, S.; Režek Jambrak, A.; Granato, D.; Montesano, D.; Bursać Kovačević, D. Novel food processing and extraction technologies of high-added value compounds from plant materials. *Foods* **2018**, 7(7), 1–16.

Qadir, M.I.; Parveen, A.; Abbas, K.; Ali, M. Analgesic, anti-inflammatory and anti-pyretic activities of *Thymus linearis. Pak. J. Pharm. Sci.* **2016**, 29(2), 591–594.

Regulation EC No 178/2002 of the European Parliament and of the Council of 28 January 2002 laying down the general principles and requirements of food law, establishing the European Food Safety Authority and laying down procedures in matters of food safety. *Off. J. Eur. Union* 2002, L31, 1–24.

Regulation EC No 1924/2006 of the European Parliament and of the Council of 20 December 2006 on nutrition and health claims made on foods. *Off. J. Eur. Union* 2006, L404, 9–25.

Rodríguez-Meizoso, I.; Marin, F.R.; Herrero, M.; Señorans, F.J.; Reglero, G.; Cifuentes, A.; Ibáñez, E. Subcritical water extraction of nutraceuticals with antioxidant activity from oregano. Chemical and functional characterization. *J. Pharm. Biomed.* **2006**, 41(5), 1560–1565.

Rodrigues, M.R.A.; Kanazawa, L.K.S.; das Neves, T.L.M.; da Silva, C.F.; Horst, H.; Pizzolatti, M.G.; … de Paula Werner, M.F. Antinociceptive and anti-inflammatory potential of extract and isolated compounds from the leaves of *Salvia officinalis* in mice. *J. Ethnopharmacol.* **2012**, 139, 519–526.

Salmalian, H.; Saghebi, R.; Moghadamnia, A.A.; Bijani, A.; Faramarzi, M.; Amiri, F.N.; Bakouei, F.; Behmanesh, F.; Bekhradi, R. Comparative effect of *Thymus vulgaris* and ibuprofen on primary dysmenorrhea: a triple-blind clinical study. *Caspian J. Int. Med.* **2014**, 5(2), 82–88.

Schaffer, S.; Eckert, G.P.; Müller, W.E.; Llorach, R.; Rivera, D.; Grande, S.; Galli, C.; Visioli, F. Hypochlorous acid scavenging properties of local Mediterranean plant foods. *Lipids* **2004**, 39(12), 1239–1247.

Scholey, A.; Gibbs, A.; Neale, C.; Perry, N.; Ossoukhova, A.; Bilog, V.; Kras, M.; Scholz, C.; Sass, M.; Buchwald-Werner, S. Anti-stress effects of lemon balm-containing foods. *Nutrients* **2014**, 6(11), 4805–4821.

Senatore, F. Influence of harvesting time on yield and composition of the essential oil of a thyme (*Thymus pulegioides* L.) growing wild in Campania (Southern Italy). *J. Agric. Food Chem.* **1996**, 44(5), 1327–1332.

Shah, M.A.; Bosco, S.J.D.; Mir, S.A. Plant extracts as natural antioxidants in meat and meat products. *Meat Sci.* **2014**, 98(1), 21–33.

Shakeri, A.; Sahebkar, A.; Javadi, B. *Melissa officinalis L.*—A review of its traditional uses, phytochemistry and pharmacology. *J. Ethnopharmacol.* **2016**, 188, 204–228.

Shen, D.; Pan, M.H.; Wu, Q.L.; Park, C.H.; Juliani, H.R.; Ho, C.T.; Simon, J.E. LC-MS method for the simultaneous quantitation of the anti-inflammatory constituents in oregano (*Origanum* species). *J. Agric. Food Chem.* **2010**, 58(12), 7119–7125.

Siebert, K.J.; Troukhanova, N.V.; Lynn, P.Y. Nature of polyphenol–protein interactions. *J. Agric. Food Chem.* **1996**, 44(1), 80–85.

Silva, F.V.; Martins, A.; Salta, J.; Neng, N.R.; Nogueira, J.M.; Mira, D.; … Palavra, A.M. (2009). Phytochemical profile and anticholinesterase and antimicrobial activities of supercritical versus conventional extracts of *Satureja montana*. *J. Agric. Food Chem.* 2009, 57(24), 11557–11563.

Singh, D.; Chaudhuri, P.K. A review on phytochemical and pharmacological properties of Holy basil (*Ocimum sanctum L.*). *Ind. Crop. Prod.* **2018**, 118, 367–382.

Stagos, D.; Portesis, N.; Spanou, C.; Mossialos, D.; Aligiannis, N.; Chaita, E.; Panagoulis, C.; Reri, E.; Skaltsounis, L.; Tsatsakis, A.M.; Kouretas, D. Correlation of total polyphenolic content with antioxidant and antibacterial activity of 24 extracts from Greek domestic Lamiaceae species. *Food Chem. Toxicol.* **2012**, 50(11), 4115–4124.

Stalińska, K.; Guzdek, A.; Rokicki, M.; Koj, A. Transcription factors as targets of the anti-inflammatory treatment. A cell culture study with extracts from some Mediterranean diet plants. *J. Physiol. Pharmacol.* **2005**, 56(1), 157–169.

Szentmihályi, K.; Csedo, C.; Then, M. Comparative study on tannins, flavonoids, terpenes and mineral elements of some *Salvia* species. *Acta Hortic.* **2004**, 629, 463–470.

Talavera-Alemán, A.; Rodríguez-García, G.; López, Y.; García-Gutiérrez, H.A.; Torres-Valencia, J.M.; Rosa, E.; … Gómez-Hurtado, M.A. Systematic evaluation of thymol derivatives possessing stereogenic or prostereogenic centers. *Phytochem. Rev.* **2016**, 10, 9412–9416.

Tapsell, L.C.; Hemphill, I.; Cobiac, L.; Sullivan, D.R.; Fenech, M.; Patch, C.S.; … Fazio, V.A. Health benefits of herbs and spices: the past, the present, the future. *Med. J.* **2006**, 21, 185–192.

Tavakolpour, Y.; Moosavi-Nasab, M.; Niakousari, M.; Haghighi-Manesh, S.; Hashemi, S.M. B., Mousavi Khaneghah, A. Comparison of four extraction methods for essential oil from Thymus daenensis subsp. lancifolius and chemical analysis of extracted essential oil. *J. Food Process. Preserv.* **2017**, 41(4).

Teixeira, B.; Marques, A.; Ramos, C.; Serrano, C.; Matos, O.; Neng, N.R.; Nogueira, J.M.F.; Saraiva, J.A.; Nunes, M.L. Chemical composition and bioactivity of different oregano (*Origanum vulgare*) extracts and essential oil. J. *Sci. Food Agric.* **2013**, 93(11), 2707–2714.

Tepe, B.; Cilkiz, M. A pharmacological and phytochemical overview on *Satureja*. *Pharm. Biol.* **2016**, 54(3), 375–412.

Tongnuanchan, P.; Bejakul, S. Essential oils: extraction, bioactivities, and their uses for food preservation. *J. Food Sci.* **2014**, 79, R1231–R1249.

Uritu, C.M.; Mihai, C.T.; Stanciu, G.D.; Dodi, G.; Alexa-Stratulat, T.; Luca, A.; … Tamba, B.I. Medicinal plants of the family *Lamiaceae* in pain therapy: A review. *Pain Res. Manag.* **2018**, ID 7801543, 1–44.

Van Wyk, B.E.; Wink, M. *Medicinal Plants of the World*; Timber Press: Portland, OR, USA, **2004**.

Van Wyk, B.E.; Wink, M. *Phytomedicines, Herbal Drugs and Poisons*; Kew Publishing, Chicago University Press: Chicago, USA, **2015**.

Venditti, A.; Frezza, C.; Celona, D.; Sciubba, F.; Foddai, S.; Delfini, M.; Serafini, M.; Bianco, A. Phytochemical comparison with quantitative analysis between two flower phenotypes of *Mentha aquatica L.*: Pink-violet and white. *AIMS Mol. Sci.* **2017**, 4(3), 288–300.

Vinceković, M.; Viskić, M.; Jurić, S.; Giacometti, J.; Bursać Kovačević, D.; Putnik, P.; Donsì, F.; Barba, F.J.; Režek Jambrak, A. Innovative technologies for encapsulation of Mediterranean plants extracts. *Trends Food Sci. Technol.* **2017**, 69, 1–12.

Vitalini, S.; Iriti, M.; Puricelli, C.; Ciuchi, D.; Segale, A.; Fico, G. Traditional knowledge on medicinal and food plants used in Val San Giacomo (Sondrio, Italy)—An alpine ethnobotanical study. *J. Ethnopharmacol.* **2013**, 145(2), 517–529.

Viuda-Martos, M.; Pérez-Álvarez, J.A.; Fernández-López, J. Laminaceae herbs: A potential ingredient to functional foods. In: Functional foods, nutraceuticals and natural products. Concepts and applications; Vattem, D.A, Maitin, V., Eds.; DESetch Publications, Inc.: Pennsylvania, **2016**; p 303–328.

Vujoššević, M.; Blagojević, J. Antimutagenic effects of extracts from sage (*Salvia officinalis*) in mammalian system in vivo. *Acta Vet. Hung.* **2004**, 52, 439–443.

Wink, M. Molecular modes of action of drugs used in phytomedicine. In: Herbal Medicines: Development and Validation of Plant-derived Medicines for Human Health; Bagetta, G., Cosentino, M., Corasaniti, M.T., Sakurada, S., Eds.; Taylor & Francis: London, UK, **2012**; pp. 161–172.

Wink, M. Modes of action of herbal medicines and plant secondary metabolites. *Medicines* **2015**, 2(3), 251–286.

Yamamoto, J.; Yamada, K.; Naemura, A.; Yamashita, T.; Arai, R. Testing various herbs for antithrombotic effect. *Nutrition* **2005**, 21(5), pp 580–587.

Yanishlieva-Maslarova, N. V. Inhibiting oxidation. In: Antioxidants in food. Practical applications; Pokorny, J., Yanishlieva, N., Gordon, M., Eds.; CRC Press: Cambridge, UK, **2001**; Vol. 1; pp 22–70.

Zheng, W.; Wang, S.Y. Antioxidant activity and phenolic compounds in selected herbs. *J. Agric. Food Chem.* **2001**, 49(11), 5165–5170.

Zivkovic, J.; Mujic, I.; Zekovic, Z.; Vidovic, S.; Mujic, A.; Jokic, S. Radical scavenging, antimicrobial activity and phenolic content of *Castanea sativa* extract. *J. Centr. Eur. Agric.* **2009**, 10, 2175–182.

Žuntar, I.; Krivohlavek, A.; Kosić-Vukšić, J.; Granato, D.; Bursać Kovačević, D.; Putnik, P. Pharmacological and toxicological health risk of food (herbal) supplements adulterated with erectile dysfunction medications. *Curr. Opin. Food Sci.* **2018**, 24, 9–15.

WHO. WHO monographs on selected medicinal plants. Essential Medicines and Health Products Information Portal. A World Health Organization resource. http://apps.who.int/medicinedocs/en/d/Js2200e/ (accessed Oct 31, 2018).

CHAPTER 9

Brief Overview of Development and Characterization of Herbal Products

ANDREW G. MTEWA[1,2*], LUCRECE AHOVEGBE[2,3], TAMIRAT BEKELE[2,4], IBRAHIM CHIKOWE[5], and FANUEL LAMPIAO[6]

[1]*Department of Chemistry, Malawi Institute of Technology, Malawi University of Science and Technology, P.O. Box 5196, Limbe, Malawi*

[2]*Pharmbiotechnology and Traditional Medicine Center of Excellence, School of Medicine, Mbarara University of Science and Technology, P.O. Box 1410, Mbarara, Uganda*

[3]*Unite de Formation et de Recherché en Pharmacie, Faculté des Sciences de la Santé, Université d'abomey-Calavi, Cotonou, Benin*

[4]*Department of Pharmacy, Ambo University, P.O. Box 19, Ambo, Ethiopia*

[5]*Department of Pharmacy, College of Medicine, University of Malawi, P/Bag 360, Chichiri, Blantyre 3, Malawi*

[6]*Africa Centre of Excellency in Public health and Herbal Medicine, College of Medicine, University of Malawi, P/Bag 360, Chichiri, Blantyre 3, Malawi*

*Corresponding author. E-mails: amtewa@must.ac.mw; andrewmtewa@yahoo.com

ABSTRACT

Herbal product development and characterization involve robust quality assurance systems and commitment. These include document and record control for all stages of development, from raw materials to the market. Various qualification tests conducted include phytochemical, bioactivity, and toxicity tests against preset standard specifications. However, large-scale production of herbal products remains a dream even after centuries of use and years or research on the same. More of these products are found with no controls, fewer regulations, and no scientific evidence of claims they carry along. Their characterization is poor largely due to lack of appropriate documentation of tests, protocols, as well as systems A large pool of herbal products is available out there, but no strong impact is felt to the point that herbal products still do not reach the prescription stage. The culture and unmethodical approach in herbal product development is found wanting in many ways, which challenges its legitimacy over the years and make it fail to grow into huge investment operations as are conventional pharmaceuticals product developments. There is a need to modify the approach taken in the practice and focus on quality and safety (Q&S) systems, as well as regulatory standard development, which will start strengthening the legitimacy of these products, which is already there by not shown off. This work has outlined basic steps for the characterization of herbal products in regard to product development. Now is time that investors relook at this potential health area and move from mere phytochemical screening to huge industrialization of herbal products. This requires a meaningful collaboration among stakeholders.

9.1 INTRODUCTION

A herbal product refers to a makeup from a plant, part(s) thereof, and/or in combination with other ingredients intended for use due to its purposeful favorable scent, flavor, medicinal properties, or a combination of these attributes. A herbal medicinal product is a herbal product that contains dietary supplementations consumed for well-being purposes such as weight loss, physical appearance modification, improved health status, and healing process initiation and its enhancement. An example of a herbal medicinal product is *Eichhornia crassipes* (Water hyacinth; *Namasipuni*) powders and extracts which are known to have dietary minerals, antimicrobial, and anticancer properties (Mtewa et al., 2018).

Herbal product development and characteristics are a set of processes undertaken in the designing, standardization, manufacturing, and valida-tion of a herbal product's readiness for the market. The development of a botanical drug involves the collection and authentication of both materials and tools used at every step of the process. These amount to the evaluation of botanical, phytochemical, pharmacological, and pharmacognostic aspects of a botanical drug product and its standardization (Lee, 2002; Sardana, 2012).

9.2 DEVELOPMENT STRATEGY

Herbal product development strategies can take various faces. One most common is to develop strategies based on the intended market level and size. Some designs cover only products that are used directly as natural herbs without formulation, without an organized consistent packaging, and usually without or with minimal marketing. These involve domestic operations by a few or a single practitioner, eying a small base of distribution. The other design covers well-formulated and packed herbal products that are processed at a higher scale and are intended for sale in a much wider market. Such designs are intended to attract the market with regard to the core competence of the facility, and all personnel work at every step of the product development process to keep improving the portfolio of the product.

9.2.1 PRODUCT FORMULATION

Formulations need to cover product formulas that are in line with the designed strategy and efficiently support product claims, and also control all product-related costs. Herbal products are prepared and sold in various formulations. These depend on different reasons influenced by culture for domestic operations and the design strategy for packed products to be sold in a wider market (Parwardhan and Mashelkar, 2009).

9.2.1.1 CULTURAL BELIEF-INFLUENCED FORMULATIONS

Culturally, some societies believe that keeping a herbal medicinal formulation in aqueous forms as practiced by early days' practitioners enhances the healing effects. Others believe that herbal medicines are only supposed to be in powder or in their natural forms, for example. Roots, bark, and leaves to be as fresh as they naturally are. Usually, cultural belief-influenced formulations are low-scale productions limited in terms of both capital investment and market base.

9.2.1.2 MARKETED FORMULATIONS

Herbal products at a professional level require some higher level of experience, professional education qualifications, and technical know-how in one or more areas of herbal medicine processing. The team make-up is usually composed of a series of stages in the manufacturing process as well as expertise in various aspects of the science behind manufacturing. Professional herbal product formulation is a venture that demands a relatively higher capital investment. This is where a more or less organized facility is maintained with all regulatory requirements being followed. Depending on the product and market strategy being followed, herbal products can be made in various forms under the professional formulation category, including capsules, tablets, solvent and excipient extracts, dried plant parts (leaves, fruits, or whole plants), teas and tinctures, among other forms.

9.3   MANUFACTURING PROCESS

The manufacturing process needs to be clearly laid down in an operation manual where every room, material coming in and out, personnel entering the facility, product safety, personnel safety, material suppliers, and every bit of activity shall be detailed with justifications, agreed upon by a facility's management team and approved by a quality control leader (Mtewa and Mtewa, 2017). This forms a basis of formulating a sound and robust quality assurance system that guarantees the wholeness of the products coming from the facility at any stage of processing. Some of the critical stages of the processing are shown in Figure 9.1.

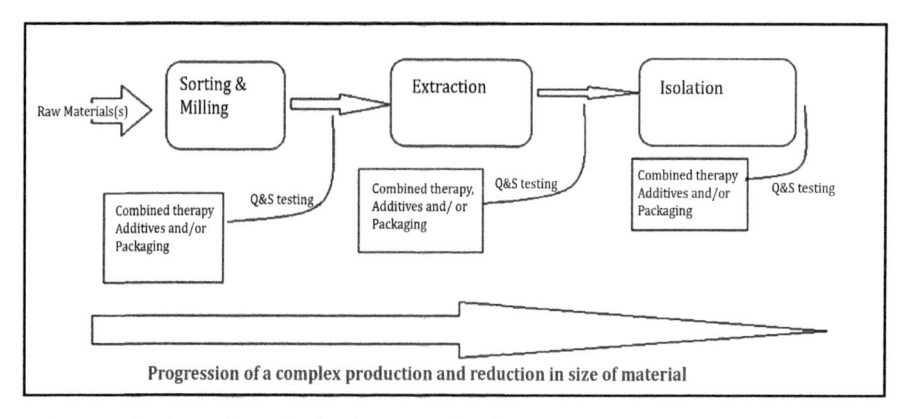

FIGURE 9.1   An outline of a simple to complex formulation development.

On the production flow, there are several products or by-product exit points, which can still serve as end points of a product themselves depending on the product development strategy. The critical point is that there needs to be clear Q&S testing at all stages of production, starting from raw materials themselves.

9.4 EXTRACTS AND THEIR ISOLATES

Solvent extracts have been in use for a long time as medicinal products. Solvents are used basically to extract molecular groups of the same or close polarity. With time and advancement in knowledge and technology, scientists have been going deeper into isolating individual compounds from various active extracts for improved and customized drug-like properties and bio-activities. Some individual molecules, including reserpine, caffeine, artemisinin, and quinine, have so far been isolated and standardized for medicinal use, replacing their respective extracts that were in use for the same purpose before (Sardana, 2012). In more recent times, some scientists have come back to combined effects of various molecules for better results. Research on combined effects shows potential improvement in activities, mainly against drug-resistant strains of microbials.

9.5 QUALITY ASSURANCE SYSTEM

Packed products often do not inform about their quality status. Mostly, the packaging outlook gives the first picture of quality the inside product may have, but it is oftentimes misleading. There are three stages at which the quality of a packed herbal product can be determined with certainty. These are the manufacturer, raw material, and packaging materials' supplier and finished product handlers. Below are some of the basic tenets for a good quality assurance system.

9.5.1 PRODUCT SPECIFICATIONS

Product specifications are test result limits that a product should meet. They may be a single-point digit for strict and realistically achievable results, but usually they come in bands of minimum to maximum allowable limits, with a fixed-point value as a target. The upper and lower limits provide for variations in the operating procedures, errors, reagent properties, and environmental variations during testing. However, they always need to be toward the target figure. The standard deviation of the test results indicates how good generally the results are in regard to the set target specification. For a processing plant, it is always good to always show continuous online results on a board with trend lines that float over these three lines on regular intervals in order to timely inform the production team how they are performing in terms of the target results.

Product specifications may be provided by specific customers, but for drugs and drug products, they are usually framed upon guidelines by the WHO. Internal facility targets can be set, which are mostly more stringent than the WHO, to avoid producing products that are out of the more flexible WHO specifications.

Product specifications are to do with all attributes that are necessary for quality, safety, customer requirements, and regulatory compliance. There needs to be specifications for materials, tools, processes, practices, and documents in the right manner.

9.5.1.1   MATERIALS

For materials, product raw materials, packaging materials, in-process materials, and finished products need to have their specific expected outcomes clearly spelt out. For example, a facility may require that all packaging materials incoming from suppliers should be white in color and always register absence for *enterobacteriaceae* on swab. This means that any packaging material that registers above zero *enterobactericeae* on swab condemns the whole or part of the whole consignment that was supplied as potentially contaminated. If the production area has a specification on raw materials that every batch of raw materials coming in should pass through a 0.5 in mesh, they should reject anything that is provided, but remaining on top of a standard 0.5 in mesh. Failing to meet specifications has penalties that may be as tough as fines, suspensions, closure of business, and ultimately, imprisonment (Wu, 2017).

9.5.1.2   PROCESSING

The processing flow should have its own guidelines clearly stated to all workers and visitors. Basic, but most important of all, is hygienic practices for drug safety. There is need to make sure that there is not potential cross-contamination from the outside environment in to the processing facility.

Care needs to be taken right from how the facility was constructed, aeration, handlers' health status (e.g., no open wounds and sneezing around), and handlers' clothing, among others. The facility needs to have appropriate dressing and hand-sanitizing points (Mtewa and Mtewa, 2017).

Critical control points in the facility need to be well identified and controlled even in the operations manual. At every regular interval, the quality control team should collect online samples for various tests in order to guide the production team as required.

9.5.1.3 FINISHED PRODUCT

Finished products need to be tested, this time with the aim of confirming the total Q&S of the product. This is a step which vindicates the processing facility's vigilance on assuring the Q&S status of the product. Each test being conducted needs to be as set in the facility's finished product sampling as well as testing protocols. The product should be tested using validated methods and positive controls as approved by authorizing official(s). Finally, the test results should be compared with finished product specifications in detail. No compromise should be made in terms of quality. It becomes a serious ethical and sometimes criminal issue when test results are compromised for a medicinal product, putting lives of many potential consumers at risks.

9.5.1.4 CUSTOMER ISSUES

Finished products that come from herbal sources face challenges in terms of consistency (Qusaj et al., 2012). For extract matrices as finished products, there is a need to standardize the formulation in terms of specific markers from every batch of raw materials coming in. Where some components in the extracts are too less or too much, there needs to be a clear protocol on the blending procedures for the facility since the product strategy already spells out what group of compounds the facility is interested in.

Where the finished herbal product is a "hit molecule," there are a lot of variations in the properties of active herbal molecules, possibly due to variations in plant growth environmental conditions, laboratory techniques and tools, as well as personalities conducting the analyses and isolation procedures. This calls for an intervention to standardize active molecules that get isolated from one herbal product to another, on a case-by-case basis. This can work best if the molecules are multiplied by means of laboratory synthesis or modified for a more stable yet more active pro-drug molecule through semisynthesis. Stable active pro-drugs can be studied for their drug-like properties from where appropriate conditions for the maintenance of "hit-molecule" status can be determined.

Labeling of a finished product is important for regulatory compliance, product safety assurance, and business profile. Basically, the labels should be clear and in a language understood by the target community. Labels should include the ingredients and any other declarations such as about allergens in the products. Regulatory approval stamps, product shelf life, and, of course, manufacturer's name and address are essential for consumers and regulators.

9.6 PRODUCT CHARACTERISTICS

The product and everything about it, from the raw material stage, need to be tested, characterized, and documented. Following are some of the most fundamental characteristics required for herbal products.

9.6.1 RAW MATERIALS

Before Q&S testing, raw materials should be well documented, starting from its botanic names, all synonyms and person who first discovered the plant for herbarium storage. The morphology of the raw materials should be well documented for easy physical description. Locations that are best known to grow the raw materials and those known to be using the same for various medicinal purposes should also be documented. Raw material suppliers and/or location from where the specific batch was collected need to be documented as well. All this information helps in the traceability of the materials in case a repeat of a specific formulation is required to be avoided or recommended, as well as for the purpose of drug substance consistency in final batch products.

9.6.1.1 ACTIVE MEDICINAL MATERIAL

Active medicinal material in raw materials needs to be both qualitatively and quantitatively analyzed at a preliminary level to determine basic phytochemical content of the material. At a more advanced stage, specific phytochemical groups can be further analyzed to identify with particular bio-activities, for example, to check whether a plant material is more anticancer or antiviral. As guided by the product development strategy, much more advanced stages can be undertaken to isolate specific individual active molecules responsible for some desired bio-activities. The levels of activities, usually identified by IC_{50}, will decide whether individual active compounds would be good enough for the said bio-activity or a particular strategic combination of molecules would be better. A robust study of individual molecules is imperative as it would point toward the usability of the drug-lead molecules as actual drugs in the human body. The molecules will be checked if they can be well absorbed, distributed, metabolized, and excreted from the body on time (Kerns and Di, 2008). Without such properties being optimal, the active drug molecules would be useless to the body and only render the whole investment to this point, a waste of efforts, energy, time, and money.

9.6.2　ANIMAL TESTS FOR SAFETY

At various stages of production, the drug material should be checked for safety on appropriate animal models way before opting for use on humans. Sometimes, a history of use of the herbal product in various communities gives confidence of safety. Variations in the dosing formulation, whether single molecule isolates or fractions are being tested, at present as opposed to crude materials being used in history should be noted. The Food and Drug Administration (FDA), for example, strictly disregards the history of use when it comes to phase III clinical trials. Another tricky part with over-reliance on the history of use is that no one would know "for sure" whether particular adverse effects (AEs) were a result of consuming a particular herbal product or other different sources all together since there had never been any clear controlled trials on the same. AEs like carcinogenicity and genotoxicity would be too complicated to measure from the history of use. Perhaps the story would change if data were available for controlled clinical trials (Chow and Pong, 2015).

9.6.3　PHARMACODYNAMICS

Nonclinical and later on clinical assessments of safety for herbal drugs need to focus on the interactions that are there between animal models and drug doses. Pharmacokinetic studies of the drug molecules assist in the interpretation of observations from clinical trial for test drugs. For isolated drug molecules, structural properties of the molecules need to be exhaustedly studied by means of testing, using the best equipments, techniques, and personnel available for the studies (FDA, 2016; Wu, 2017). Once all stages are approved for human clinical trials, the test drug needs to be tested on various categories of people, including the aged, children, hepatic challenged, those with renal conditions, races, among other variations, in order to collect enough data on whom the drug works best and on whom they show AEs so that further modifications and/or prescription limitations be made based on evidence (Patwardhan et al., 2008).

9.6.4   *TOXICITY STUDIES*

Toxicity studies should encompass all possible protocols for a test drug, particularly if the test drug is to get into phase III of clinical trials. The FDA recommends that the maximum dose used in the test should, at best, show some AEs on animal models to give a good picture of the best dose formulation for a drug for humans (FDA, 2016; Li, 2002). Apart from general toxicology, reproductive toxicology and carcinogenicity and genotoxicity studies should form part of the toxicity studies for a test drug.

All chemistry and pharmacology data obtained should be well documented, and secured, with all the documents and records control procedures being adhered to accordingly. These form the basis for assuring that everything is being done as planned and according to regulatory requirements; in doing so, product development guarantees safe and quality herbal products to consumers.

9.7   CHALLENGES IN THE ADVANCEMENT OF HERBAL PRODUCT DEVELOPMENT

Although herbal medicine and traditional medicines have been in practice for many centuries, they are failing to break the ceiling of taking position as the first line of medication or practice for medical care. The world is still at a stage where herbal medicines are finding it very hard to make it to the level of prescription drugs by physicians (Sardana, 2012). This is partly due to "politics in the pharmaceutical industry," "geopolitics," and challenged credibility of herbal medicine products and their practices on many fronts. It is very hard for some to believe that traditional and herbal medicines can stand alone and so effectively. It is believed that the status of herbal medicines without being coupled with some conventional standards and the way they are used may not take them as far as everyone would want them to go. This is the case due to some deficiencies they suffer in their core traditional forms. This includes the lack of robust, if at all, quality assurance systems at all stages of the preparation, storage, use, and monitoring of their drug forms.

9.7.1 ANECDOTAL DATA

Over reliance of practitioners and patients in subjective data and claims makes the management and promotion of herbal medicines very difficult to penetrate some societies. The notion that a particular herbal medicine works against some ailment, to some, is the same old story that promoters of herbal medicines and business operatives gimmick from their personal biased positions to coax new users and believers in the medicines. There is usually little or no effort at all amongst its traditional users and promoters to go deeper into the verifiable science of the medicines and base their practices on the same. There are sentiments that go sympathetically with lack of scientific evidence for most of the herbal medicines in use. Some quarters of the society think that it is too early for herbal medicine practice to be based on grounded scientific methods, techniques, and evident data because research in the area is not too old. However, one would also argue that traditional herbal medicine practice is centuries older than conventional medicine, and the latter has seen most robust and vigorous scientific research, claiming huge banks of data today, and more in progress of development. Something serious needs to be done, collaboratively from all fields of scientific and artistic research, to penetrate societies with data that has an objective, reliable and verifiable findings, to substantiate any claims on herbal medicines.

9.7.2 VAGUE THERAPEUTIC AND SAFETY CLAIMS

Traditional herbal medicines go with many claims of healing. History of use and folklore seem to be enough evidence for therapeutic claims. Sometimes, these claims go as far as generalizing that all herbal medicines can heal literally all diseases. On safety, there is a general claim that goes around even in scientific publications that being natural botanic products, herbal medicines are free from side effects. With the little scientific research already portrayed so far, these two claims seem only to have been misrepresentations of facts about herbal medicines. Today, many plants are known to not only be unsafe for use, but also toxic in nature despite being natural botanics. It is from the same botanic products that we get toxins for insect and animal repellents through various mechanisms of actions, natural anesthetics, psychotic compounds, and many others which we would consider as unsafe or/and toxic, depending on the context of use and definitions of doses. It is therefore not true to generalize the claim that natural herbal medicines are wholesomely without side effects (Patwardhan et al., 2008).

One the part of claims for universal therapeutic potential of herbal medicines by some, this also requires robust investigation into chemical composition, pharmacological make up, and mechanisms of actions toward various biologics. The authors of this chapter fully understand the possibility of every plant to potentially have therapeutic compounds for as wide a range of biological targets as there could be, only limited by quantity, synergistic, and antagonistic interactions within the plant matrices. Note needs to be taken that the previous statement is yet another claim by the authors that requires in-depth analysis and scrutiny. However, even this could be the case, it is misleading to report as if there is not net bioactivity on biological targets, which is affected by additive as well as counteractive mechanisms within the plant. It is prudent and ethical to report activity with a qualitative term as "claim" or "as studied by … or in a particular set up of testing." The more individual test results of claims from different setup conditions accumulate, the better the scientific evidence for a particular claim the science community as well as the general public shall benefit for the betterment of traditional herbal medicine practices.

It is not expected to always have matching test results for batches under study; rather, more research studies, even if it means repeating the same method over and over to substantiate some earlier findings, need to be undertaken. This practice is usually termed "re-inventing the wheel" amongst scientists who feel it is a waste of time and other resources to repeat analytical tests that were done by another earlier in the same or a different test setting. If resources are available, science should encourage repetitions, over and over, which is one of the reasons there are methods published along all research findings. If they cannot be repeated, then why do we have these in place? Science should be able to give a platform of honest debate over both methods and findings among other sections of reporting.

Much as the claims are there, they should only form a solid basis for repeatable scientific research today to either substantiate the claims or to bring in a new school of thought which can have evidence that can be subjected to scrutiny using agreed methods of science.

9.7.3 LACK OF QUALITY ASSURANCE SYSTEMS

Most practitioners of herbal medicines do not have any system to guide their operations and trade. There is no piece of document one can find on how prescription is being conducted, how many patients were given what the previous week, and what herbal formulation led to the deaths of patients. All such data are, deliberately or not, not being kept. In the Chinese herbal medicine practice, some basic systems are being practiced, unlike in the African societies where no one has an opportunity to learn from their own mistakes. There is a chance to keep repeating the same mistake over and over, risking people's lives just because no one is keeping records of observations (Mtewa and Mtewa, 2017). This keeps the practice in Africa at the same level, and its growth hardly celebrated as an achievement. The lack of systems can also be attributed to the culture of secrecy, where no one wants to share knowledge. This is proving to be retrogressive, and most of the valuable knowledge there was two centuries ago went along with those who bore them; the efficiency of passing on from father to son could never be 100%. In Africa, one country that is most familiar with and other practitioners changed such, their herbal medicine practice would be better than one can imagine.

9.7.4 THIN LINE BETWEEN SPIRITUALISM AND HERBALISM

Some people still equate herbal medicines to spiritual mediation. This could be so because in the near past, most traditional medicine practices were being championed by herbalists/spiritualists who practiced both trades simultaneously. Due to religious beliefs against spiritualists associated with witchcraft, the practice of herbal medicine has been condemned, demonized, and ignored by many. It is only in recent times that some people have started realizing that the two are completely different. This perception has greatly contributed to the lack of credibility of herbal medicines in many societies.

9.8 CONCLUSION

The increase in the use of herbal medicine has encouraged the concept of herbal medicine production largely at a very small scale. The production of herbal products at present is chaotic with unclear or no regulations that makes its up-scaling very challenging. Lack of proper quality assurance and safety measures in the practice belittles the rich potential that the practice has to grow into a huge venture it should be, as most products can hardly be characterized with proper medicine guidelines. Mostly, there are no history documentations, laboratory test results, safety tests data, and owing to poor characterization of documents, this field faces a lot of challenges in terms of claims thereof as well as legitimacy. Being products that have survived centuries, today herbal medicine is expected to be flourishing at par or even far ahead of conventional medicine. This is about time when research goes much far beyond phytochemical screening to the industrialization of herbal medicine. This cannot happen if everyone works individually. There is a need to put the practice in order to encourage flexibility and objective collaboration among all stakeholders for a meaningful end.

KEYWORDS

- **product development**
- **herbal products**
- **herbal-product characteristics**
- **pharmacokinetics**
- **herbal regulations**
- **drug approvals**
- **drug patents**

REFERENCES

Chow, S. C.; Pong, A. Scientific issues in botanical drug product development. *Ann Biometrics Biostatist.* 2015, 2(1): 1–12.

Kerns, E. H.; Di, L. Drug-like properties. In: *Concepts, Structure Design and Methods.* 1st ed.; New York, NY: Elsevier, 2008.

Lee, K. M. Overview of drug product development. *Curr Protoc Pharmacol.* 2002, 15(1): 7.3.1–7.3.10.

Li, W. Botanical drugs: A future for herbal medicines. *J Contemp Health Law Policy.* 2002, 19(1): 117–149.

Mtewa, A.; Mtewa, T. *Quality Assurance Principles for Processors in the Developing World.* 1sted.; Beau Bassin: Lambert Academic Publishers, 2017.

Mtewa, A. G.; Deyno, S.; Ngwira, K.; Lampiao, F.; Peter, E. L.; Ahovegbe, L. Y.; Ogwang, P. E.; Sesaazi, D. C. Drug-like properties of anticancer molecules elucidated from *Eichhornia crassipes. J Pharmacogn Phytochem.* 2018, 7(5): 2015–2079.

Parwardhan, B.; Mashelkar, R. A. Traditional medicine-inspired approaches to drug discovery: Can Ayurveda show the way forward? *Drug Discov Today*. 2009, 14: 804–811.

Patwardhan, B.; Vaidya, A. D. B.; Chorghade, M.; Joshi, S. P. Reverse pharmacology and systems approaches for drug discovery and development. *Curr Bioact Compound*. 2008, 4(4): 1–11.

Qusaj, Y.; Leng, A.; Alshihabi, F.; Krasniqi, B.; Vandamme, T. Development strategies for herbal products reducing the influence of natural variance in dry mass on tableting properties and tablet characteristics. *Pharmaceutics*. 2012, 4: 501–516.

Sardana, S. Herbal drug development from natural sources. *J Adv Pharm Technol Res*. 2012, 3(2): 82.

Wu, C. Recent developments of marketing authorization of botanical drugs in the USA. *Botanical Review Team Office of Pharmaceutical Quality*, CDER/FDA. EMA Seminar, September 14, 2017.

FDA. Botanical drug development guidance for industry. Pharmaceutical Quality/CMC Revision, 2016, 1:1–34.

CHAPTER 10

Aromatic Medicine

SAKSHI BAJAJ[1*] and HIMANGINI BANSAL[2]

[1]*Department of Pharmacognosy and Phytochemistry,
Delhi Institute of Pharmaceutical Sciences and Research, Sec-III,
Pushp Vihar, M.B. Road, Delhi 110017, India*

[2]*Department of Pharmaceutical Chemistry,
Delhi Institute of Pharmaceutical Sciences and Research,
Sec-III, Pushp Vihar, M.B. Road, Delhi 110017, India*

Corresponding author. E-mail: sakshibajaj84@gmail.com

ABSTRACT

Aromatherapy, likewise alluded to as essential oil (EO) therapy, can be characterized as the craftsmanship and science of utilizing naturally extricated aromatic essences from plants to harmonize, balance and promote the soundness of body, spirit, and mind. It seeks to unify psychological, physiological, and spiritual processes to enhance a person's natural healing process. Essential oils are natural and contain the genuine aromatic fragrance and other naturally useful properties of the plant the aromatic oil was distilled form. Essential oils have properties of both aromatic and therapeutic. Remedially, aromatherapy is a minding hand on treatment which seeks to induce relaxation physically, mentally, and spiritually domain of an individual's response to EO. Today it is being acknowledged world over and is being utilized in homes, facilities, and emergency clinics for different applications.

10.1 INTRODUCTION

Aromatic medicine, otherwise called scientific aromatherapy or aromatology or essential oil (EO) treatment, alludes on a very basic level to the utilization of EOs, administered by means of a scope of clinical applications, to impact physiological well-being. Experts of integrated aromatic medicine anyway interface two corresponding paradigms, specifically comprehensive aroma-based treatment and aromatic science, which traverses the two methodologies, grasping the best of the two methodologies in order to start mending (Stea et al., 2014).

It is the art and science of using naturally extracted aromatic fragrance from plants to harmonize, balance, and promote the well-being of body, psyche, and soul. It seeks to bind together the physiological, physiological, and spiritual realm of the individual's reaction to aromatic extricates as well as to observe and upgrade the person's inalienable mending process. As an all-encompassing practice, aromatherapy is both a safeguard approach just as a functioning technique to use during extreme and unending periods of ailment or disease.

In aromatherapy, treatment is done using fragrance or scents that are completely natural and are derived from flowers, resins, wood, roots, fruits, herbs, leaves, seeds, and furthermore distilled from resins. The aroma of a plant or any of its parts may be because of the EO which from time to time exists in a free state as in the case of rose flower. Most edible fruits have a pleasant flavor which is due to the presence of aromatic substances in them. The EOs are formed in special cells, glands or duct either in one particular organ or distributed in many parts of the plant (Vankar, 2004)

Aromatherapy is the most commonly applied topically or through inhalation. When applied topically, the oil is usually added to carrier oil and used for massage. Essential oils can be breathed in through a humidifier or by soaking gauze and placing it near the patient (Boehm et al., 2012). Olfactory and tactile sensory stimulation produced by these oils can enhance normal human activities such as eating, social connection, and sexual contact (Cino, 2014)

In a general sense, aromatic medicine is underpinned by science. However, the very nature of aromatic oil molecules, with their perplexing complex structure, accomplishes both the unpretentious effects—profusely described in contemporary aromatherapy literature, simultaneous with clinical impacts validated by current scientific research.

10.2   EVOLUTION OF AROMATHERAPY

The roots of aromatherapy can be traced back more than 3500 years before the birth of Christ, to a time when the use of aromatics was first recorded in human history. However, among the numerous countries that have a documented history of utilizing aromatic plants in their healing traditions are China, Egypt, France, Greece, India, Iraq, Syria, Switzerland, Tibet, UK, and the USA (Buckle, 1997).

Many antiquated civilizations like Egypt, China, and India have utilized this as a prominent complementary and alternative therapy at least for 6000 years (Krishna, 2000).

Literature survey uncovers that this therapy has picked up a great deal of consideration in the late 20th century and is prominent in the 21st century as well, and because of its significance, popularity and widespread use, it is perceived as aroma science therapy (Esposito et al., 2014). The renaissance of aromatherapy began in France. It was a clinical affair and initially involved only three people: a chemist (Gattefosse), a physician (valnet), and a nurse (Maury). The expression "Aromatherapie" was authored in 1928 by René Maurice Gattefosse, a French chemist, to portray the therapeutic utilization of aromatic oils. Albeit present day aromatherapy practice initially encompassed human pathology and the use of EOs in the treatment of different physical and emotional conditions, it has thusly formed into an increasingly comprehensive practice, which grasps all parts of being—body, brain, and soul—to encourage physical, mental, and enthusiastic prosperity.

Initially aromatherapy picked up fame in English-speaking nations like the UK, USA, Australia, and South Africa, basically as "feel better," against stress, or beauty/cosmetic treatments. These "soft" techniques of delivery usually refer to massage application of EOs that have been significantly diluted in vegetable oils, creams, etc., just as encompassing fragrancing by means of vaporizing lamps and air sprays. However, the expanding acknowledgment by orthodox and integrated medication practitioners, of the therapeutic ability of EOs, is gradually changing the discernment that this is the norm. This advancement is attributed to the impact of the European methodology, which features clinical use of EOs, supported by scientific rigor. There are various erudite health professionals and scholastics from nations, for example, France, Germany, Belgium, and Turkey, among others, who have spearheaded this development. Aromatherapy along these lines implies various things to various people. Whichever technique for practice is preferred; good quality EOs are imperative (Boesl and Saarinen, 2016).

10.3   CLASSIFICATION OF AROMATHERAPY

10.3.1   COSMETIC AROMATHERAPY

This therapy uses certain aromatic oils for skin, body, face, and hair restorative products. The EOs utilized on the body as a cleanser, moisturizer, skin, and hair-care products go under this category. These items are utilized for their different impacts as purifying, moisturizing, drying, and toning. A sound skin can be gained by utilization of EOs in facial products.

On a personal level, cosmetic aromatherapy of full body or foot shower will be a straightforward and an effective way to revitalize and rejuvenate the body other than considering the essential cleaning, conditioning, moisturizing, and defensive properties of EOs for different skin and hair types. Thus, few drops of proper oil give a reviving and rejuvenating experience (Ziosi et al., 2010).

10.3.2   MASSAGE AROMATHERAPY

Essential oils can be ingested through the skin by scouring or topical application. This advances an all-encompassing recuperating of the entire body by going through the bloodstream and influencing various organs of the body. A portion of these oils are likewise potent antiviral, antifungal, and antiseptic in nature. The utilization of grape seed, jojoba, or almond oil in unadulterated vegetable oil during massage has been appeared to have amazing effects. This is additionally called as recuperating touch of massage therapy (Soden et al., 2004).

10.3.3   MEDICAL AROMATHERAPY

The founder of modern aromatherapy-based treatment Rene-Maurice Gattefosse has utilized EOs to rub patients amid surgery; thus utilizing medical aromatherapy information of the impact of EOs on advancing and treating clinically diagnosed restorative tribulations (Maeda, 2012).

10.3.4   OLFACTORY AROMATHERAPY

In this technique, EOs are apperceived through the feeling of smell to give a remedial benefit. This is done by a direct or indirect inhalation or aeronautical dispersion of EOs. Olfactory fragrance-based treatment is helpful because the mind is molded from memory for different sorts of smells. This is made synchronize and manage the regular powers of the body to set up an inborn equalization and harmony. In olfactory aromatherapy, where basic inhalation has brought about upgraded enthusiastic well-being, smoothness, unwinding, or revival of the human body. The arrival of stress is welded with pleasurable aromas which unlock odor memories. Essential oils are supplemented to remedial treatment and can never be taken as a substitution for it (Price, 1993).

10.3.5 PSYCHO-AROMATHERAPY

In psycho-aromatherapy, certain states of conditions of dispositions and feelings can be gained by these oils giving the delight of unwinding, empowerment, or a wonderful memory. The inhalation of the oils in this treatment is immediate; however, the imbuement in the room of a patient. Psycho-aromatherapy and aromacology, both manage the investigation and impacts of smell be it common or synthetic. Psycho-aromatherapy has constrained itself with investigation of natural EOs (Perry and Perry, 2006). The author of present-day fragrance-based treatment Rene-Maurice Gattefosse has utilized EOs to rub patients amid surgery, along these lines using the therapeutic aromatherapy learning of the impact of EOs on advancing and treating clinically analyzed restorative ailments.

10.4 HOW AROMATHERAPY WORKS

For centuries, EOs have discovered their significance as an aroma with a recuperating potential on the body, mind, and soul. These aroma molecules are amazingly serious natural plant synthetic substances that make the environment free from disease, bacteria, virus, and fungus (Baratta et al., 1998b). Their versatile character of antibacterial, antiviral, anti-inflammatory nature alongside invulnerable booster body with hormonal, glandular, enthusiastic, circulatory, quieting impact, memory, and sharpness enhancer, is well documented by numerous researchers (Svoboda and Hampson, 1999). Many pilot undertakings and studies have been guided on humans to unravel their tendency and role with disease and disorder (Liu et al., 2013). These oils are known for their vitality explicit character, as their strength is not lost with time and age. The incitement properties of these oils lay in their structures which are intently in similarity with actual hormones. It has been proposed that volatile oils, either inhaled or applied to the skin, demonstration by methods for their lipophilic fraction responding with the lipid parts of the cell membranes, and accordingly, modify the activity of the calcium ion channels. They can associate with the cell membranes by means for their physiochemical properties and molecular shapes, and can affect their enzymes, carriers, receptors, and ion channels (Buchbauer and Jirovetz, 1994).

The penetration ability of oils to achieve the subcutaneous tissues is one of the critical characters of this therapy. Their effects are also complex and subtle due to their intricate structure and chemical properties. The mechanism of their action includes combination of EOs into an organic signal of the receptor cells in the nose when inhaled in. These fuse brain incitement, nervousness assuaging sedation and antidepressant activities, as well as increasing the cerebral blood flow. The fragrance compounds are absorbed by inward breath and can cross the blood–brain barrier and interface with receptors in the central nervous system. The signals are transmitted to limbic and hypothalamus parts of the brain through olfactory bulb. These signals cause brain to release neuro messengers like serotonin, endorphin etc., to connect our nervous and other body frameworks guaranteeing a desired change and to give a sentiment of relief. Serotonin, noradrenalin, and endorphin are discharged from calming oil, euphoric, and stimulating oil, respectively, to give expected impact on psyche and body (Buchbauer and Jirovetz, 1994).

10.5 METHODS OF APPLICATION OF AROMATHERAPY TREATMENT

Various methods are used to apply the treatment in aromatherapy. The most usual methods are as follows:

- A diffuser, usually powered by electricity, giving out a fine mist of the EO.
- A burner, with water is added to the fragrance to prevent burning of the EO. About 1–4 drops of EO are added to about 10 mL water. The burner can be warmed by candles or electricity. The latter would be safer in a hospital or a children's room or even a bedroom.
- A steaming shower with included drops of EO. This outcomes in the moderate volatilization of the EO, and the scent is breathed in by

means of the mouth and nose. Any impact is not probably going to be through the absorbtion of the EO through the skin as expressed in aromatherapy books, as the EO does not blend with water. Droplets either form on the surface of the water, regularly coalescing, or else the EO adheres to the side of the bath. Pouring in an EO blended with milk fills no helpful purpose as the EO still does not mix with water, and the premixing of the EO in a carrier oil, concerning rub, just outcomes in a dreadful oily scum around the bath.

- A bowl of high temp water with drops of EO, regularly utilized for splashing feet or utilized as a bidet. Again, the EO does not blend with the water. This is, nonetheless, a helpful technique for breathing in EOs in respiratory conditions and colds. The EO can be taken in when the head is set over the container and a towel set over the head and container. This is a set-up strategy for treatment and has been utilized effectively with obas oil, Vicks Vaporub and Eucalyptus oils for a long time, so it is not amazing that it works with aromatherapy EOs.

- Metal rings and ceramic placed on an electric light bulb with a drop or two of aromatic oil. This outcomes in a quick burnout of the oil and goes on for a brief timeframe because of the fast volatilization of the EO in the warmth.

- Massage of hands, feet, back, or everywhere throughout the body utilizing 2–4 drops of EO (single EO or mixture) diluted in 10 mL carrier oil (fixed, sleek), for instance, almond oil, jojoba oil, or grapeseed, wheatgerm oils, thus on. The massage applied is usually by delicate efleurage with some petrissage (kneading), with and without some cases, lymph waste sometimes, and is vigorous, as indicated by the aroma specialist's aptitudes and conviction.

- Compresses utilizing EO drops on a wet cloth, either hot or cold, to calm inflammation treat wounds, thus on. Again, the EO cannot blend with the water and can be concentrated in a couple of territories, making it a conceivable health hazard.

- Oral admission is more like regular than "alternative" use of EOs. Although it is practiced by various aromatherapist, this is not to be overlooked except if the aromatherapist is medically qualified. Essential oil drops are "mixed" in a tumbler of high temp water or exhibited on a sugar cube or "mixed" with a teaspoonful of honey and taken internally. The inability of the EO to mix with aqueous solutions presents a health hazard, as do different techniques, as such strong concentrations of EOs are included.

10.6   SOME PLANTS USED IN AROMATHERAPY

Numerous plants have been accounted for to use in the aromatherapy due to the presence of essential or volatile oils in various plants materials like barks, flowers, leaves, stem, roots, fruits etc. Some of the plants utilized in aromatherapy are abridged below.

10.6.1   CLARY SAGE

Clary sage (*Salvia sclarea* Linn.) belongs to the family of Lamiaceae. The leaves are large purple tinted furry and green in color. It is the principal source of EO in clary sage, perennial herb. It is extraordinary from, *Salvia officinalis* or a typical sage. Further, it tends to be differentiated by its size of leaves which are much larger than the regular one and its shading is somewhat blue white in late summer. It contains predominantly linalyl acetate, linalool, germacrene D, alpha-terpineol, geranyl acetate, and caryophyllene (Dogan et al., 2015).

Clary sage is calming to the nervous system, particularly in case of depression, insomnia, stress, and deep-seated tension. It besides is a good tonic for the womb and female functions, in general, such as painful periods, meager menstruation, and relaxation during labor; thus encouraging a less excruciating birth. It helps in controlling the sebum production; hence, can be used for both dry and oily skin, along with acne, wrinkles and for controlling cellulite (Svoboda and Hampson, 1999; Szentmihályi et al., 2001). Clary sage oil's greatest benefit lies in its calming and sedating influence on the nerves, emotions, female functions, kidneys, and digestive system. In ongoing investigations, this oil is observed to be exceptionally effective in controlling cortisol levels in women alongside its antimicrobial activity (Lee et al., 2014).

10.6.2   EUCALYPTUS

Eucalyptus [*Eucalyptus globulus* Labill (*E. globulus*)] belonging to the family of Myrtaceae, is a long evergreen plant with a tallness up to 250 feet. It is known for its constituents like Eucalyptol (cineole), D-limonene, α-pinene, terpinene, alloaromadendrene, *p*-cymene, α-phellandrene, and carvone (Hillis, 1967). Its oils have been used to control and initiate the different system like nervous system for neuralgia, cerebral pain, and debility. The immune system supports the invulnerability against measles, influenza, cold, and chickenpox. Leucorrhea and cystitis of genitourinary system can likewise be very much treated with it. Throat diseases, cough, catarrh, bronchitis, asthma, and sinusitis related with respiratory system have been dealt with by EOs of this plant. In addition, skin problems like cuts, wounds, burns, herpes, lice, insect repellent, and insect bites can be treated with it.

The treatment of rheumatoid joint pain, muscle and joint torments and aches is well reported from EOs of this plant (Mulyaningsih et al., 2011; Sadlon and Lamson, 2010). Eucalyptus oil has exhibited its, anti-inflammatory, antioxidant, antiproliferative, and antibacterial activities and scientists have demonstrated its adequacy certain in the treatment of different metabolic and irresistible diseases. The results are promising and can be utilized for the treatment of multifactorial infections of different origin in human (Aazza et al., 2014; Mulyaningsih et al., 2011).

10.6.3 GERANIUM

Geranium (*Pelargonium graveolens* L' Herit) belongs to the family of Geraniaceae. It is a plant of choice of EO. A perennial shaggy bush local of South Africa, up to one meter in stature, likewise found and developed in France, Spain, Italy, Egypt, Central America, Japan, and Congo. Eugenol, citronellol, linalool, geraniol, citronellyl formate, citral, myrtenol, terpineol, and sabinene are the chemical constituents of its EO (Rana et al., 2002). A standout among the best common fragrance, complete is geranium oil, for the most part utilized in soaps and detergents since its one of a kind is never tried with alkalinity of soaps. In this manner, this oil is commonly used to control the feelings in aroma therapy. It is utilized in eczema, dermatitis, maturing skin, some fungal infections, alongside stress, and nervousness-related problems. The oil has some antibacterial activity and is a significant ingredient for endometriosis treatment. This oil is additionally utilized for its calming properties, nerve tonic, in throat infection, to rectify the blood disorder diabetes and for menopausal-related problems. A couple of reports are there about its supportive therapy in uterine and breast malignant growth, and it additionally positively can help the patient in adapting to the pain. Individuals have used this as an enhancing specialist for nourishment stuff alongside alcoholic and nonalcoholic beverages. It is a compelling creepy crawly repellant (Tisserand and Young, 2002). In addition, this oil is picking up notoriety as antidiabetic, anticancer, antibacterial, and antimicrobial agent (Ben Hsouna and Hamdi, 2012).

10.6.4 LAVENDER

Lavender (*Lavandula officinalis* Chaix.) belonging to the family of Lamiaceae, is a lovely herb of the garden. The lavender EO is synthetically made out of camphor, terpinen-4-ol, linalyl acetate, linalool,, beta-ocimene, and 1,8-cineole (Price, 1993). Among the previously mentioned chemical constituents, linalool and linalyl acetate demonstrated a sedative effect and marked narcotic actions, respectively. Moreover, linalool and linalyl acetate derivation have extraordinary absorbing properties for skin amid massages. In addition, lavender EO was all around recorded for the treatment of scraped areas, burns, stress, headaches, skin problems, muscular pain, and boosting the insusceptible system (Kim et al., 2011).

Lavender oil demonstrates its antibacterial and antifungal properties against numerous types of bacteria, especially when antibiotics neglect to work, yet the exact mechanism are yet to be established. In aromatherapy, it is very much archived for the treatment of abrasions, burns, stress, headaches, in advancement of new cell growth, skin problems, agonizing muscles, and boosting an invulnerable system. This oil is utilized in the treatment of essential dysmenorrheal (Koulivand et al., 2013).

10.6.5 LEMON

Lemon [*Citrus limon* Linn. [(*C. limon*)]] belongs to the family of Rutaceae. *C. limon* long trees grow up to the 15 feet height and bear rich scented lemon fruits all year round. Its oil constituents are copious in the terpenes, limonene, and D-limonene, together forming about 90% of the main part of the oil. Traces of pinene, phellandrene, and sesquiterpene are likewise present (Price, 1993). Essential oil of lemon is able to accelerate the production of white blood cells, fortify the immune system, and help in the assimilation processes. The main constituents of lemon oil have exhibited antiseptic, astringent, and detoxifying properties for imperfections related with slick skin (Tisserand and Young, 2002). Lemon EO is mainly used to boost the immune system and to accelerate the white corpuscles generation along with counteracting acidity and ulcers through citric acid, which helps digestion, by forming carbonates and bicarbonates of potassium and calcium (Tisserand and Young, 2002).

10.6.6 PEPPERMINT

Peppermint [*Mentha piperita* Linn. (*M. piperita*)] belongs to the family of Lamiaceae. Till date, all the 600 sorts of mints are raised from 25 well-characterized species. The two most significant are peppermint (*M. piperita*) and spearmint (*Mentha spicata*). Spearmint bears the strong fragrance of sweet character with a sharp menthol undercurrent. Its oil constituents incorporate carvacrol, menthol, carvone, methyl acetate, menthone, and limonene. Spearmint bears the strong smell of sweet character with a sharp menthol undercurrent. Its oil constituents include carvacrol, menthol, carvone, methyl acetate, limonene, and menthone.

Peppermint oil has been intensively contemplated for its anti-inflammatory, anti-infectious, antimicrobial, and fungicidal impact as well as anti-septics and carminative properties. It is seen that the single constituents of peppermint can calm numerous bacterial, parasitic, and viral contaminations when breathed in or applied as vapor demulcent. On the other side, Ali et al. reported that menthol, the essential constituent of peppermint oil, is responsible for pharmacological activity. It is seen that it can alleviate numerous bacterial, fungal, and viral infections when breathed in or applied in the form of vapor emollient. Sinus and lung blockage are likewise known to be cleared from this oil. Much have been said and examined about the mentha oil by numerous researchers for its different activities yet its utilization in aromatherapy needs more endeavors (Tassou et al., 1995).

10.6.7 ROSEMARY

Rosemary (*Rosmarinus officinalis* Linn.) belonging to the family of Lamiaceae bears little light blue blossoms in pre-summer/late-spring and grows up to the tallness of 90 cm. It has three assortments (silver, gold, and green stripe). It is the green assortment that is utilized for its therapeutic properties. This plant is rich in bitter principle, tannic acid, resin, and volatile oil. The active chemical constituents are bornyl acetate, borneol along with other esters and, special camphor like that possesses by the myrtle, cineol, pinene and camphene (Svoboda and Deans, 1992). Its oil has a marked action on the digestive system, with diminishing the symptoms of indigestion, flatulence, and stomach cramps. It works as liver and gallbladder tonic. The oil additionally has some great action on the cardiovascular system. It regularizes the blood pressure and retards the solidifying of arteries. The advantages of rosemary EO in treating respiratory issues are unmatched. The aroma of the oil gives help from throat blockage. In most recent human trials, aromatherapy is a strong nonpharmacological treatment for dementia and may have some potential for improving psychological capacity, particularly in Alzheimer's disease patients, because of its free radical scavenging activity (Atsumi and Tonosaki, 2007). Brilliant skin tonic properties, a relieving, beneficial outcome on menstrual issues, for hair development are a portion of the other significant properties of this oil. Different advantages of rosemary incorporate a stimulant for the scalp empowering hair development and giving treatment to dandruff and oily hair (Al-Sereiti et al., 1999).

10.6.8   TEA TREE

Tea tree (*Melaleuca alternifolia* Cheel) belonging to the family of Myrtaceae, with yellow or purple bloom and needles like leaves is a bush of mucky zone. Because of its commercial esteem, it is cultivated on estates. The principal constituent of its oil is terpinen-4-ol, an alcoholic terpene with a clean smelly smell. The antiviral action is expected to α-sabine with antibacterial and antifungal impacts. It is an immune booster due to terpinen-4-ol while cineole is responsible of its antiseptic character (Hammer et al., 2003). The tea tree itself has antibacterial, antiinflammatory, antiviral, insecticidal, and immune stimulant properties. The aromatherapy uses the blend of lemon, blue gum, clary sage, eucalyptus, lavender, rosemary, ginger, and Scotch pine for the treatment of various infirmities. The oil is utilized in herpes, blister acne, burns, cold sores, insect bites, abscess, mouth blisters, dandruff, and sleek skin. Further, in the treatment of respiratory-related issues it has been utilized for tuberculosis, bronchitis, cough, asthma, catarrh, and whooping cough. Additionally, it is utilized in females for vaginitis, cystitis, and pruritus treatment. Cold, fever, influenza, and chickenpox have required its use. Well-characterized studies have been done on *Melaleuca alternifolia* on herpes through clinical trial endeavors with a promising consequence of this plant (Pazyar et al., 2013).

10.7   ESSENTIAL OILS

Essential oils are the gathering of hydrophobic and highly volatile secondary plant metabolites that can be extracted from plants and are utilized as fragrances, flavorings, and alternative medicine techniques, for example, aromatherapy. While a solitary compound may have a particular smell, most

EOs are really many intensifies that, when consolidated, make the smell related with that specific plant (Dhifi et al., 2016)

Essential oils are commonly derived from at least one plant parts, for example, blooms (e.g. rose, jasmine, carnation, clove, mimosa, rosemary, lavender), leaves and stems (e.g., geranium, patchouli, petitgrain, verbena, cinnamon), leaves (e.g., mint, Ocimum spp., lemongrass, jamrosa), roots (e.g., angelica, sassafras, vetiver, saussurea, valerian), bark (e.g., cinnamon, cassia, canella), wood (e.g., cedar, shoe, pine), seeds (e.g., fennel, coriander, caraway, dill, nutmeg), organic products (bergamot, orange, lemon, juniper), rhizomes (e.g., ginger, calamus, curcuma, orris) and gums or oleoresin exudations (e.g., balsam of Peru, *Myroxylon balsamum*, storax, myrrh, benzoin).

Essential oils are "the steam distillate of aromatic plants" (Tisserand and Balacs, 1995). They have been portrayed as "the volatile, natural constituents of fragrant plant matter and add to both flavor and aroma and are extricated either by distillation or by distillation or by cold pressing (expression)" (Tisserand and Balacs, 1995). The historical backdrop of distillation of EOs began with the medieval doctor Avicenna (Abu Ali al-Hussein Ibn Abdallah Ibn Sina 980-1037) from Persia who imagined the process of distillation and was presumably the first to distil oil of the rose plant (Tisserand, 1988). Today, roughly 40 unique oils got from plants are utilized in aromatherapy. Lavender, rosemary, eucalyptus, chamomile, marjoram, jasmine, peppermint, lemon, ylang ylang, and geranium are the absolute generally prevalence.

10.7.1 CHEMICAL COMPOSITION

The essential objective of plants, like all living beings, is to develop and reproduce. Most of the metabolites delivered by plants, along these lines, are polysaccharides and proteins that give the plants structure and function. Plants likewise produce little measures of secondary metabolites: exacerbates that are not straightforwardly identified with development or reproduction. Many of these auxiliary metabolites are in all respects economically profitable and some have complex chemistry. Most of the plant mixes utilized in scents, flavors, and natural medicines are secondary metabolites.

10.7.1.1 TERPENES AND TERPENOIDS

One of the key secondary metabolites building blocks is a five-carbon particle called isoprene (or, more formally, 2-methyl-1,3-butadiene) The isoprene units are commonly connected with each other through C_1 and C_4 position (Figure 10.1). Along these lines of linkage of isoprene units is alluded to as the head-to-tail association. The previous structural correlation is known as the *Isoprene rule.*

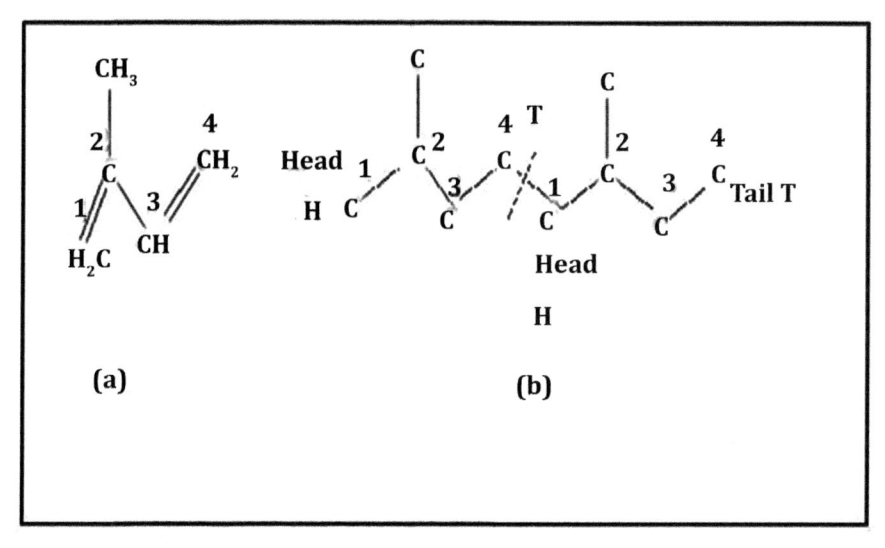

(a)   (b)

FIGURE 10.1   (a) Isoprene (2-methyl-1,3-butadiene), (b) carbon skeleton showing 1,4-union of two isoprene units.

Isoprene is utilized by plants to produce terpenes, molecules made from two or more isoprenes, and terpenoids, terpenes that have slight chemical alterations, particularly terpene alcohols. Terpenes and terpenoids are the synthetic compounds responsible for many plant smells and flavors. The carbon skeleton of terpenoids are distinct into five carbon isoprene (2-methyl-1,3-butadiene) units. Consequently, terpenoids are additionally mentioned as isoprenoids.

Essential oils formed by terpene subsidiaries are normally made from mono and sesquiterpenes, which might be hydrocarbons or oxygenated. Be that as it may, they can also be derived from aliphatic or alicyclic derivatives and some may have sweet-smelling structures. Examples of EOs from this broad gathering incorporate turpentine (Pinus spp.) with α and β pinene, geraniol from damask rose (Rosa damascene), tea tree oil (*Melalevca alternifolia*) with terpinen-4-ol, coriander (*Coriandrum sativum*) with linalool, peppermint (*M. piperita*) with menthol, caraway (*Carum carvi*) with carvone, wormwood (*A absinthium*) with thujone, (*Eucalyptus globus*) with cineol, and

(*Chenopodium abrosioides*) with ascaridole as instances of monoterpenes. Examples of sesquiterpenes incorporate sandalwood (*Santalum album*) with Santalol, german chamomile (*Maticaria recutita*) with bisabolol, and ginger (*Zingiber officinale*) with zingiberol.

10.7.2 ABSORPTION OF ESSENTIAL OIL

- *Essential oil components can be absorbed by four routes*: (Buckle, 1997)
- *Topical*: via external skin massage, compress, or bath.
- *Inhaled*: Directly or indirectly, with or without steam using diffusers, aromastones, fans, humidifiers, aromasticks individual patches, individual packets or nostril clips.
- *Internal*: Via internal skin using mouthwashes, gargles, douches, pessaries or suppositories.
- *Oral*: Using gelatin capsules or diluted in honey, alcohol, or dispersant. Each method of application has its own physiologic process, advantages and disadvantages.

10.7.3 TECHNIQUES FOR EXTRACTION OF ESSENTIAL OILS

Plants owe their scent to the presence of traces of EOs in various parts. Various fragrance materials are available in flowers, root, barks, stems, leaves, leafy foods, resins, and oleoresins are additionally important crude materials for fragrances by uprightness of their tenacious but soft odor. Essential oil can be obtained by means of water distillation, water, and steam distillation, or steam distillation alone. In fact, these are the most traditional and ordinarily utilized techniques. In this protocol, to keep away from the loss of hydro dissolvable compounds, the condensed water is returned from the separator back to the still. Different procedures for obtaining EOs incorporate maceration and enfleurage, also increasingly present-day strategies, for example, extraction with solvents or supercritical liquids. Maceration can be utilized when the yield from distillation is poor, while enfleurage and dissolvable extraction is reasonable for costly, fragile, and thermally unsteady materials (Ester et al., 2012).

One alternative to distillation that avoids chemical alternation to the final product is supercritical CO_2 extraction. This procedure will in general leave labile mixes and a wide scope of different standards altered (Herzi et al., 2013). Several procedures like cold expression, distillation, microwave-assisted

TABLE 10.1 List of Some Constituents Existing in Common Aromatic Drugs

Sr. No	Common Aromatic Drug	Constituents
1.	Dill	Sabinene, α-Thujene, Myrcene, *p*-Cymene, *trans*-β-Ocimene, β-Pinene, β-Phellandrene, α-Pinene, α-Phellandrene, Carvone
2.	Parsley	Sabinene, Myrcene, *p*-Cymene, *trans*-β-Ocimene, β-Pinene, γ-Terpinene, *cis*-3-Hexenyl acetate, β-Phellandrene d, α-Terpineol, *trans*- Caryophyllene
3.	Caraway	Sabinene, α-Thujene, Myrcene, *p*-Cymene, *trans*-β-Ocimene, β-Pinene, β-Phellandrene d, Linalool, Limonene, Bornyl acetate, *cis*-Pinane, *trans*-Caryophyllene, *p*-Cymene, *o*-Cymene, *trans*-Pinocamphone, Bornyl acetate, Isobornyl acetate, b-Elemene, b-Caryophyllene
4.	Basil	Sabinene, Myrcene, *trans*-β-Ocimene, β-Pinene, β-Phellandrene d, Linalool, Limonene, Terpinolene, *cis*-Carveol, α-Pinene, β-Caryophyllene, *cis*-Pinane, *trans*-Caryophyllene, 1,8-Cineole, *o*-Cymene, Camphor, *cis*-Linalool oxide, *iso*-Pinocamphone, Myrtenol, Terpinen-4-ol, b-Cedrene, b-Elemene, a-Humulene, a-Copaene, b-Caryophyllene
5.	Majoram	Sabinene, α-Thujene, Myrcene, Camphene, *p*-Cymene, *trans*-β-Ocimene, β-Pinene, γ-Terpinene, Terpinene, β-Phellandrene d, Linalool, Limonene, Terpinolene, Bornyl acetate, α-Pinene, α-Phellandrene, β-Caryophyllene, α-Terpinene, 1,8-Cineole, *p*-Cymene, D3-Carene, *o*-Cymene, Camphor, *iso*-Pinocamphone, *iso*-Borneol, Thymol, Terpinen-4-ol, Bornyl acetate, Isobornyl acetate, b-Cedrene, b-Elemene, a-Humulene, a-Copaene, b-Caryophyllene
6.	Sage	Sabinene, α-Thujene, Myrcene, Camphene, *p*-Cymene, β-Pinene, γ-Terpinene, Linalool, Limonene, *trans*-Carveol, α-Pinene, α-phellandrene, dodecanal, β-Caryophyllene, *trans*-Caryophyllene, 1,8-Cineole, *p*-Cymene, *o*-Cymene, (–)-Citronellal, Camphor, *iso*-Pinocamphone, Myrtenol, Terpinen-4-ol, *trans*-Pinocamphone, Bornyl acetate, Isobornyl acetate, b-Cedrene, a-Humulene, b-Caryophyllene, Caryophyllene oxide
7.	Balm	α-Thujene, Myrcene, *p*-Cymene, β-Pinene, γ-Terpinene, α-Terpinene, β-Phellandrene d, Linalool, Limonene, α-Pinene, β-Caryophyllene, *trans*-Caryophyllene, 1,8-Cineole, *o*-Cymene, (–)-Citronellal, Camphor, Terpinen-4-ol, *trans*-Pinocamphone, b-Cedrene, b-Elemene, a-Humulene, b-Caryophyllene, Caryophyllene oxide

TABLE 10.1 *(Continued)*

Sr. No	Common Aromatic Drug	Constituents
8.	Oregano	α-Thujene, Myrcene, Camphene, *p*-Cymene, γ-Terpinene, α-Terpinene, Linalool, Limonene, Terpinolene, α-Pinene, α-phellandrene, β-Caryophyllene, *cis*-Pinane, *trans*-Caryophyllene, α-Terpinene, 1,8-Cineole, *p*-Cymene. D3-Carene, *o*-Cymene, (−)-, *iso*-Pinocamphone, Thymol, Terpinen-4-ol, Citral, b-Cedrene, a-Humulene, a-Copaene, b-Caryophyllene, Caryophyllene oxide
9.	Mint	Myrcene, Camphene, *p*-Cymene, *trans*-β-Ocimene, γ-Terpinene, α-Terpinene, cis-3-hexenyl acetate, Limonene, nonanal, Carvomenthyl acetate, *trans-p*-Mentha-2.8- dien-1-ol, bornyl acetate, *cis*-Carveol, *trans*-Carveol, α-Pinene, a-Humulene, *trans*-Caryophyllene
10.	Fennel	Myrcene, *p*-Cymene, β-Pinene, γ-Terpinene, β-Phellandrene d, α-phellandrene, *p*-Cymene, D3-Carene, *cis*-Linalool oxide
11.	Lavender	α-Thujene, Myrcene, Camphene, *p*-Cymene, *trans*-β-Ocimene, γ-Terpinene, β-Phellandrene d, Linalool, Limonene, bornyl acetate, α-phellandrene, β-Caryophyllene, *cis*-Pinane, *trans*-Caryophyllene, *p*-Cymene, D3-Carene, *o*-Cymene, Camphor, *iso*-Pinocamphone, Myrtenol, Terpinen-4-ol, Bornyl acetate, Isobornyl acetate, b-Cedrene, a-Humulene, b-Caryophyllene, Caryophyllene oxide
12.	Vervain	*trans*-β-Ocimene, γ-Terpinene, Linalool, Limonene, α-Pinene, β-Caryophyllene, *trans*-Caryophyllene, 1,8-Cineole, *o*-Cymene, Terpinen-4-ol, Citral, b-Elemene, a-Humulene, a-Copaene, b-Caryophyllene

extraction, ultrasonic-assisted extraction, and solvent extraction are accessible for the extraction of fragrance standards are discussed in the following sections.

10.7.3.1 COLD EXPRESSION

Expression or cold pressing is the oldest extraction method and is utilized only for the generation of citrus EOs from the peels of citrus plants, for example, orange, lemons, etc. This technique alludes to any physical procedure amid which the EO organs in the peel and cuticles are broken all together for the oil to be discharged. This procedure results in the creation of a watery emulsion which is along these lines centrifuged to separate out the EO (Bousbia et al., 2009).

Expression or cold squeezing, as it is additionally known, is just utilized in the production of citrus oils. The term expression indicates to any physical procedure where the EO glands in the peel are squashed or broken to release the oil. One technique that was drilled numerous years back, especially in Sicily (spugna strategy), initiated with splitting the citrus natural product pursued by mash expulsion with the guide of honed spoon-blade (known as a rastrello). The oil was expelled from the peel either by squeezing the peel against a hard object of baked clay (concolina) which was set under an enormous characteristic wipe or by bowing the strip into the wipe. The oil emulsion consumed by the wipe was expelled by crushing it into the concolina or some other compartment. It is accounted for that oil created along these lines contains a greater amount of the organic product smell character than oil produced by some other method. It is no uncertainty conceivable to extract the EO from citrus peels by steam distillation, yet the oil has been found to come up short on the freshness of fragrance which is normally connected with hand squeezed oils which bring better cost.

10.7.3.2 DISTILLATION

The customary methods for secluding volatile compounds as EOs from plant material are steam distillation, hydro distillation, and hydro-steam distillation (Essentielles, 2000). Amid all distillation procedure, fragrant plants presented to bubbling water or steam release their EOs through evaporation. Steam and EO vapors are condensed, both are gathered and isolated in a vessel generally called the "Florentine flask" (Dugo and Di Giacomo, 2002).

10.7.3.2.1 *Steam Distillation*

A standout amongst the most widely recognized is steam distillation since it takes into consideration the separation of somewhat unpredictable, water-immiscible substances by methods for low temperature distillation, being of specific use when the components boil at high temperature (higher than 100 °C) and are susceptible to decay beneath this temperature. Even though this technique exhibits a few points of interest, it is important to endure at the top of the priority list that it is not only a straightforward steam hauling business. The release of the segments present in the stomas is brought about by cell-wall rupture because of the higher weight and the oil content extension of the cell generated by heat. The steam stream gets in through the stomas, breaks them and in the end hauls the EO (Başer and Buchbauer, 2010).

Steam distillation (Figure 10.2) comprises of steaming because of a straight present of steam water, which warms the blend just as it diminishes the bubbling temperature as a result of the higher steam pressure inborn in water to those of volatile components in essential oils. The steam originating from the distillator gets cold in a condenser and, at long last, the immiscible blend gets isolated in a clarifier or Florentine jar. This philosophy is more helpful than organic solvents extraction or straight distillation as water steam has a lower cost contrasted with organic solvents. Likewise, it avoids oil warming or the utilization of sophisticated equipment. Regardless, the extraction technique depends, among different elements, on the sort of material to be handled and the area of the parts inside the vegetable structure as indicated by the species and botanical family (Bandoni, 2000).

Steam is infused through steam tubes underneath the charge and the pressure inside the distillation vessel is controlled according to the nature of material being distilled. This technique is productive and gives higher yields.

10.7.3.2.2 *Hydrodistillation*

To confine EOs by hydrodistillation, the fragrant plant material is pressed in a still and an adequate amount of water is added and heated to the point of boiling; on the other hand, live steam is injected into the plant charge. Because of the impact of boiling water and steam, the EO is liberated from the oil organs in the plant tissue. The vapor blend of water and oil is consolidated by indirect cooling with water. From the condenser, distillate streams into a separator, where oil isolates consequently from the distillate water. Hydrodistillation framework, however the most established, is yet being broadly drilled for oil extraction. The plant material is in direct contact with bubbling water in an unrefined metallic distillation outfit. Orange bloom and flower petal oil units utilize this method (Reyes-Jurado et al., 2015)

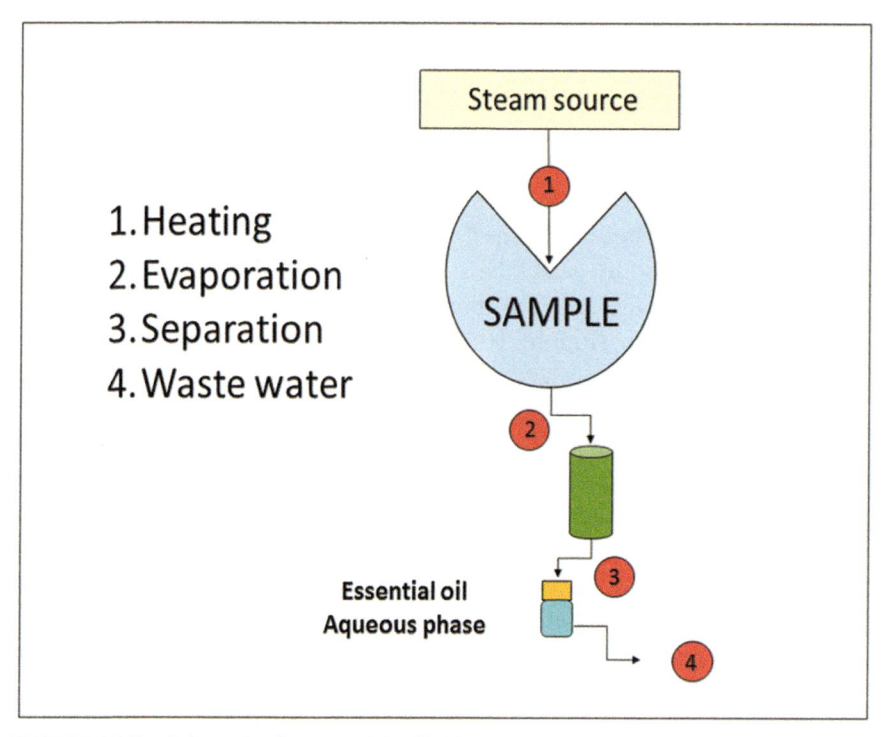

FIGURE 10.2   Schematic diagram of the distillation.

10.7.3.2.3 Hydro-steam Distillation

In hydro-steam distillation, the steam can be created either in a satellite kettle or inside the still, albeit isolated from the plant material. Like water distillation, water and steam distillation is generally utilized in rustic regions. Additionally, it does not require significantly more capital use than water distillation. Additionally, the gear utilized is commonly like that utilized in hydro refining; however, the plant material is bolstered over the bubbling water on a punctured framework. Truth be told, usually people performing water refining inevitably advancement to water and steam distillation.

Hydro-steam distillation is utilized where the perfumery material is helpless against direct steam. Therefore, the plant material is bolstered on a punctured network or screen embedded at some separation over the base of the still. The lower some portion of the still contains water up to level just underneath the matrix. The measure of oil created relies upon length of refining time, temperature, weight, and sort of plant material (Naves, 1974).

10.7.3.2.4 Direct Steam Distillation

As the name proposes, direct steam distillation is the way toward distillation plant material with steam created outside the still in a satellite steam generator for the most part alluded to as an evaporator. As in water and steam refining, the plant material is upheld on a punctured matrix over the steam channel. A genuine preferred position of satellite steam generation is that the measure of steam can be promptly controlled. Since steam is created in a satellite evaporator, the plant material is warmed no higher than 100 °C and, subsequently, it ought not to experience warm debasement. Steam distillation is the most generally acknowledged process for the generation of EOs on enormous scale. All through the flavor and scent supply business, it is a standard practice.

10.7.3.3 SOLVENT EXTRACTION

Solvent extraction can be used to extract EOs that are thermally labile. This process is generally used in recovering oils having a delicate flowery note, which may ordinarily be destroyed under steam distillation. The main advantage of extraction over distillation is that a lower temperature is used during the process. Therefore, reducing the risk of chemical changes due to high temperature, which are used during distillation. The solvent commonly employed is petroleum ether of low boiling point. The solvent is boiled in a separate vessel, and its vapors can enter a rotary drum in which baskets, containing the flowers to be extracted, are fitted. These baskets are pierced so that the vapors pass freely through the flowers as well as from one basket to another; by keeping the drum in rotation, fresh solvent comes at regular intervals in contact with flowers. There is also another method in which the solvent is allowed to pass through a series of vessels and extraction is carried out by the counter current system in which fresh solvent is allowed to pass through a series of vessels and extraction is carried out by the counter current system in which fresh solvent is allowed to pass over the flowers which have already given up most of their perfume by previous extraction.

The saturated solvent is evaporated and an "absolute" of the flower perfume is obtained.

10.7.3.4   ENFLEURAGE

Despite the introduction of the cutting-edge procedure of extraction with volatile solvents, the outdated strategy for enfleurage, as passed on from dad to child and culminated over the span of ages, still assumes a significant job. Enfleurage on an enormous scale is today done uniquely in the Grasse area of France, with the conceivable exemption of isolated instances in India where the procedure has remained primitive.

It is another conventional extraction strategy utilized for extracting the oil from blossoms of jasmine, tuberoses, cassia and so on. Amid this method (Figure 10.3) a purged, saturated, and scentless cold fat is spread on the plant material. The fat is meagerly layered on the two sides of a glass plate bolstered on a rectangular wooden edge or body. A few pursues are put one over the other sandwiching the bloom between two layers of fat. Spent blooms are expelled (defleurage) and new charge is made. Turning around of the glass slab is called patage. Patage is completed a few times to get most extreme aroma absorption. Furrows are made with combs to increase assimilation surface. The process of defleurage, crisp charging, and patage is kept on getting fat of the ideal aroma quality. The fat adsorbs the odoriferous bodies present in the flowers and when it is saturated with the fragrance of the flowers, it is shaken with alcohol at low temperature which breaks down the fragrance however for all intents and purposes no fat.

10.7.3.5   SUPERCRITICAL FLUID EXTRACTION

Supercritical liquid extraction (SFE) is an intriguing procedure for the extraction of flavoring compounds from vegetable materials. It can comprise a modern option in contrast to dissolvable extraction and steam distillation processes (Anitescu et al., 1997; Baysal and Starmans, 1999). The utilization of supercritical liquids, particularly carbon dioxide, in the extraction of plant volatile compounds has expanded amid the two last decades because of the expected advantages of the supercritical extraction process.

Due to the low solubilizing intensity of vapors and gases, investigation has been completed on gas phase extractions performed at raised weights and temperatures. When a gas is compressed and heated, it can achieve a condition of collection at which no refinement between the gas and fluid can be observed. Such a supposed supercritical gas has improved solubilizing properties that are generally equivalent to those of fluids, yet has a very high dissemination coefficient, looking like that of a characteristic gas. A super-critical gas additionally has a thickness like that of a vaporous stage, which implies that the contacting area between the plant material and a supercritical liquid is additionally comparative (Paulaitis et al., 1983).

FIGURE 10.3 Showing Enfluerage method.

CO_2 is the supercritical solvent of choice in the extraction of fragrance compounds. It behaves like a lipophilic solvent but, compared to liquid solvents, it has the advantage that its selectivity or solvent power is adjustable and can be set to values ranging from gas like to liquid like (Stahl et al., 1983).

There are five intentions in picking supercritical carbon dioxide as the extraction medium:

1. It has a basic temperature of 31 °C; this implies extractions can be led at temperatures that are low enough not to harm the physicochemical properties of the extract.

2. It is dormant in nature; in this manner, there is no danger of side responses, for example, oxidation.
3. It is nontoxic; carbon dioxide is an innocuous material that is frequently utilized in beverages. It has been acknowledged by most European food and drugs acts as an extraction medium for the isolation of food related compounds.
4. It has a low polarity; the extremity of carbon dioxide is near that of pentane and hexane, which are solvents commonly utilized in fluid extraction procedures. In this manner, a comparative scope of mixes can be extricated utilizing both techniques.
5. It permits fractionated separation; by basically picking distinctive temperature and weight conditions for various consecutive separator vessels, a fractionated separation of the organic compounds can be achieved.

10.7.3.6   *AROMA EXTRACTION FROM SEEDS AND OTHER FOODS*

Terpene hydrocarbons found in the aromatic oils of plants are among the most significant flavor and aroma mixes and offer a wide assortment of lovely and flower aromas. Notwithstanding their use in fragrances and perfumed items, these hydrocarbons can be utilized to season sustenance's and beverages, underlining their incentive in advanced chemistry.

Blossoms, herbs, and fruits are typical sources of scent and fragrance compounds. When aromatic compounds are incorporated into processed foods, by utilizing hydrodistilled aromatic oils or by including solvent extracted oleoresins, they don't completely take after the regular fragrances. This might be the aftereffect of warmth and water incited changes occurring amid the distillation procedure or the presence of solvent residue in the oleoresin.

However, by utilizing extraction with supercritical liquids, at moderate temperatures, scents and smells can be isolated with no procedure-initiated changes. Current improvements in the extraction of terpene hydrocarbons concern the isolations (Miyake et al., 2000) and enantiospecific portrayal of limonene subordinates and higher terpenes, for example, sesquiterpenes (Castaneda-Acosta et al., 1995) and triterpenes (Sewram et al., 1995). The enantiospecific characterization of terpenes can be connected as an analytical tool to recognize natural and nature-indistinguishable flavors and fragrances.

10.7.3.7 FRACTIONATION OF LOW VAPOR PRESSURE OILS

Generally, EOs segregated from plant material contain a mixture of fascinating compounds. The EOs of plant material, for example, ginger, pepper, and cinnamon contain nonvolatile, pungent compounds that decide to a significant degree of overall flavor profile. Anyway, in EOs of other plant material, for example, the members from the Umbelliferae family (e.g., celery, fennel, caraway, anise, dill, coriander) elevated amounts of scentless triacylglycerol oils are present. The important organic compounds in EOs can be isolated from such contaminants by passing the supercritical carbon dioxide extract through at least two exclusively controlled separator vessels, each worked at a specific pressure and temperature. The subsequent stage-wise reduction in dissolvability of the multicomponent extract subsequently prompts the detachment of pure secondary metabolites at low temperatures. Other authors have revealed the enhanced isolation of palmitate by fractionation of rice-wheat oil (Zhao et al., 1987), and the explained counter current decontamination of the generally inedible lampante olive oil.

10.7.3.8 MICROWAVE-ASSISTED EXTRACTION

Microwave energy is notable to significantly affect the rate of different processes in the food and chemical industry. All the revealed applications have demonstrated that microwave-assisted extraction is an option in contrast to ordinary methods for such matrices. The principal advantages are decline of extraction time and solvents used. The advantages of utilizing microwave energy, which is a noncontact heat source, for the extraction of EOs from plant materials, include: progressively viable warming, quicker energy transfer, decreased thermal gradients, specific heating, reduced equipment size, quicker reaction to process warming control, quicker start-up, expanded production, and elimination of procedure steps. Extraction procedures performed under the activity of microwave radiation are accepted to be influenced to a limited extent by polarization, volumetric, and particular heating (Chema and Lucchesi, 2006).

The treatment of plant material with microwave irradiation previously as well as amid an extraction procedure can result in an improved recuperation of secondary metabolites and aromatic compounds. Much consideration has been given to the utilization of microwave dielectric heating for the extraction of normal items that regularly required hours or days to achieve fulfillment with traditional methods. Utilizing microwaves, full reproducible extractions would now be able to be completed in a seconds or minutes with high reproducibility, decreasing the utilization of dissolvable, simplifying manipulation and work-up, giving higher immaculateness of the final product, eliminating posttreatment of waste water and consuming just a portion of the energy typically required for a customary extraction technique, for example, steam distillation or dissolvable extraction.

10.7.3.9 *MICROWAVE-ASSISTED HYDRO DISTILLATION*

The constrained warming of water (Figure 10.4) in the center of the material can cause the steam-actuated opening of the external layers of the plant material. Such opening (or puffing) of the lattice material effectively abbreviates the way of dissemination of the secondary metabolites, and in this manner advances a progressively fruitful extraction of the material. The removed mixes are taken up by a reasonable encompassing medium to encourage the detachment from the rest of the plant material. This encompassing medium can be either a fluid or a gas (Craveiro et al., 1989). In case of a liquid, to some extent progressively expand division step is important to acquire the pure compounds, though because of a gas, just a straightforward buildup step is sufficient. The last decision depends on the simplicity of volatilization of the ideal compounds.

10.8   MEDICAL USES OF ESSENTIAL OILS IN HUMAN DISEASES

10.8.1 *ANTIOXIDANT*

The EO from seeds of *Nigella sativa* L. is a potent antioxidant in vitro, with compelling hydroxyl radical rummaging action. Kanuka (*Kunzea ericoides*), Manuka (*Leptospermum petersonii*), and (*Leptospermum scoparium*) have good antioxidant and antibacterial action properties. The EO from the *M. armillaris* has stamped antioxidant potential; it adjusts the parameters of superoxide dismutase, improves Vitamin E and Vitamin C concentrations (Baratta et al., 1998a). Dorman et al. (1995) screened Pelargonium sp., *Monarda citriodora* var. *citriodora, Myristica fragrans, Origanum vulgare* ssp. Hirtum and *Thymus vulgaris* for their antioxidative impact utilizing a thiobarbituric acid (TBA) test. The oils indicated active antioxidant activity at incredibly low dimensions of weakening. Rosemary has for some time been perceived as having antioxidant molecules and these have been distinguished as carnosic acid, carnasol, carsolic acid, rosmaridiphenol, and rosmarinic acid, found in ethanol-solvent part. Antioxidant properties are likewise found in the volatile oil fraction. In any case, it is essential to understand that in certain cases, antioxidants can be ace oxidant and can invigorate free radical responses (Damien Dorman et al., 1995).

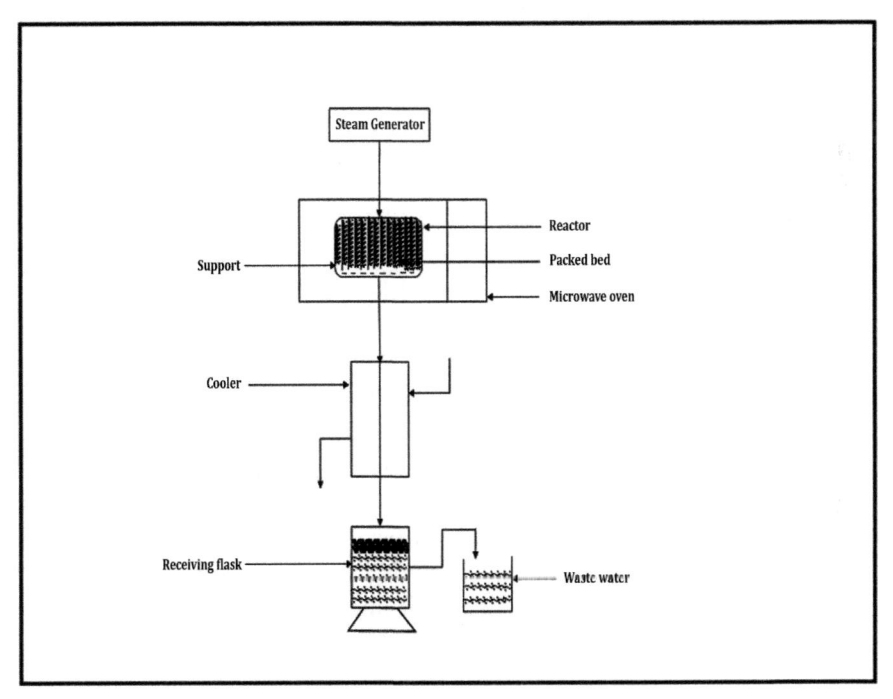

FIGURE 10.4   Schematic diagram of the microwave hydro-distillation process.

10.8.2 ANTI-INFLAMMATORY

Inflammatory conditions are related with torment, redness, and swelling, prompting loss of imperative capacities. Essential oils have been utilized for quite a few years to soothe torment and aggravation. Generally, EOs have more effective and pain-relieving properties than numerous pharmaceutical analgesics. The utilization of EOs has numerous advantages in the treatment of inflammation since it has fewer reactions than numerous synthetic and conventional medications. An audit of the therapeutic properties of chamomile reported that plants contain a larger number of flavonoids with anti-inflammatory properties than others do. These inflammation decreasing compounds can enter the skin effectively and lessen inflammation. Tea tree oil has been appeared to increment monocytic differentiation in vitro and lessen irritation, in this way helping the mending of incessant injuries. Other promising applications have been proposed for *Helichrysum italicum* (Roth) G. Don as antispastic (Rigano et al., 2013), *Pelargonium roseum* L'h´er. as anti-inflammatory (Boukhatem et al., 2013), and *O. majorana* L. as antimutagenic agent (Mossa et al., 2013).

10.8.3 ANTICARCINOGENIC

Essential oils demonstrated a potential anticancer activity against liver, lung, colon, and prostate malignancy, for example, *Artemisia lavandulaefolia* DC and its principal single constituent 1,8-cineole against a subline of the pervasive KERATIN-shaping tumor cell line HeLa which derived from an epidermal carcinoma of the mouth (Cha et al., 2010). Further examinations exhibited that some single constituents like carvacrol (Jayakumar et al., 2012), limonene, and citral (Chaouki et al., 2009) and thymol (Deb et al., 2011), have appeared encouraging cytotoxic impact against various human malignancy cell lines essentially because of the enlistment of the mitochondrial dysfuction. Elshafie et al. (2017) have pointed out the adequacy of *O. vulgare* EO and its fundamental single constituents (carvacrol, thymol, citral, and limonene) on tumor liver cell line (HepG2) following the MTT viability assay.

10.8.4 ANTIMICROBIAL ACTIVITY

10.8.4.1 ANTIBACTERIAL

Elshafie et al. (2016) assessed the in vitro antibacterial action of three EOs extricated from *Schinus terebinthifolius Raddi* (leaves and natural products) against two strains of Gram positive bacteria (*Bacillus megaterium* de Bary and *Clavibacter michiganensis* Smith) and four strains of Gram negative microbes (*E. coli, Xanthomonas campestris* Pammel, *Pseudomonas savastanoi* Janse, and *P. syringae pv. phaseolicola* Van Hall) contrasted with the synthetic antimicrobial tetracycline. They brought about the way that the previously mentioned EOs had the option to essentially repress the development of tested bacterial strains particularly against Gram positive bacteria.

10.8.4.2 ANTIFUNGAL

Melaleuca alternifolia (tea tree) oil tried positive for its all constituents for in vitro antifungal activity but beta-myrcene. Hammer et al. recognized that the majority of the components of tea tree oil had wide scope of fungicidal potential, particularly against dermatophytes and filamentous organisms (Hammer et al., 2003). Elshafie et al. assessed the antifungal effect of the single segments of *O. vulgare* EO (carvacrol, thymol, linalool, and trans-caryophyllene) against the postharvest pathogens *M. laxa, M. fructigena,* and *M. fructicola.* The fungicidal activity assay of the over four single components has been done in vitro.

In a similar report, *Elshafie* et al. reported that the bioactive medicines which displayed the most noteworthy in vitro activity were chosen for further in vivo experiments against the equivalent postharvest pathogenic growths furthermore, inferred that carvacrol and thymol demonstrated a promising hindrance of the brown rot disease of peach fruit caused by *M. laxa, M. fructicola,* and *M. fructigena.* The antifungal effect is chiefly credited to the inhibition of both mycelium development and spore germination. This theory proposes that obstructing the initial infection and the ensuing mycelial spread beyond the infection site will restrict the expression of disease (Elshafie et al., 2015).

10.8.5 ANTIDANDRUFF

In a single blind and parallel-group study, it was seen that shampoos which contain 5% tea tree oil were compelling and very much endured by patients having gentle to moderate dandruff and at least 41% improvement was observed (Satchell et al., 2002). Not much have been investigated on the antidandruff capability of plant products, and particularly on volatile products, a few endeavors have been made by Anjum et al, yet the outcomes are not promising (Anjum et al., 2014).

10.8.6 HORMONAL ACTION

Geranial, neral, geraniol, nerol, and trans-anethole are well established for their stimulation of estrogenic response, when compared to eugenol which has antiestrogenic activity. Citral, that is, the combination of geraniol, nerol, and eugenol was effective in replacing [3 H]17b-estradiol from the estrogen receptors in recombinant yeast cells.

Geranial, geraniol, neral, nerol, and trans-anethole are entrenched for their incitement of estrogenic reaction, when contrasted with eugenol which has antiestrogenic activity. Citral, that is, the combination of geraniol, nerol, and eugenol was effective in supplanting [3 H]17b-estradiol from the estrogen receptors in recombinant yeast cells (Grunfeld and Gresty, 1998; Howes et al., 2002).

10.9 CONCLUSION

Our overview suggests that aromatherapy, which generally delivered with massage therapy, may initiate relaxation which, thus, may improve pain and psychological health. Aromatherapy is characteristic and noninvasive endowment of nature for humans. It is not only the ailment manifestations which are eradicated but also the whole body is revived by utilization of aroma. Aromatherapy manages the physiological, profound, and mental upliftment for the new time of life.

Essential oils can be a useful nontherapeutic choice or can likewise be joined with conventional care for some prosperity conditions, gave security, and quality issues are considered. The tilt of the scientific community toward alternative and complementary medicine has given the new hope to reduce the undesirable effects of present day medicine by these EOs and if appropriately investigated to their maximum capacity, this treatment can be an aid not exclusively to the patients yet notwithstanding a typical man. Plant EOs referenced in the parts are viewed as promising characteristic options in contrast to customary pharmaceutical drugs.

Aromatic medicine does not purport to replace emergency medicine and conventional allopathic medicine. It presents both curative and preventative competencies for acute and chronic conditions. Pharmaceutical industries are going after for ecological cordial, alternative and natural medicine for disease related with pathogens and metabolism. There might be a plausibility of upgrading the rate of reaction and bioavailability of medications from the utilization of these EOs.

KEYWORDS

- **essential oil**
- **aromatic essences**
- **physiological**
- **psychological**
- **spiritual**
- **therapeutic**
- **domain**

REFERENCES

Aazza, S., Lyoussi, B., Megías, C., Cortés-Giraldo, I., Vioque, J., Figueiredo, A.C., Miguel, M.G., 2014. Anti-oxidant, anti-inflammatory and anti-proliferative activities of Moroccan commercial essential oils. Nat. Prod. Commun. 9, 587–94.

Al-Sereiti, M.R., Abu-Amer, K.M., Sen, P., 1999. Pharmacology of rosemary (Rosmarinus officinalis Linn.) and its therapeutic potentials. Indian J. Exp. Biol. 37, 124–30.

Anitescu, G., Doneanu, C., Radulescu, V., 1997. Isolation of coriander oil: Comparison between steam distillation and supercritical CO_2 extraction. Flavour Fragr. J. 12, 173–176. https://doi.org/10.1002/(SICI)1099-1026(199705)12:3<173:AID-FFJ630>3.0.CO;2-1

Anjum, F., Bukhari, S.A., Shahid, M., Bokhari, T.H., Talpur, M.M.A., 2014. Exploration of nutraceutical potential of herbal oil formulated from parasitic plant. African J. Tradit. Complement. Altern. Med. AJTCAM 11, 78–86.

Atsumi, T., Tonosaki, K., 2007. Smelling lavender and rosemary increases free radical scavenging activity and decreases cortisol level in saliva. Psychiatry Res. 150, 89–96. https://doi.org/10.1016/j.psychres.2005.12.012

Bandoni, A.L. (Ed.), 2000. Los Recursos Vegetales Aromáticos en Latinoamérica. Argentina.

Baratta, M.T., Dorman, H.J.D., Deans, S.G., Biondi, D.M., Ruberto, G., 1998a. Chemical composition, antimicrobial and antioxidative activity of laurel, sage, rosemary, oregano and coriander essential oils. J. Essent. Oil Res. 10, 618–627. https://doi.org/10.1080/10412905. 1998.9700989

Baratta, M.T., Dorman, H.J.D., Deans, S.G., Figueiredo, A.C., Barroso, J.G., Ruberto, G., 1998b. Antimicrobial and antioxidant properties of some commercial essential oils. Flavour Fragr. J. 13, 235–244. https://doi.org/10.1002/(SICI)1099-1026 (1998070)13:4<235:AID-FFJ733>3.0.CO;2-T

Başer, K., Buchbauer, G., 2010. Handbook of Essential Oils: Science, Technology, and Applications, 2nd edition, CRC Press/Taylor & Francis, Universitat Wien, Austria. https://doi.org/doi:10.1201/9781420063165-c3

Baysal, T., Starmans, D.A., 1999. Supercritical carbon dioxide extraction of carvone and limonene from caraway seed. J. Supercrit. Fluids 14, 225–234. https://doi.org/10.1016/S0896-8446(98)00099-0

Ben Hsouna, A., Hamdi, N., 2012. Phytochemical composition and antimicrobial activities of the essential oils and organic extracts from Pelargonium graveolens growing in Tunisia. Lipids Health Dis. 11, 167. https://doi.org/10.1186/1476-511X-11-167

Boehm, K., Büssing, A., Ostermann, T., 2012. Aromatherapy as an adjuvant treatment in cancer care--a descriptive systematic review. African J. Tradit. Complement. Altern. Med. AJTCAM 9, 503–18.

Boesl, R., Saarinen, H., 2016. Essential oil education for health care providers. Integr. Med. (Encinitas). 15, 38–40.

Boukhatem, M.N., Kameli, A., Ferhat, M.A., Saidi, F., Mekarnia, M., 2013. Rose geranium essential oil as a source of new and safe anti-inflammatory drugs. Libyan J. Med. 8, 22520.

Bousbia, N., Abert Vian, M., Ferhat, M.A., Petitcolas, E., Meklati, B.Y., Chemat, F., 2009. Comparison of two isolation methods for essential oil from rosemary leaves: Hydrodistillation and microwave hydrodiffusion and gravity. Food Chem. 114, 355–362. https://doi.org/10.1016/j.foodchem.2008.09.106

Buchbauer, G., Jirovetz, L., 1994. Aromatherapy—use of fragrances and essential oils as medicaments. Flavour Fragr. J. 9, 217–222. https://doi.org/10.1002/ffj.2730090503

Buckle, J., 1997. The evolution of aromatherapy, in: Clinical Aromatherapy: Essential Oils in Healthcare. Churchill Livingstone, Philadelphia, USA, pp. 2–14.

Buckle, J., 1997. Overview: How essential oil works, in: Clinical Aromatherapy: Essential Oils in Healthcare. Churchill Livingstone, Philadelphia, USA, pp. 16–36.

Ester, C., Silvia, N., Walter, M.G., Sequeria A.F. 2012. Study of the chemical composition of essential oils by gas chromatography, in: Salih, D.B. (Ed.), Gas Chromatography in Plant Science, Wine Technology, Toxicology and Some Specific Applications. InTech, pp. 307–324. https://doi.org/10.5772/33201

Castaneda-Acosta, J., Cain, A.W., Fischer, N.H., Knopf, F.C., 1995. Extraction of bioactive sesquiterpene lactones from magnolia grandiflora using supercritical carbon dioxide and near-critical propane. J. Agric. Food Chem. 43, 63–68. https://doi.org/10.1021/jf00049a013

Cha, J.-D., Kim, Y.-H., Kim, J.-Y., 2010. Essential oil and 1,8-cineole from Artemisia lavandulaefolia induces apoptosis in KB cells via mitochondrial stress and caspase activation. Food Sci. Biotechnol. 19, 185–191. https://doi.org/10.1007/s10068-010-0025-y

Chaouki, W., Leger, D.Y., Liagre, B., Beneytout, J.-L., Hmamouchi, M., 2009. Citral inhibits cell proliferation and induces apoptosis and cell cycle arrest in MCF-7 cells. Fundam. Clin. Pharmacol. 23, 549–556. https://doi.org/10.1111/j.1472-8206.2009.00738.x

Chemat, F; Lucchesi, M., 2006. Microwave-assisted Extraction of Essential Oils, in Microwaves in organic synthesis. Anal. Bioanal. Chem. 385, 753–759.

Cino, K., 2014. Aromatherapy hand massage for older adults with chronic pain living in long-term care. J. Holist. Nurs. 32, 304–313. https://doi.org/10.1177/0898010114528378

Craveiro, A.A., Matos, F.J.A., Alencar, J.W., Plumel, M.M., 1989. Microwave oven extraction of an essential oil. Flavour Fragr. J. 4, 43–44. https://doi.org/10.1002/ffj.2730040110

Damien Dorman, H.J., Deans, S.G., Noble, R.C., Surai, P., 1995. Evaluation in vitro of plant essential oils as natural antioxidants. J. Essent. Oil Res. 7, 645–651. https://doi.org/10.108 0/10412905.1995.9700520

Deb, D.D., Parimala, G., Saravana Devi, S., Chakraborty, T., 2011. Effect of thymol on peripheral blood mononuclear cell PBMC and acute promyelotic cancer cell line HL-60. Chem. Biol. Interact. 193, 97–106. https://doi.org/10.1016/j.cbi.2011.05.009

Dhifi, W., Bellili, S., Jazi, S., Bahloul, N., Mnif, W., 2016. Essential Oils' Chemical Characterization and Investigation of Some Biological Activities: A Critical Review. Med. (Basel, Switzerland) 3, 25–41. https://doi.org/10.3390/medicines3040025

Dogan, G., Hayta, S., Yuce, E., Bagci, E., 2015. Composition of the essential oil of two Salvia taxa (Salvia sclarea and Salvia verticillata subsp. verticillata) from Turkey. Nat. Sci. Discov. 1, 62–69. https://doi.org/10.20863/nsd

Dugo, G., Di Giacomo, A., 2002. Citrus: The Genus Citrus, 1st. edition, CRC Press, London, Newyork.

Elshafie, H., Armentano, M., Carmosino, M., Bufo, S., De Feo, V., Camele, I., 2017. Cytotoxic activity of origanum vulgare L. on hepatocellular carcinoma cell line HepG2 and evaluation of its biological activity. Molecules 22, 1435–51. https://doi.org/10.3390/molecules22091435

Elshafie, H.S., Camele, I., 2017. An overview of the biological effects of some mediterranean essential oils on human health. Biomed Res. Int. 2017, 1–14. https://doi.org/10.1155/2017/9268468

Elshafie, H.S., Ghanney, N., Mang, S.M., Ferchichi, A., Camele, I., 2016. An *in vitro* attempt for controlling severe phytopathogens and human pathogens using essential oils from mediterranean plants of genus *Schinus*. J. Med. Food 19, 266–273. https://doi.org/10.1089/jmf.2015.0093

Elshafie, H.S., Mancini, E., Sakr, S., De Martino, L., Mattia, C.A., De Feo, V., Camele, I., 2015. Antifungal activity of some constituents of *Origanum vulgare* L. Essential oil against postharvest disease of peach fruit. J. Med. Food 18, 929–934. https://doi.org/10.1089/jmf.2014.0167

Esposito, E.R., Bystrek, M. V, Klein, J.S., 2014. An elective course in aromatherapy science. Am. J. Pharm. Educ. 78, 79. https://doi.org/10.5688/ajpe78479

Grunfeld, E., Gresty, M.A., 1998. Relationship between motion sickness, migraine and menstruation in crew members of a "round the world" yacht race. Brain Res. Bull. 47, 433–6.

Hammer, K.A., Carson, C.F., Riley, T. V, 2003. Antifungal activity of the components of Melaleuca alternifolia (tea tree) oil. J. Appl. Microbiol. 95, 853–60.

Herzi, N., Camy, S., Bouajila, J., Destrac, P., Romdhane, M., Condoret, J.-S., 2013. Supercritical CO_2 extraction of Tetraclinis articulata: Chemical composition, antioxidant activity and mathematical modeling. J. Supercrit. Fluids 82, 72–82. https://doi.org/10.1016/j.supflu.2013.06.007

Hillis, W.E., 1967. Polyphenols in the leaves of eucalyptus: A chemotaxonomic survey—II.: The sections renantheroideae and renantherae. Phytochemistry 6, 259–274. https://doi.org/10.1016/S0031-9422(00)82772-7

Howes, M.-J.R., Houghton, P.J., Barlow, D.J., Pocock, V.J., Milligan, S.R., 2002. Assessment of estrogenic activity in some common essential oil constituents. J. Pharm. Pharmacol. 54, 1521–1528. https://doi.org/10.1211/002235702216

Huiles essentielles, 2000. Monographies relatives aux huiles essentielles (Tome 2). AFNOR, Afnor, Paris.

Jayakumar, S., Madankumar, A., Asokkumar, S., Raghunandhakumar, S., Gokula dhas, K., Kamaraj, S., Josephine Divya, M.G., Devaki, T., 2012. Potential preventive effect of carvacrol against diethylnitrosamine-induced hepatocellular carcinoma in rats. Mol. Cell. Biochem. 360, 51–60. https://doi.org/10.1007/s11010-011-1043-7

Kim, S., Kim, H.-J., Yeo, J.-S., Hong, S.-J., Lee, J.-M., Jeon, Y., 2011. The effect of lavender oil on stress, bispectral index values, and needle insertion pain in volunteers. J. Altern. Complement. Med. 17, 823–826. https://doi.org/10.1089/acm.2010.0644

Koulivand, P.H., Khaleghi Ghadiri, M., Gorji, A., 2013. Lavender and the nervous system. Evid. Based. Complement. Alternat. Med. 2013, 681304. https://doi.org/10.1155/2013/681304

Krishna, A., Tiwari R., Kumar S., 2000. Aromatherapy-an alternative health care through essential oils. J. Med. Aromat. Plant Sci. 22, 798–804.

Lee, K.-B., Cho, E., Kang, Y.-S., 2014. Changes in 5-hydroxytryptamine and cortisol plasma levels in menopausal women after inhalation of clary sage oil. Phyther. Res. 28, 1599–1605. https://doi.org/10.1002/ptr.5163

Liu, S.-H., Lin, T.-H., Chang, K.-M., 2013. The physical effects of aromatherapy in alleviating work-related stress on elementary school teachers in Taiwan. Evid. Based. Complement. Alternat. Med. 2013, 853809. https://doi.org/10.1155/2013/853809

Maeda K, Ito T, S.S., 2012. Medical aromatherapy practice in Japan. Essence (Downsview). 10, 14–16.

Miyake, M., Shimoda, M., Osajima, Y., Inaba, N., Ayano, S., Ozaki, Y., Hasegawa, S., 2000. Extraction and Recovery of Limonoids with the Supercritical Carbon Dioxide Micro-Bubble Method, in: Citrus Limonoids. ACS Symposium Series, pp. 96–106. https://doi.org/10.1021/bk-2000-0758.ch007

Mossa, A.-T.H., Refaie, A.A., Ramadan, A., Bouajila, J., 2013. Antimutagenic effect of *Origanum majorana* L. essential oil against prallethrin-induced genotoxic damage in rat bone marrow cells. J. Med. Food 16, 1101–1107. https://doi.org/10.1089/jmf.2013.0006

Mulyaningsih, S., Sporer, F., Reichling, J., Wink, M., 2011. Antibacterial activity of essential oils from *Eucalyptus* and of selected components against multidrug-resistant bacterial pathogens. Pharm. Biol. 49, 893–899. https://doi.org/10.3109/13880209.2011.553625

Naves, Y.R., 1974. Technologie et chimie des parfums naturels; essences concrètes, résinoïdes, huiles, et pommades aux fleurs, 1st. edition, Masson, Paris.

Paulaitis, M.E., Krukonis, V.J., Kurnik, R.T., Reid, R.C., 1983. Reviews in chemical engineering., Reviews in Chemical Engineering. Freund Publishing House.

Pazyar, N., Yaghoobi, R., Bagherani, N., Kazerouni, A., 2013. A review of applications of tea tree oil in dermatology. Int. J. Dermatol. 52, 784–790. https://doi.org/10.1111/j.1365-4632.2012.05654.x

Perry, N., Perry, E., 2006. Aromatherapy in the management of psychiatric disorders: clinical and neuropharmacological perspectives. CNS Drugs 20, 257–80.

Price, S., 1993. The aromatherapy workbook: A complete guide to understanding and using essential oils, 1st. edition, Thorsons, London, UK.

Rana, V.S., Juyal, J.P., Amparo Blazquez, M., 2002. Chemical constituents of essential oil of Pelargonium graveolens leaves. Int. J. Aromather. 12, 216–218. https://doi.org/10.1016/S0962-4562(03)00003-1

Reyes-Jurado, F., Franco-Vega, A., Ramírez-Corona, N., Palou, E., López-Malo, A., 2015. Essential oils: antimicrobial activities, extraction methods, and their modeling. Food Eng. Rev. 7, 275–297. https://doi.org/10.1007/s12393-014-9099-2

Rigano, D., Formisano, C., Senatore, F., Piacente, S., Pagano, E., Capasso, R., Borrelli, F., Izzo, A.A., 2013. Intestinal antispasmodic effects of Helichrysum italicum (Roth) Don ssp. italicum and chemical identification of the active ingredients. J. Ethnopharmacol. 150, 901–906. https://doi.org/10.1016/j.jep.2013.09.034

Sadlon, A.E., Lamson, D.W., 2010. Immune-modifying and antimicrobial effects of Eucalyptus oil and simple inhalation devices. Altern. Med. Rev. 15, 33–47.

Satchell, A.C., Saurajen, A., Bell, C., Barnetson, R.S., 2002. Treatment of dandruff with 5% tea tree oil shampoo. J. Am. Acad. Dermatol. 47, 852–855. https://doi.org/10.1067/mjd.2002.122734

Sewram, V., Nair, J.J., Mulholland, D.A., Raynor, M.W., 1995. Open tubular supercritical fluid chromatography of triterpene acids from Dysoxylum pettigrewianum (Meliaceae). J. High Resolut. Chromatogr. 18, 363–366. https://doi.org/10.1002/jhrc.1240180608

Soden, K., Vincent, K., Craske, S., Lucas, C., Ashley, S., 2004. A randomized controlled trial of aromatherapy massage in a hospice setting. Palliat. Med. 18, 87–92. https://doi.org/10.1191/0269216304pm874oa

Stahl, E., Quirin, K.-W., Gerard, D., 1983. Solubilities of soybean oil, jojoba oil and cuticular wax in dense carbon dioxide. Fette, Seifen, Anstrichm. 85, 458–463. https://doi.org/10.1002/lipi.19830851202

Stea, S., Beraudi, A., De Pasquale, D., 2014. Essential oils for complementary treatment of surgical patients: state of the art. Evid. Based. Complement. Alternat. Med. 2014, 726341. https://doi.org/10.1155/2014/726341

Svoboda, K., Hampson, J., 1999. Bioactivity of essential oils of selected temperate aromatic plants: antibacterial, antioxidant, antiinflammatory and other related pharmacological activities, in: Proceedings NAHA, 25–28 September. St. Louis Missouri, USA, pp. 105–127.

Svoboda, K.P., Deans, S.G., 1992. A study of the variability of Rosemary and Sage and their volatile oils on the British market: Their antioxidative properties. Flavour Fragr. J. 7, 81–87. https://doi.org/10.1002/ffj.2730070207

Szentmihályi, K., Forgács, E., Hajdú, M., Then, M., 2001. In vitro study on the transfer of volatile oil components. J. Pharm. Biomed. Anal. 24, 1073–80. https://doi.org/10.1016/S0731-7085(00)00542-2

Tassou, C.C., Drosinos, E.H., Nychas, G.J., 1995. Effects of essential oil from mint (Mentha piperita) on Salmonella enteritidis and Listeria monocytogenes in model food systems at 4 degrees and 10 degrees C. J. Appl. Bacteriol. 78, 593–600.

Tisserand, R. 1988. Essential oils as psychotherapeutic agents. In S. Van Toller & G. H. Dodd (Eds.), Perfumery: The psychology and biology of fragrance (p. 167–181). Chapman & Hall/CRC.

Tisserand, R., Balacs, T., 1995. Essential Oil Safety: A Guide for Health Care Professionals, Illustrate. ed. Churchill Livingstone.

Tisserand, R., Young, R., 2002. Essential oil safety: a guide for health care professionals, 1st cdition cd. Churchill Livingstone, Edinburgh.

Vankar Padma S, 2004. Essential oils and fragrances from natural sources. Resonance 30–41.

Zhao, W., Shishkura, A., Fujimoto, K., Arai, K., Saito, S., 1987. Fractional extraction of rice bran oil with supercritical carbon dioxide. Agric. Biol. Chem. 51, 1773–1777. https://doi.org/10.1080/00021369.1987.10868305

Ziosi P, Manfredini S, Vertuani S, Ruscetta V, Radice M, Sacchetti G, 2010. Evaluating essential oils in cosmetics: Antioxidant capacity and functionality. Cosmet. Toilet. 125, 32–40.

CHAPTER 11

Understanding Classical Naturopathy: The Hippocratic Way of Healing

SRIJAN GOSWAMI[1*] and USHMITA GUPTA BAKSHI[2]

[1]*Department of Biochemic Medicine, Indian School of Complementary Therapy and Allied Sciences, Ichapur 743144, Kolkata, West Bengal, India*

[2]*Department of Biochemic Medicine, Indian School of Complementary Therapy and Allied Sciences, Ichapur 743144, Kolkata, West Bengal, India*

Corresponding author. E-mail: srijangoswamiigmgs@gmail.com

ABSTRACT

Naturopathy is an art of treatment, completely based on scientific principles, governed by natural laws. But due to the lack of proper knowledge of its concepts, this scientific practice is being represented as pseudoscience, nonevidence based, and are constantly being defamed. Thus, the current chapter is dedicated to understanding the modus operandi underlying this promising discipline. In this chapter, the authors discussed the postulates given by the Great Hippocrates about the art of healing which clearly indicates that the system of medicine that he taught people about is the naturopathic way of treatment. The sections of the chapter explain some of the postulates of the Great Hippocrates: how they form the basis of the Naturopathy system of medicine and how they are governed by the natural laws. In some instances, the chapter also provides a comparative view of how conventional medicine differs from the Naturopathy in their approach of dealing with patients and how integrating these two systems together will lead to a personalized healthcare system.

11.1 INTRODUCTION

Knowledge advances not by repeating known
facts but by refuting false dogmas.
—Karl Popper, n.d.

The author begins the chapter with the above quote because society is flooded by false beliefs and cultural myths. Cultural myths have been embedded deeply into the fabric of our society by industry-controlled media. The thing about the cultural myth is that one does not realize that they are in the grasp of a cultural myth, when in reality they are in a cultural myth—it is a reality by consensus. So when one looks around in the 21st century, the only medicine that is practiced in hospitals, the only medicine the insurance pays for, the only medicine they make television shows about, and the only medicine people are taught about in schools and that is provided with huge research grants is the conventional, industrially synthesized chemical drug-based system, known as Allopathy or Western Medicine. And whenever other types of medicines are talked about, the talk is disparaging. One such system which has been ignored by the media and is the victim of cultural myth is Naturopathy. The present chapter is dedicated to understanding the modus operandi underlying this promising field.

Naturopathy is an ancient and holistic form of treatment. But in the present age of scientific and technological advancements, this very promising system of healing has been marked as unscientific, pseudoscientific, and nonevidence-based procedures. And the medical professionals practicing Naturopathy are termed as "Quacks." Several reputed websites and professionals' claims that naturopathy practitioners are against conventional medical practices and protocols, vaccination, and surgery, and instead the Naturopaths depends on unscientific perceptions, often leading to diagnoses and treatments are not genuine. These are ad-hominem arguments—the arguments that do not argue the facts and disparage people. These types of arguments try to damage the prestige of their competitors, and this is exactly what happened to Naturopathy and other alternative medical practices for over the past 100 years.

In the present chapter, the authors pointed out the postulates given by the Great Hippocrates on the art of healing, which clearly indicates that the system of medicine that he taught people about is the naturopathic way of treatment. The chapter also covers the controversial historical events that took place under the direction of Carnegie and Rockefellers, which led to the rise of modern medicine and defamation and illegalization of all other medical practices. The following sections explain some of the postulates of the Great Hippocrates: how they form the basis of the naturopathic system of medicine and how they are governed by natural laws. In some instances, the chapter also provides a comparative view of how conventional medicine differs from the naturopathic system in their approach of dealing with patients. And since Naturopathy is a wholistic treatment approach, whenever the word "wholistic" appears in the chapter, it will indicate the naturopathy or naturopathic system of medicine.

11.2 HISTORICAL PERSPECTIVES

When it comes to treatment of diseases, it is believed that conventional medicine is the only system which is evidence based and scientific. This belief is not completely true. This is because people are taught only a certain portion of scientific history from which money-making industries can easily control and manipulate masses (Angell, 2005). In the book titled *The Medical Mafia*, authored by Dr Ghislaine Lanctot, it is clearly described how pharmaceutical industries control medical education, research, and delivery, and how they control medical universities, journals, and professionals for their personal profit by legalizing medical practices that in reality are not helping the society at all (Lanctot, 1995).

People have to realize that medical institutions, all the media, medical journals, the newspapers, television, and radio are totally controlled by Big-Money (Lanctot, 1995). In the early 1990s, natural systems of medicine were very popular. Physicians used to heal patents and treat complicated diseases using different kinds of remedies available in nature such as herbs, exercises, dietary regulations, energy therapies, lifestyle modifications, and counselling. Several types of medical practices existed quite harmoniously. All forms of medical systems were equally practiced, acknowledged, and respected. There were mutual respect and harmony among practitioners belonging to different ideologies. But this harmony was disrupted permanently after the approval of Flexner Report (Lanctot, 1995). In 1907, American Medical Association made an elaborate survey of medical education systems of United States and Canada. This survey was conducted with the help from the Carnegie Foundation and was named Bulletin Number 4. During this time period, industrialists such as Rockefeller and Carnegie found out ways to isolate and patent petrochemical compounds that could be used as drugs for treatment of diseases. They planned to make those petrochemical-based patented drug molecules mainstream. For achieving that goal, they appointed a person named Abraham Flexner who was neither a physician nor a scientist.

He was given the authority to visit all the medical schools for inspection and create a report based on that. The purpose of the report published in 1910 was to create medical establishment that promotes petrochemical-based patented drug molecules as mainstream medical practice. According to the Flexner report, those medical schools that were fit for and agreed to teach and promote the petrochemical-based patented chemicals as drugs were legalized as scientific, and all other medical practices that did not agree were illegalized. So, all the medical schools except those that were running in terms of the Flexner continued to grow, whereas the other medical schools were shut down overnight. In addition to that the Carnegies and Rockefellers gave millions of dollars of grants to the hospitals and medical schools that practiced allopathic medicine. They also glad-handled legislations to pass laws that made it illegal to practice medicine unless they were an MD practicing petrochemical-based allopathic medicines, thereby delivering patented pharmaceuticals. And finally, the topics on natural methods of healing and details about herbal medicines were completely omitted from the medical courses so that people do not get to learn anything about natural modes of healing; they should only get to know about chemical drugs, reactions, etc. designed to prescribe more and more pills (Glidden, 2010). All the natural systems of medicine such as Naturopathy, Herbalism, Homoeopathy, and Chiropractic were demoted as quackery. In 1997, the World Health Organization obtained full control over medicine as the validity of the Flexner report was extended worldwide (Glidden, 2010). This was the beginning of the end of all natural systems of medicine, and this is why all other medical practices currently are referred to as alternatives, not because they are substandard, but because the big pharmas have occupied the first place seat for so long that this is just what we have become used to (Glidden, 2010).

11.3   DIFFERENT TYPES OF MEDICINES PRACTICED IN THE WORLD

The science of medicine is vast as an open sky. There exist different types of medical practices all around the world, which are completely scientific. It is not possible for a single person to be a master of medicine. Similar to the conventional physicians with the Master of Medicine (MD) degree are specialized in a single form of medicine known as Allopathy, the physicians who have qualified in an equivalent MD degree and received a practice license from an accredited medical school on Naturopathy, uses "ND" after their names. Figure 11.1 gives an overview of different forms of medical practices that are practiced worldwide in a hierarchical manner.

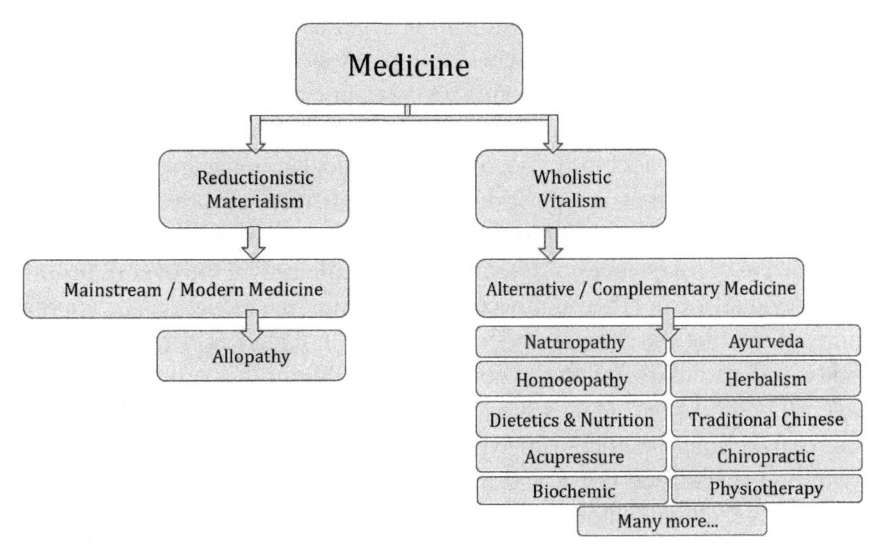

FIGURE 11.1   Different systems of medicine.

The Allopathy or Western Medicine is known to all. There are other scientific medical practices that coexist and are recognized by the World Health Organization for treatment, education, and research; they include naturopathy, homoeopathy, Ayurveda, biochemic medicine, traditional Chinese medicine, acupressure, acupuncture, pranayama, yoga, nutrition and dietetics, physiotherapy, chiropractic, and herbalism. All these medical practices are classified into two broad medical philosophies: the reductionism and the wholism. Now let us understand the concept of reductionism and wholism with reference to medical science.

> *Any understanding of human illness or wellness has*
> *to be in totality and not in bits and pieces.*
> —Hedge, 2014

> *Human body does not work in bits and pieces but functions as a whole.*
> —Hedge, 2014

Thus, when it comes to the approach of treating diseases, two completely different ideologies exist: (a) reductionistic approach and (b) wholistic/holistic approach.

Conventional medical science follows reductionistic ways, but if one uses their common sense, he/she can easily realize that the universe is working as a whole and not in bits and pieces. The following section explains the concept of reductionism and wholism in relation to human health in brief.

In the reductionistic approach, or Allopathy, every organ or components of a body are viewed as discrete units working in isolation with other organs or components of that body. Sir Isaac Newton's postulates make the foundation for reductionistic ideology. So, when a particular organ gets affected, the treatment procedures are performed and medicines are prescribed with an intention to providing relief to the particular organ under consideration. The effects and consequences of treatment and medicines on the overall health of an individual are not thoroughly considered. Thus, the reductionistic approach aims at treating the disease under consideration that is evident in a physical body. It does not deal with the consequences it imposes on other organs and overall homeostasis. This is the reason behind specialized disciplines under the conventional medicine in which every discipline is concerned strictly with one specialized organ and their issues. According to the reductionistic perspective, a human body is considered a machine made out of parts, and if anything goes wrong, the only way to manage the situation is by administering industrially synthesized chemical drugs or surgery. Conventional medicine provides excellent results in the case of emergency and critical care where the vital force is so weak that immediate quick fixes are required—also in surgical cases and trauma care (Glidden, 2010).

On the contrary, in the wholistic system of treatment, for example, in naturopathy or naturopathic medicine, every single component of the body is regarded as a functional whole. The body components are functionally and structurally interconnected with each other at a subatomic level (Jain, 2015). In this approach, a body is considered as complex combinations of multiple specialized systems functioning at different levels (Jain, 2015). These systems carry out their individual functions and also operate in synchronization with other components of the body, thereby making the life possible (Jain, 2015). The maintenance of homeostasis is the ultimate objective of these systems. As long as the homeostasis is maintained, the body is protected from diseases. While addressing a certain disease, the effort is made to determine the cause that is responsible for the derangement of homeostasis. The treatment regimen is prescribed keeping in mind that for curing a disease, the cause must be identified and addressed. Special care is taken to ensure that the prescribed treatment regimen does not derange the overall homeostasis. Naturopathic system treatment provides excellent results in chronic and lifestyle diseases and stimulating quick recovery after surgical interventions (Lipton, 2016).

Conventional medical science (reductionism) believes that everything that exists and happens around us can only be explained through concepts of Newtonian Physics. However, in the wholistic system, there exists a natural healing force/vital force/vital energy, which cannot be explained using Newtonian Physics. The presence of this vital energy drives life and is responsible for maintaining homeostasis.

The reductionistic or mechanist view has predominated in the 20th century because virtually all information gathered from observation and experiment has agreed with it. But over the past few years, lots of evidence has been published and accepted worldwide on the authenticity of the concept of Vitalism. It was shown by an Alternate Nobel Laureate, Sir Hans Peter Durr, the Director of Max Plank Institute, in his scholarly work titled "Matter is not made out of matter." In his scholarly work, it was stated that to actually understand the authenticity of the wholistic view of life, one need to understand the laws of quantum physics—the physics based on which the entire universe actually works. The authenticity of Vitalism was also proved by renowned scientists around the world, but among all, the two most notable works were the book written by Dr Bruce Lipton, a well-known cell biologist and professor of medicine, *The Biology of Belief*, and another book written by Dr Amit Goswami, a well-known quantum physicist, *The Quantum Doctor* (Hegde, 2014). One must not forget what the Father of Medicine the Great Hippocrates himself postulated and the author(s) quotes

Everyone has a doctor in him or her, we just have to help it in its work. The Natural Healing Forces in each of us are the greatest Force in getting well. Nature itself is the best Physician. (Hippocrates, n.d.)"

The wholistic approach of healing, such as Naturopathy, strictly agrees with the above lines postulated by the Great Hippocrates about the inherent wisdom of human body or consciousness. The human race would have been extinct from the face of the Earth, if the statement that drugs, preventive screening of apparently healthy population, advanced medical intervention, or vaccinations were necessity for keeping human alive. The Mother Nature has provided human beings with her inbuilt doctor that in the modern day known as the human consciousness (Lipton, 2016). The human body is a colony of approximately 100 trillion individual cells working in tandem with each other. Each and every cell has its own intelligence, gathered from its ancestors based on their environment and associated evolutionary changes. Evolution of human does not follow Darwinian laws, because there is natural evidence that the environment is actually responsible for control of gene expression (Lipton, 2016). Human cells have survived on the planet for millions of years and possess a huge amount of information about the nature around them. Each and every cell that makes up a human body has its own individual consciousness that can be termed as the intelligence of the body. So, each and every cell's own wisdom is much greater than the amount of knowledge gathered by human race from literatures and media. The human body knows how to defend itself from threats found in nature and how to adjust itself according to the environment, and it knows very well how to heal itself (Bollinger, 2015). The naturopathic physicians are of the belief that nature has provided human body with inherent consciousness that are not measurable with instruments available to mankind, but its presence and effects can be felt in daily life. So the human body at any moment in time, wants to fix itself, knows how to fix itself, and is constantly trying to fix itself (Glidden, 2010). The above statement has been represented in Figure 11.2.

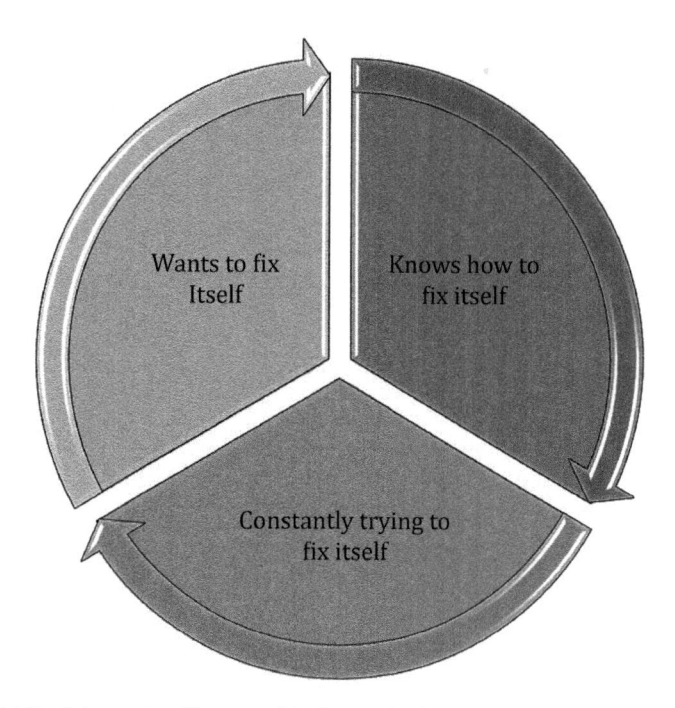

FIGURE 11.2   Inherent intelligence of the human body.

11.4 RESTORATION OF HEALTH: THE NATUROPATHIC APPROACH

According to the World Health Organization, HEALTH is defined as

"A state of complete physical, mental and social well-being and not just the absence of disease or infirmity."

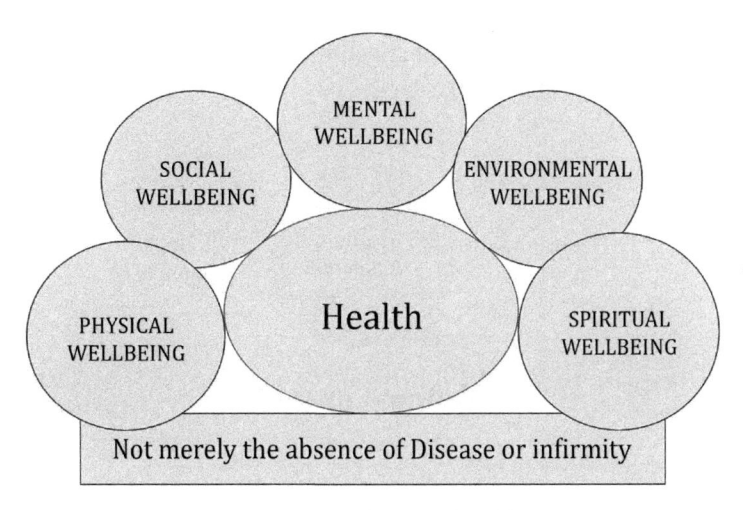

FIGURE 11.3   Pictorial representation of definition of health.

If the definition of health is analyzed in a proper manner, then it will be clear that for maintaining proper health one needs to consider physical, mental, social, environmental, and spiritual aspects of the patient, and not just the disease or infirmity that is clinically observable. So, in order to restoring a patient's health, a medical professional should consider all the given aspects as a whole. But if one carefully observes and analyzes the approach of the conventional system in chronic and lifestyle diseases, he/she can clearly see that it only takes into consideration the physical well-being along with the presence or absence of any disease or infirmity. Conventional medicine does not consider mental, social, environmental, or spiritual well-being when it comes to healing a patient. And in some rare cases where they consider the mental aspect along with the physical one, they end up charging medications for brain, on the basis of a misunderstanding that mind is in the brain (Lipton, 2016).

Now, let us understand what Naturopathy is, how naturopathic medicines or naturopathy approaches in order to address chronic illnesses or lifestyle diseases and restore health.

Naturopathic physicians treat the patient and not the disease. Naturopathy believes that the intelligence of the patient's body heals itself only with the components that are natural to them. The naturopathic system believes that health is composed of eight interconnected dimensions, namely physical, intellectual, environmental, emotional, financial, social, spiritual, and occupational. The eight dimensions of health as per the naturopathic system are depicted in Figure 11.4. According to Naturopathy, a physician must consider all the eight aspects that constitute health while restoring the health of the patient (Glidden, 2010).

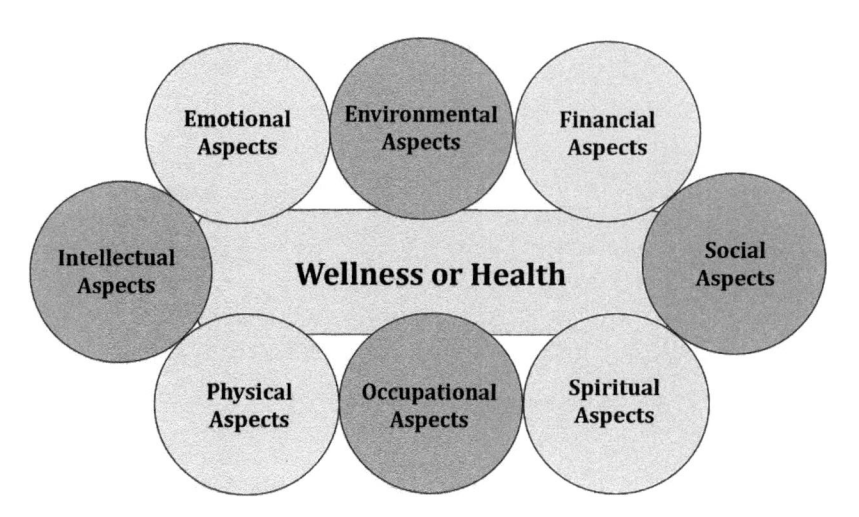

FIGURE 11.4 Naturopathic concept of health.

Since the health is a multidimensional phenomenon, the ideal approach for restoration of health should also be multidimensional. Treatment procedures should be able to address all the eight dimensions of health in an appropriate manner. To address these multidimensional aspects of health, the naturopathic system of medicine is equipped with a diverse range of treatment modes and strategies that are completely natural to the human race. Naturopathy implements biologically based approaches (herbal remedies, diet and nutrition, and lifestyle management, etc.), manipulative and body-based therapies (physiotherapy, acupressure, acupuncture, chiropractic, etc.), mind–body interventions (yoga, spiritual training, pranayama, etc.), alternative medicines (Ayurveda, Homoeopathy, Biochemic, etc.) and energy-channeling therapies (Qigong, magnetic therapy, meditation, etc.); all these aspects are represented in Figure 11.5.

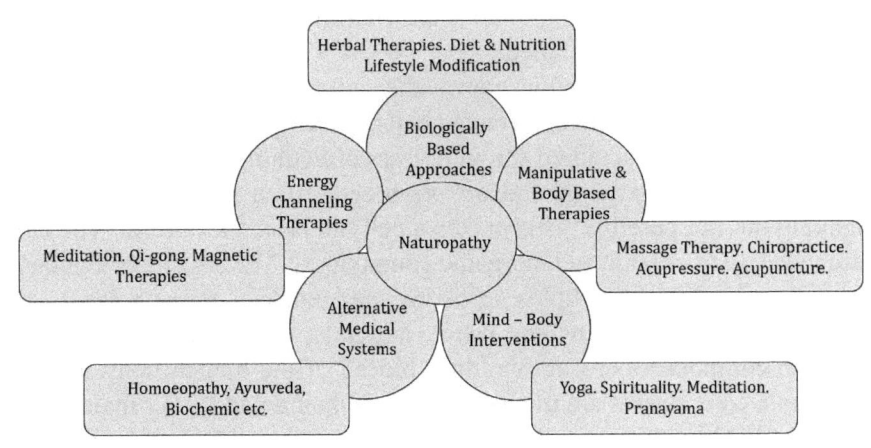

FIGURE 11.5   Treatment modes under naturopathy.

So, from the above discussion, it is clear that the naturopathic system of medicine possesses techniques that efficiently addresses multidimensional aspects of health restoration as compared to the single-dimensional approach taken by the conventional system of medicine in treating chronic illnesses and lifestyle diseases.

11.5   FUNCTIONS OF A HUMAN BODY AND THE NATUROPATHIC VIEW OF DISEASE DEVELOPMENT

> *Everyone has a doctor in him or her, we just have to*
> *help it in its work. The natural healing for within each*
> *one of us is the greatest force in getting well.*
> —Great Hippocrates, n.d.

> *Natural forces within us are the true healers of the disease.*
> —Hippocrates, n.d.

In the above statements, the Great Hippocrates mentioned that a natural healing force or a vital force exists that drives life and keeps human beings healthy. When in disease, this inherent natural energy heals all the derangements that occurred as a consequence of the disease. He clearly explained that the human body is endowed with built in capability to protect itself from diseases. The natural healing forces described by the Great Hippocrates and the vital force postulated by Sir Samuel Hahnemann was in reality the same concept. The natural healing force is the vital force. This is the very basis of the naturopathic system of medicine.

As per the naturopathic view, the human body is composed of approximately 100 trillion individual cells. Each of these cells possesses its own consciousness acquired through years of evolution. The cells perform their individual function and also work together in a synchronous manner, making the life possible. This makes a cell the basic unit of life. The cells are made up of organic and inorganic components. The organic components include carbohydrates, lipids, proteins, and nucleic acids, whereas the inorganic components include water, minerals, vitamins, and gases. The organic components are the building blocks of the human body, and the inorganic components are the workers of the human body. For maintaining absolute health, a proper balance between the organic (the building blocks) and inorganic (the workers) components are crucial. The proper balance and synchronized interaction among these components, guided by the natural forces, are the key to the maintenance of homeostasis or state of health. Figure 11.6a is a schematic representation of the maintenance of homeostasis or health.

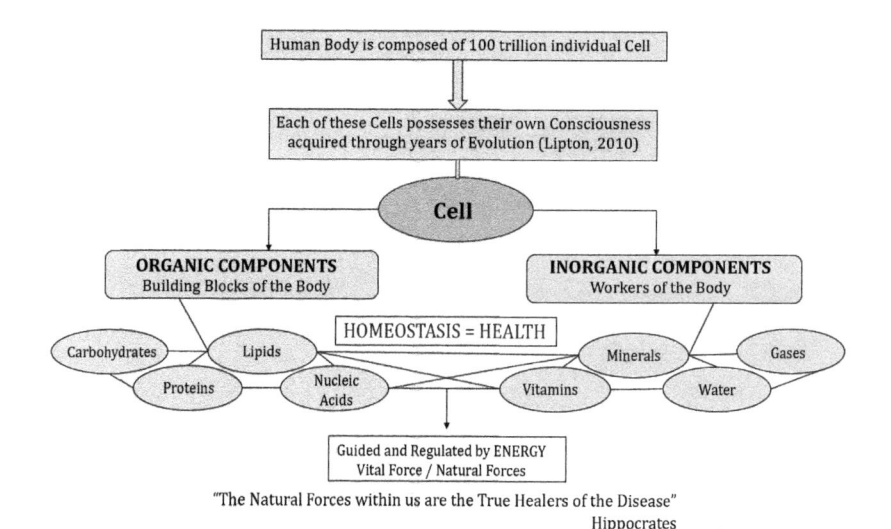

FIGURE 11.6a   Naturopathic view of state of health.

These organic and inorganic components work by interacting with each other and not in an isolated manner. When the balance and synchronization among the building blocks and the workers of the body gets disturbed, the homeostasis gets disrupted. Disturbed homeostasis is disease, and whatever the disease may be, from common cold to cancer, originates at this very point (marked with the cross in Figure 11.6b).

Just as food causes chronic illness, it can be the most powerful cure.
—Great Hippocrates, n.d.

Naturopathic physicians only provide the components that the human body needs to perform the defense and the healing mechanisms in an efficient manner. These components that the Naturopaths provide to their patients are organic or inorganic constituents that the human body is made up of, and sometimes the appropriate ENERGY that controls the molecular motion of these organic and inorganic components and compels them to act. This is the basis of Dr Hahnemann's vital force theory. The natural healing force proposed by the Great Hippocrates and the vital force proposed by the Great Hahnemann are in reality the same concepts.

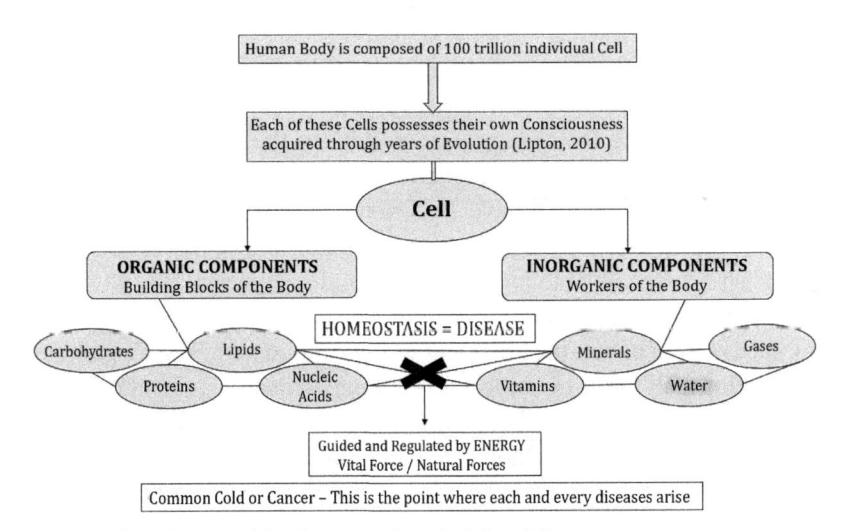

FIGURE 11.6b Naturopathic view on point of origin of disease state.

The greatest medicine of all is teaching people how not to need it.
—Great Hippocrates, n.d.

The naturopathic physicians believe, in fact, that natural healing forces within each of us are the greatest force in getting well, so they believe in spreading awareness among people that not all diseases need chemical pills or surgeon's knife. No physicians have ever cured any diseases. The healing, if occurred, was completely done by the intelligence of the human body itself. Naturopathic physicians just provide them with the organic substances that are the building blocks of the body, the inorganic components that are the workers of the body, and appropriate energies that compel the workers of the body to manage and channelize the building blocks in response to healing mechanisms. This is why a majority of the naturopathic practices such as physiotherapy, chiropractic, and acupressure do not use any medicines. Simple lifestyle modification and healthy diet habits can provide protection from several critical diseases. Naturopathy encourages people to spend their life in synchronization with nature because human beings are the intricate part of this nature. As long as human beings lead their life based on the laws of nature, the health is maintained. As the great Father of Medicine said:

> *Illness do not come upon us out of the blue. They are developed from small daily sins against Nature. When enough sins have accumulated, illness will suddenly appear.*
>
> —Hippocrates, n.d.

Based on the above saying of the Great Hippocrates, Naturopaths believe that living in synchronization with nature is the fundamental aspect of health, and any disturbance in these process results in the deterioration of health. Let us understand the above statement from the naturopathic point of view. Naturopathic physicians believe that none of the chronic diseases occur overnight but follow a progressive growth pattern. The human body always gives us stress signals in the form of signs and symptoms. It is a tendency of most of the people to ignore these mild stress calls given by the body. This ignorance is what makes the treatment more complicated. When a pathological state develops, all forms of tissue injury initiate with molecular alteration in the cells. The normal cell is biochemically restless and is constantly modifying its structure and function in response to changing demands and stresses present in its internal as well as external environment. In a normal condition, the cells tend to maintain an equilibrium by maintaining the fluid balance, temperature balance, glucose balance, amino acid balance, fatty acid or glycerol balance, vitamin balance, pH balance, electrolyte balance, and oxygen–carbon dioxide balance. This state of equilibrium is termed as homeostasis. As long as homeostasis is maintained, the body is protected from disease. When the cell encounters excessive physiologic or pathologic stresses, it undergoes the adaptation process and achieves an altered but steady state while maintaining the homeostasis despite the continued stress. But in the conditions where no adaptive response is possible or cell's adaptive capability is exceeded, the homeostasis gets disturbed and cell injury develops. Up to a certain point, cell injury is reversible, but with severe or persistent stress, the cell suffers irreversible injury and is fated to die (Abbas et al, 2004). This is what happens in a majority of chronic cases. The patient visits a physician when it is too late to reverse the pathologic damage. All of the diseases that the modern society have are the direct result of bad lifestyle choices.

11.6 NATUROPATHY AND USE OF DRUGS

It is a common belief among people that naturopathy does not support the use of drugs or does not believe in administering drugs when needed. This is a false belief. Naturopathy is not against drugs. The naturopathic system uses drugs and very well acknowledges the role of drugs in restoration of health. For understanding when and why naturopaths use and prescribe drugs, the reader should understand the function of drugs in the human system and how does a drug achieve its function.

Let us first understand how drugs actually function in the human system. It is our common belief that drugs are the component that cures any disease state. If that were the case, then why a patient is prescribed more and more types of chemical drugs as the time progresses and not less and less? The reason behind this is our half-hearted understanding of drug activity. Figure 11.7 explains the function of drugs in a human body in a concise manner.

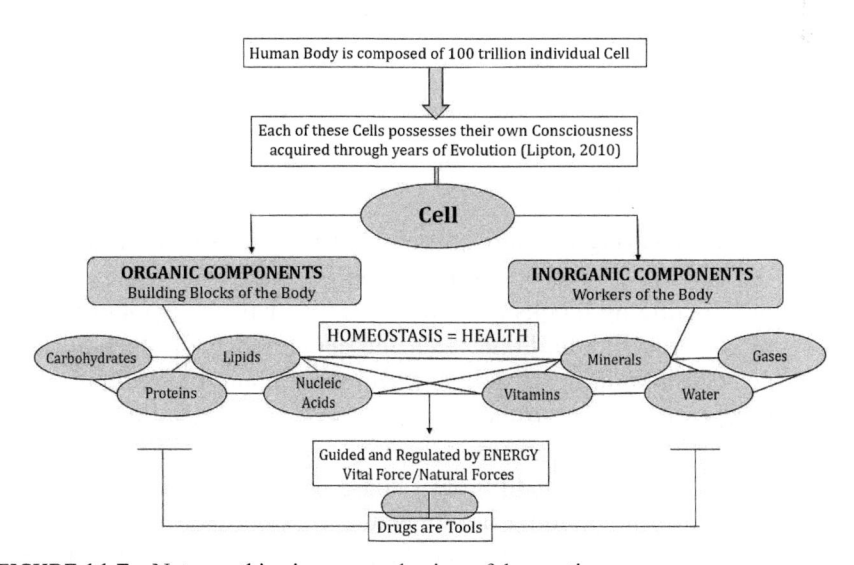

FIGURE 11.7   Naturopathic view on mechanism of drug action.

The mother nature has provided human beings with in-built doctors that in modern day are known as the immune system or healing mechanisms of the body. A human body is a colony of 100 trillion individual cells working together in a synchronized manner. As explained in an earlier section, each and every cell has its own intelligence, gathered from its ancestors based on its environment and associated evolutionary changes (Lipton, 2016). Human cells have survived on the planet from millions of years and possess a huge amount of information about the nature around them. Each and every cell that makes up a human body has its own individual consciousness known as the intelligence of the body. So, a cell's wisdom is much greater with respect to all kinds of knowledge gathered by human from literatures and media combined. Our body knows how to defend itself from threats found in nature, how to adjust itself according to the environment, and how to heal itself (TTAC, 2016). The human body is well equipped to protect itself from the diverse array of stresses. But there comes a situation when the healing ability of the body needs some extra push. This is where outside intervention in the form of drugs (natural or industrial origin) is required. As per naturopathy, drugs are neither the workers nor the building blocks of the human body. Drugs are the just tools utilized by the workers (inorganic components) and building blocks (organic components) of the body for achieving the healing process in an efficient manner in cases where the natural healing force of the patient is not powerful enough to take over the disease force all by itself. Drugs are tools that help in achieving the healing processes by up-regulating or down-regulating certain biochemical pathways. In such cases, drugs should be used only to reverse the pathologic state by stimulating natural healing mechanisms of the body.

For the proper understanding of this concept, let us consider an analogy. Suppose for constructing a building, certain amount of building blocks and workers are required.

- While repairing a small broken house, if the situation arises that the number of workers is more than required, whereas the amount of the building blocks needed is less than half the amount required to construct the building, then the construction of the building is not possible until the deficiency in the amount of building blocks is replenished even if there exists more than the required number of workers.
- While repairing a small broken house, if the situation arises that the amount of the building blocks is in excess but there are no workers who know how to use the building blocks to construct the building,

then the construction of the building is not possible in the absence of proper workers.

- Now, suppose if it is required to repair a high-rise building, even if an adequate amount of building blocks and more than required amount of workers are provided, there will still be a need of proper tools that the workers will be needing to put the building blocks back in the places where needed to complete the restoration. It is wise to use the tools only when needed. But it should also be realized that without the building blocks and the dedicated workers who know how to use the tool, the tool itself is of no use because the tool itself is not going to perform the function of the workers or building blocks all by itself.

Similarly, it should be noted that in the case of any disease, providing only the drugs to the patients will not solve the issue. One needs to provide appropriate types of inorganic and organic components along with the drugs in a proper amount, form, and potency because without proper building blocks and proper workers, the tools are of no use. This is the basis of use and selection of drugs as per the naturopathic systems of medicine. Naturopaths mainly use herbal and mineral substances that are natural in origin so that the components can get assimilated in the human system readily and without any adverse effects.

11.7 EFFICIENCY AND SAFETY OF HERBAL MEDICINES PRESCRIBED IN THE NATUROPATHIC SYSTEM AS COMPARED TO ALLOPATHIC DRUGS

People are made to believe that herbal drugs or herbal products are not real medicine and are not evidence based. But this kind of thinking is not at all correct. It was proved in a study performed by a professor of Genetics, Dr D. C. Wallace, in his Mitochondrial Chip experiment that each and every industrially synthesized chemical drug, from simple aspirin to complex statin, when enters the human body, the human body system recognizes that as foreign. The intelligence of the human body after recognizing them as foreign transports them to the liver. The liver tries its best to destroy the chemical drug that is thrown to it, and in that process the liver itself gets damaged. There comes a point when the liver becomes so damaged that it fails to neutralize the drug molecule completely. This is one of the major causes of non-alcoholic cirrhosis of liver (Hegde, 2014). The drug molecules that remain, escape the liver and start getting mixed with blood, and through blood, those drug molecules are deposited on the tissues. This phenomenon is called the first pass effect. The gathering of the drug molecules in the tissues and the bloodstream disrupts the normal functioning of the body and thus leading to adverse drug reactions. However, when the same experiment was performed with a wide variety of herbal drugs or natural supplements, the outcome was astonishing. The body recognized each and every herbal drug as a known substance, and the natural components readily got absorbed into the blood stream and was transported to the tissues requiring the remedies. And no adverse reactions were seen with natural remedies (Wallace, 2008). The entire process is schematically represented in Figure 11.8.

From the experiment conducted by Dr Wallace, it was evident that since the human race and the substance that occurs in nature have coexisted for a long time and undergone evolutionary changes together, the consciousness of human cells acquired through time and experience has certain memory that recognizes the naturally occurring nutrients as safe. On the other hand, the counterpart of the same chemical, when synthesized artificially, is recognized as foreign, even if the artificially created compound exactly possesses the same structure and function as their natural counterpart. Although more research needs to be performed in this regard, but the Mitochondrial Chip experiment made it clear that herbal remedies are much more effective, efficient, and safe than their synthetic counterparts.

11.8 NATUROPATHY AS A PERSONALIZED SYSTEM OF MEDICINE

> *It is more important to know the person who has the*
> *disease, than it is to know the disease the person has.*
> —Hippocrates, n.d.

> *Positive health requires knowledge of man's primary constitution, and the*
> *powers of various foods, both natural to them and those resulting from*
> *human skills. But eating alone is not enough for health. There must also*
> *be exercise. The combination of these two things makes regimen.*
> —Hippocrates, n.d.

According to the personalized system of treatment, it is considered that each and every individual is unique at the proteomic level. The genotype of an individual is determined by the diverse set of genes present in his/her body. But the presence of the set of genes does not ensure their expression in a person's lifetime. Instead the phenotype of an individual is determined by the products of the genes or proteins. The expression of genes is directly regulated by the surrounding environment a person lives in, his/her lifestyle, as well as his/her thought processes (epigenetics).

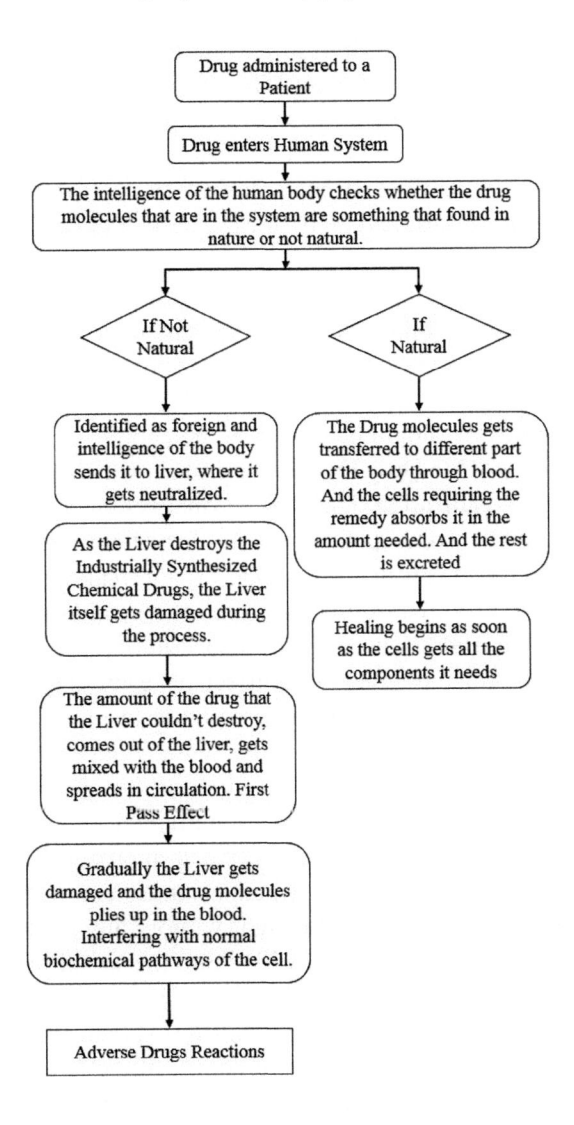

Mitochondria as Chi (Wallace DC, 2008).

FIGURE 11.8 The fate of different drugs inside human system as performed by Dr Wallace.

So, the genotype of individuals may be the same, but every individual becomes unique at the proteomic level. The word "proteome" refers to the set of all the gene products or proteins present in an individual. Individuals' susceptibility to a disease and their response to a particular drug is influenced by their proteome. This defines the difference in the constitution of a patient. So, in the case of a particular disease, the efforts are made to determine the treatment procedures, and medications are prescribed that are compatible to one's constitution. This system aids in the selection of the appropriate remedy for the patient under consideration, without any side effects (Jain, 2015).

Naturopathy is a personalized system of medicine because it considers the constitution of the patient while selecting a treatment regimen. Personalization of treatment is not a new concept. The Great Hippocrates postulated that the practice of medicine and treatment of patient should be personalized, and he also postulated how it should be done. On the contrary, the conventional medical science takes "One Drug Fits All" approach, which is exact opposite of the concept of personalization. The conventional system of medicine treats the disease, whereas the naturopathic system of medicine treats the patient who has the disease. Naturopathic physicians do not treat the symptoms but aim at addressing the root cause that triggered the symptoms. One should understand that each and every person is constitutionally different. One particular drug will not have the same effect on each and every person. Because every person is different at the proteomic level and this is where the drugs of conventional medicines act, the "One Drug Fits All" approach can only deliver symptomatic relief and palliation but can never cure the disease under consideration. Since every individual is constitutionally different, his/her response to a certain stress and approach to cope with it is also different. For example, when a patient under conventional system complains about pain, vomiting, and cough, without any delay, the patient will be provided medications that act by blocking the signal pathway that is triggering the symptoms. In this way, the body's ability to generate the stress call is kept deactivated with medication, and the patient and their family live with the illusion that the problem is cured. But if the cases are carefully studied in detail, it will be observable that every patient complaining about having pain reacts to it and deals with it in a different manner, and this is not coincidence, but because of the fact that each and every individual is constitutionally different. The name of the symptom they are having may be the same but their physiological, psychological, and even social responses to the symptom under consideration are diverse. This is because they are constitutionally different. And consequently, according to the naturopathic system, the approach of dealing with each and every clinical case should be patient-centric and not disease-specific.

11.9 HIPPOCRATES, NATUROPATHY, AND THE ART OF HEALING

> *If we could give every individual the right amount of*
> *nourishment and exercise, not too little and not too much,*
> *we would have found the safest way to health.*
> —Hippocrates, n.d.

> *Leave your drugs in the chemist's pot if you*
> *can heal your patient with food.*
> —Hippocrates, n.d.

In the first statement, the Great Hippocrates clearly stated that the right amount and type of diet along with appropriate forms of exercise are the safest way to keep a person healthy. He emphasized on the concept of diet and nutrition therapy and exercise therapies (physical exercises, yoga, pranayama, massage, etc.). In the second statement, the Great Hippocrates clearly stated that a physician must try to heal the patient with food and nutritional substances first and not to use chemical drugs if the patient is curable through natural substances and dietary management. But unfortunately, when a naturopathic physician tries to heal a patient with nutritional management, he/she is tagged as an "unscientific practitioner. It is believed that foods cannot be medicines, and the only substances that can restore heath are industrially synthesized chemical drugs. On the contrary, the naturopathic system of medicine firmly believes that *we are made up of what we eat*, because it is from the food itself the human body gets its building blocks and nutrients required for normal functioning. As the great Father of Medicine also stated:

> *Each of the substances of a man's diet acts upon his body and changes it*
> *in some way and upon these changes his whole life depends.*
> —Great Hippocrates, n.d.

The body of person who will have poor nutrition or wrong dietary choices will be deprived of its components required to heal the daily wear and tear that occurs in the body. And consequently, the person will die, whereas the exact opposite will be the case in those who are under proper and healthy dietary choices. In naturopathy, maintaining a healthy dietary regimen as per the need of an individual is key to health and can protect the body from severe illnesses.

*While selecting a remedy the attention should be given
to the season of the year, changes in the wind, and the
age of the individual, situation of his home.*

—Hippocrates, n.d.

If one analyzes the above statement, it will be evident that the Great Hippocrates clearly stated that the factors such as the weather condition, season of the year, social behavior, age, occupation, and the mode of adobe of the patient should be considered while choosing a remedy. Complete knowledge about these aspects is important because a patient is the innate part of the environment he/she lives in. These environmental conditions influence and drive the life force of the patient and by extension possesses the ability to interfere with the drug action.

This ideology is one of the pillars of naturopathic treatment. In naturopathy, it is important for physician to consider all the aspects that influence the day-to-day activities of the patient under consideration. This wholistic approach helps a naturopathic physician to design an effective and personalized treatment regimen for the patient under consideration. Furthermore, naturopathy uses various natural diagnostic methods to pinpoint the type and progression of the pathologic state, for example, pulse diagnosis, facial diagnosis, nail diagnosis, tongue diagnosis, acupressure reference points on palms of the hand, and soles of feet.

The Great Hippocrates also stated that understanding the nature of spine is very important. Due to wrong body posture or certain day-to-day activities, the normal alignment of the vertebrae gets disturbed, thereby resulting in serious health complications and even death. In such cases, the proper adjustment of the spine is important for addressing those complications. This statement is an indication that the Great Hippocrates presented the idea of chiropractic. Chiropractic is a form of treatment in which healing is achieved by the manipulation of the musculoskeletal system, especially spine. The major goals of these kinds of manipulations are to provide relief from pain, functional improvement of body parts, and triggering body's inherent healing ability.

Mainstream scientific communities believe that energy-based therapies or energy-channeling therapies are unscientific and lack clinical relevance. This kind of thinking is not correct. The Great Hippocrates said that there were many experienced physicians who mastered the art of healing patient by placing their hand on the affected part and transferring the healing energy to the patient. The Great Hippocrates stated that when he was healing his patients, he felt that there were remarkable property in his hands that could remove the impurities or complications from the affected part just by laying the hand upon the place.

The statement indicates the Great Hippocrates himself performed and provided the concept of energy-channeling therapy like Reiki and other energy therapies. But still these practices have been tagged as unscientific, and physicians practicing these energy therapies are called "Quacks." The major reason behind this is our lack of knowledge and the illusion that we know everything about how nature functions. There is a need for understanding the basic principles based on which the energy-channeling therapies work. One should realize that creating energy-channeling therapists in medical schools or universities like other medical professionals are created is not an easy task. People may have medical degrees in energy therapies, but genuine energy therapists are hard to become. Energy therapies are based on the concepts of quantum physics. It is not possible to explain the physics behind the mechanism of action of energy therapies in a single chapter. The actual mechanism of action calls for a detailed chapter of its own. But for the in-depth understanding, it is recommended that the readers should follow the research works and evidence presented by Dr Larry Dossey, Dr Hans Peter Durr, Dr Bruce Lipton, Dr Amit Goswami, and Dr Candice Pert, details of which are given in the "References" section (documentaries) of this chapter. According to these concepts, the matter is not of particle nature (reductionism) but is of energies or waves (wholism). Unlike particles, the waves cannot be separated, and thus we are not particles colliding with each other but energy waves interacting with each other (quantum entanglement). And there is either positive energy or negative energy. All atoms absorb energy and all atoms give off energies, and when two energies interfere, one can change the power from destructive interference (state of disease) to constructive interference (state of health), and this is the physical foundation of hands on healing (Lipton, 2016). And according to the second law of thermodynamics, energy flows from its higher state to its lower state until it reaches equilibrium. This is exactly what happens in the case of energy-channeling therapies. The professionals performing energy therapy should themselves possess a high level of positive energy (a state of consciousness which is not easy for a person to achieve) and establish an equilibrium with the patient whose body is lacking the appropriate energy level. When the healer with positive life force in him touches the patient with weak life force, the energy gets transferred from a higher state to a lower state until the energy level reaches equilibrium. This process is not easy as it sounds. The proficiency in these scientific practices can only be achieved by a person who has achieved a higher spiritual (not to be confused with religion) equilibrium. Spirituality has nothing to do with religion. If the energy level of the healer is not spiritually high, he/she will never be able to heal his/her patients even if he/she possesses a degree in hands on healing. This is an art of healing which cannot be learned through textbooks. This is the toughest of all the methods of healing. And the major reason behind the failure of all the energy therapy clinical trials is the person performing the energy therapies has not attained that superior level of spirituality or excellence.

> *When Nature opposes, everything else is in vain; but when nature leads the way to what is more excellent, the instruction in the art takes place. Physician treats, Nature heals.*
>
> —Hippocrates, n.d.

In this statement, the Great Hippocrates indicated that the nature heals the disease, not the physician or his/her pills. So, if the nature stops healing processes, a physician or his/her pills cannot do anything. Naturopathy firmly believes that the mother nature created us, nurtures us, provides us with all those things we need to survive in our surrounding environment and also provides us with an in-built healing mechanism to protect ourselves from stresses that we call diseases. So, as long as the natural healing mechanism is guiding one's life, the person is alive. But when the natural healing force stops performing its function, no kind of drugs or surgery can heal a patient. This is why the Great Hippocrates said, "Physician treats but Nature heals."

11.10   LIMITATIONS OF THE NATUROPATHIC SYSTEM OF MEDICINE

Even though the naturopathic system provides promising results in addressing chronic illnesses and lifestyle diseases, there are some limitations of this system that need to be considered.

- The first one is that the entire treatment procedure is mainly dependent on the body's inherent capability to heal itself. So this system will not be effective in cases where the vital force of the patient is too weak. In such cases, the conventional system should be preferred.
- The naturopathic system is not suitable for emergency and critical care cases.
- It should be realized that one cannot address advanced pathological states that require surgical interventions with only naturopathic treatments.

Thus, it is important that one must know the limitations and weaknesses of every treatment method. It should be realized that it is impractical to believe that a single medical system can address all kinds of health issues that prevail. Every physician should possess in-depth knowledge and critical analytic ability about pathological states and must be able to distinguish the line when he/she should refer the cases to the conventional system where surgical interventions or modern critical care medicines are required. One must be wise enough to determine which patient requires naturopathic treatment and which patient requires conventional treatment.

11.11 FUTURE PERSPECTIVES

Naturopathy is currently at a critical stage of evolution. There are several challenges that exist in the way of the naturopathic system, some of which are addressed in the present section.

So, from the discussions presented in the above sections, it is evident that:

- Naturopathy views the human body as a dynamic and interconnected system of mind, physical body, and universal consciousness, whereas the conventional system views the human body as a machine composed of a discrete system reduced into its constituent parts.
- In naturopathy, emphasis is placed on the identification of the areas of imbalance or dysfunction in normal physiology, whereas in the conventional system, emphasis is placed on the identification of a disease or pathological tissue change.
- In naturopathy, diagnosis integrates data from many different systems and methods (patient-centric), whereas in the conventional system, diagnosis is system-specific (disease).
- Naturopathy is based on the ideology of the personalized system of medicine, whereas the conventional system is based on the "One Drug Fits All" approach.
- Naturopathy addresses the underlying source (cause) of the dysfunction, whereas the conventional system addresses the result of the dysfunction, that is, the symptoms. Thus, the conventional system addresses the result of disease, whereas naturopathy treats the underlying cause.
- In naturopathy, emphasis is placed on both the subjective and objective pieces of information which are gathered based on the concept of optimal physiological function, whereas in the conventional system, emphasis is placed on on how the patient is doing based on charts, statistics, and test results that are measured against statistical normal population.

- As per naturopathy, health is measured along the wellness spectrum, which covers all the minute aspects in between health and disease state. As a consequence, interventions can be designed at every phase of the spectrum for restoring health.

The major concern with naturopathy is the lack of professionally trained naturopathic practitioners who can take of the education, research, and further treatment. So proper education and systematic training and research are essential for naturopathy. The natural therapies that already exist and are in practice are very cheap, easily available, and easily affordable. When these will be manufactured under large industries and will be sold and controlled like a pharmaceutical product, there will be an exponential rise in its cost and lots of people will not be able to afford it.

The best solution is to educate people and create awareness about the alternative natural therapies approved by the World Health Organization for treatment purposes and make the systematic and harmonic utilization of different medical practices. Mainstream medicine should be implemented once the natural systems of treatments have failed. This way the patient is protected from unnecessary toxic effects of the drug materials of synthetic origin. As the great Father of Medicine Hippocrates himself said:

"Keep the drugs in chemist's pot, if you can heal your patients with food."

Last but not least, it is evident that naturopathy strictly follows and satisfies the postulates and treatise of the great Hippocrates in relation to the art of medicine. So the naturopathic system of medicine and the practitioners of naturopathy should be given the position and honor they deserve for the service to the mankind and for following the path shown by the great Father of Medicine himself in serving those who are sick.

11.12 CONCLUSION

Thus, from the instances presented in the chapter, it is clear that the naturopathic system of medicine is not a pseudoscience but is completely based on scientific principles that are guided by the natural laws. The only difference is their approach or point of view toward addressing a clinical condition. The naturopathic system contains a diverse array of medical practices; some of them use drugs to treat diseases, whereas others implement energy-channeling processes for healing. The father of Medicine the Great Hippocrates included naturopathic practices among the art of healing. Being a follower of science, it is our duty to keep an open mind toward various scientific ideologies, and criticizing a scientific discipline just because it does not fit into our inherent set of knowledge is not a good scientific practice and is not an attribute of a man of science. Finally, the author wishes the readers to understand the fact that both conventional and naturopathic practices are an important aspect of the healthcare system. One system possesses the potential to overcome the limitations faced by others. The conventional system is excellent in addressing emergency and critical care cases and the cases that require surgery, but it does not do well while dealing with chronic and lifestyle diseases. Whereas naturopathy does not do well in addressing emergency and critical care cases and the cases that require surgery but is excellent in healing and preventing chronic illnesses and lifestyle diseases. People should acknowledge that both the systems are important, and it will be wise to combine the best from the two systems and create an integrated system of medicine, which will be a judicial combination of both conventional and naturopathic practices. Integrating the wholistic approach of healing along with modern medical practices will bring promising results. The integration of the two ideologies will make the healthcare system more personalized and ultimately will lead to a healthy future of mankind.

KEYWORDS

- hippocrates
- natural remedies
- vital force
- vitalism
- reductionism
- wholism
- energy medicine
- pseudoscience
- monopoly medicine
- personalized medicine
- herbal remedies
- natural supplements
- herbs

REFERENCES

Abbas, A.K.; Aster, J.C.; Kumar, V.; *Robbins and Cotran Pathologic Basis of Disease*, 7th ed. Philadelphia, PA: Elsevier Saunders, 2004.

Angell, M.; *The Truth about the Drug Companies: How They Deceive us and What to Do About It*, Reprint edition; New York, NY: Random House Publisher, 2005.

Bollinger, T; *The Truth About Cancer: A Global Quest Episodes 1-9*, 2015. Retrieved from https://www.youtube.com/watch?v=KqJAzQe7_0g&t=6s. (accessed on 2nd May 2020).

Dossey, L.; Healing Words: The Power of Prayer and the Practice of Medicine, 1st ed. Harper One, 1995.

Dossey, L.; Reinventing Medicine: Beyond Mind-Body to a New Era of Healing, 1st ed. Harper One, 2000.

Durr, H. P.; Quantum Physics and Creativity (Part 1/2), 2014. Retrieved from https://www.youtube.com/watch?v=WUaT3IoCvkA (accessed on 2nd June 2020).

Durr, H. P.; Quantum Physics and Creativity (Part 2/2), 2014. Retrieved from https://www.youtube.com/watch?v=KPOJEik5BhU (accessed on 2nd June 2020).

Glidden, P.; *The MD Emperor Has No Clothes: Everybody Is Sick and I Know Why*, 1st ed., Peter Glidden BS, ND, 2010.

Goswami, A.; Consciousness, Quantum Physics and Being Human, 2012. Retrieved from https://www.youtube.com/watch?v=bnQ63AOrs6 (accessed on 2nd June 2020).

Goswami, A.; Quantum Doctor: A Quantum Physicist Explains the Healing Power of Integral Medicine (Camino Guides) Paperback, 2011 Revised Edition. Hampton Roads Publishing Co, 2011.

Hegde, B.M.; *What Doctors Don't Get to Study in Medical School*, 4th ed. Telangana: Paras Medical Publisher, 2014.

Jain, K.; *Textbook of Personalized Medicine*, 2nd ed. Totowa, NY: Humana Press, 2015.

Lanctot, G.; *The Medical Mafia: How to Get Out of It Alive and Take Back Our Health & Wealth*, (Paperback). Here's The Key Inc., 1995.

Lipton, B.; *The Biology of Belief*, 1st ed. New Delhi: Hay House Publishers, 2016.

Lipton, B.; Biology of Belief Full Lecture, 2016. Retrieved from https://www.youtube.com/watch?v=82ShSNuru6c&t=81s (accessed on 2nd June 2020)

Pert, C.; Molecules of Emotion, 1999. 1st Touchstone ed. NY: Simon & Schuster.

Wallace, D.C.; Mitochondria as Chi; *Genetics*. 2008, 179(2):727–735.

Naturopathy

BHUSHAN R. RANE[1*], SANDIP A. TADAVI[2], and RAJ K. KESERWANI[3]

[1]*Shri D.D. Vispute College of Pharmacy and Research Center, Panvel, Raigad 410206 (Affiliated to University of Mumbai), Maharashtra, India*

[2]*PSGVPM's College of Pharmacy, Shahada 425409 (Affiliated to Kavayitri Bahinabai Chaudhari North Maharashtra University), Jalgaon, Maharashtra, India*

[3]*School of Pharmaceutical Sciences, Rajiv Gandhi Proudyogiki Vishwavidyalaya, Airport Bypass Road, Bhopal 462036, Madhya Pradesh, India*

Corresponding author. E-mail: rane7dec@gmail.com

ABSTRACT

Naturopathy is based on the science of origin of nature. The naturopathy/ nature cure is the ancient/oldest method among all the prevalent healing sciences. Many principles, theories, and concepts used in naturopathy are based on old literature, that is, Vedas. In Vedas, it was reported that water can be used in therapies, that 50 days of yearly fasting is useful in cancer treatment as our body destroys cancer cells during fasting, and that fasting is advantageous to cure other diseases. Since many years in history, fasting and water therapy are being used by Indians to cure many diseases. Yoga, which comes from ancient science "Vedas," is also part of treatment to cure various diseases. Yoga is the systematic practice of exercise of the body and mind in the presence of fresh air; it can motivate us, is responsible for the even functioning of the human body, increases blood circulation, boosts body metabolism, and strengthens immunity. Naturopathy helps selecting the right foods for maintaining the best health. However, maintaining the highest amount of nutrients in our daily diet is totally dependent on the food cooking method. Physiotherapy is one of the best and appropriate methods to treat injury and pain. Acupuncture is also one of the ancient techniques used to treat various health-related problems. In acupuncture, some points of the body are stimulated by inserting fine needles with an aim to stimulate various sensory receptors and transmitting the impulses to the brain.

12.1 INTRODUCTION

Every living (animals, plants, and human being) and/or nonliving organisms are part of nature. All the living systems have their own structure and physiology, and they vary from other species. Likewise, human beings also have different physiology but have something special and important, that is, a well-developed brain. During normal functioning, a human body requires some nutrients.

Nature/creature provides us all the essential nutrients through various plants, herbs, and natural resources. Theses naturally available things are used in the preparation and manufacturing of various medicinal products for the treatment and management of a variety of diseases and disorders (Sharma, 2002).

Approximately 70% of our body comprises water, and we can never imagine a life without water. In Vedas, it is reported that water is the best medicine for curing of various disorders and diseases. A normal adult requires 4-5 l of water daily to keep the body healthy and free from toxins.

Human cannot survive for more than 8-10 days without water and will die within 5 min without air. Natural cures are based on some of the elements, with air being one of the most important elements among them. Air is an essential component of the respiratory system where oxygen is taken by lungs and transported to all the cells and tissues through blood. Naturopathy is the best approach to treat various diseases and disorders by natural means to have a complete mental, healthy, and social well-being. It is not only a simple practical approach but is also based on scientific theories which lay a complete foundation to health of human being in a natural way. Naturopathy is the most economical approach to treat any disease or disorder.

Naturopathy is a drug-free therapy that involves the uses "Panchmahabhutas," that is, five elements of nature [Earth (*Prithvi*), Water (*Jal*), Fire (*Agni*), Air (*Vayu*), and Ether (*Aakash*)]. All these five elements are responsible to renovate, make, and maintain the health; thus, naturopathy is called the science of healthy living and a drugless system of healing based on well-known philosophy.

Most of the nutrients, nutraceuticals, foods, dietary supplements, herbal products, and processed products such as cereals and soup are used as diet during naturopathy treatments.

Naturopathy is a system of human building in harmony based on the combined principles of nature on physical, mental, moral, and spiritual grounds. Naturopathy promotes health, prevents diseases, and has ability to cure diseases as well (Rastogi, 2012).

According to public declaration (manifesto) of British Naturopathic Association, "Naturopathy is way of treatment which identifies the existence of the important curative force within the body." Therefore, it is a helping aid to the human physiology to remove the basic cause of diseases. For treating diseases, through unusual natural toxins, nonessential matters are expelled from the human body (Figure 12.1).

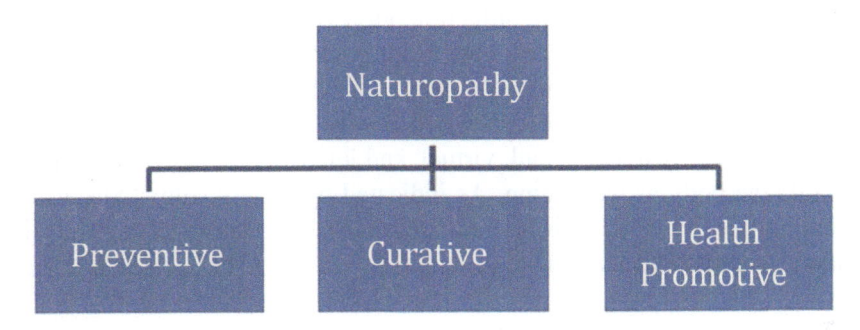

FIGURE: 12.1 Potential of naturopathy.

The nature fix is better than all strategies of well-being sciences because of the aforementioned five components, which do not react nor have any delayed consequence. Treatment through yoga is likewise one of the strategies. It covers the air treatment through breathing exercises. Body stances are utilized in yoga treatment to mend or fix numerous ailments.

Naturopathy reveals why infection is created? Whenever we have to find out the reason of infection or disease we must have to update the information about it.. Only updated knowledge is utilized to provide information about what should be required by body to re-establish to maintain the balance.

People can improve their well-being by taking own prepared nourishment enriched food. Another reason of increasing trends of nutraceutical is awareness in society, renewable sources, use of best cultivation processing methods, and its availability at neighborhood site.

The fundamental tenets of naturopathy are as follows:

All ailments, their drivers, and treatment methods are similar. Except for vulnerable and other conditions the main cause of all disease is one and it is deposition of unwanted substances in body. The best treatment of all ailments is removal of toxic substances from body The primary root cause of illness is deposition of toxic substances. Bacteria and viruses enter and alive in body when they have suitable environment; such environment provided by deposition of unwanted substances in body. Thus, the main driver of infection is bleak issue and also microbes. All the smaller scale living being is not destructive to person. Short life ailments are restoring endeavors of the body. Henceforth, they are not unsafe, these are our companions. Transient infections are a direct result of wrong treatment and concealment of the intense malady. Nature is the best mending force. The human body can shield itself from infection and recoup well-being if grim.

Nature cure treats not only infection but it also cures complete system of patient which is caused and recouped Naturopathy fixes patient experiencing endless illness effectively. In naturopathic framework, fix the stifled sickness brought to surface and are evacuated totally. Naturopathy system treats complete body by treating soul, virtual, and intellectual aspect. Naturopathy regards the body as well-being. As indicated by Naturopathy, "Sustenance is eating like drug," there is no compelling reason to take prescription (i.e., called as nutraceuticals)

12.2 THREE BASIC PRINCIPLES

The application of naturopathic system is performed to repair, to maintain the health of human being for the treatment of physical disorders containing physical, chemical, biological, spiritual, and mental.

Therefore, natural treatment has three basic principles: elimination, accumulation, and cure within blending of human with nature.

1. The first and most significant tenet of the characteristic fix is that all types of ailments are because of the same underlying drivers—to be specific, all ailments are caused by the holding of waste material and eating undesirable sustenance and because the body denies the framework. These waste materials in a sound individual are discharged from the body framework through the end organs. In individual patient, there is constant supply of unwanted things to body due to bad living habits (such as wrong diet, improper care, worry, stress, and extra work)

leads to moral and mental weakness in individual and nervousness. If somebody has follows the first important rule of elimination of toxic substances from body then it is a best method to cure all aliments.

2. The second essential standard of naturopathy is connected to every single instance of infection, for example, fever, cold, and irritations. In the case of stomach-related unsettling influences and skin issues, the body makes an endeavor and evacuates all the aggregated lethal waste materials. The ceaseless concealment of intense infection may prompt constant sickness, for example, coronary illness, diabetes, hypertension, stiffness, asthma, kidney issue, and malignant growth through destructive techniques, for example, drugs, antibodies, opiates, and removal of organs.

3. The third rule of the nature fix is that the body contains a mind-boggling recuperating instrument that has the ability to arrive at the typical state of well-being by utilizing the right techniques that empower it to do as such. As it were, the intensity of restoring all the presence sickness is inside the body itself and not in the hands of the specialist (i.e., resolution).

There are two natural media: an outside world that impacts the living being and an inside domain that operates the human organs. The body's inward nature improves the health condition and treatments illnesses itself. The objective of the treatment here is not to decline the condition of the patient (Hamidli et al., 2010).

2.3 SCIENTIFIC PRINCIPLES

The recuperating intensity of nature is the famous piece of naturopathic drug. Improving the recuperating capacity or by applying applicable fundamental learning to every living framework advances, keep up, and reestablish typical elements of individual. Naturopathic doctors advance or support, encourage, and evoke recuperating by using techniques and treatment with the regular procedure (Downey, 2000; Pizzorno, 1996).

12.3.1 FIVE SCIENTIFIC PRINCIPLES (Downey, 2000; Pizzorno, 1996):

1. Do not mischief to tolerant (Do not cut): Means utilizing the least intrusive—first, utilize most delicate treatment, so as to stay away

from the concealment of indications, support or advance the patient's own self-mending process, and limit the opportunity of symptom.

2. Identify and treat (To address the major reason for malady): This happens when an illness condition, for example, physical, mental pressure, passionate pressure, or otherworldly convictions may lead to the disturbance of well-being. Indications are distinguished as side effects. The body makes an endeavor to oppose, protect, adjust, or mend itself when face with such difficulties. Identifying the actual problem in patient is the main hurdle during health treatment. There is regularly more than one reason recognized and controlled.

3. Treat (To co-work with recuperating forces of nature): The entire individual mirrors the multidimensional parts of the individual and condition. Every patient is one of a kind, and the numerous ideas of ailment require a far-reaching, singular methodology utilized to analyze treatment. Naturopathic doctors urge to the patient to connect with both of their inner assets (e.g., profound convictions and processing) and outside help (e.g., family, companions, and dietary admission) to give a strong establishment from whichever or push ahead into a more noteworthy condition of well-being.

4. Well: This implies more than the basic nonappearance of sickness. Surveying danger factors and hereditary inclination helps naturopathic doctors make proper intervention and guide patients to build up equalization and ideal well-being. What's more, through instruction and advancement of good way of life propensities or sound living, persistent learns approach to dodge malady and improve well-being.

5. Doctor as Teacher: To show the standards of solid living and preventive medications. This helps outline strategies by which naturopathic doctors attempt to set up an agreeable specialist relationship, so as to teach about their well-being and to energize patient's self-obligation through their treatment methodologies.

12.4 COMMON PRINCIPLES OF NATURAL CURES

1. Sickness occurs because of numbness of nature, and following characteristic laws again serves to recoup well-being. In the event that somebody pursues common laws, there are extremely less odds of occurring of infections. All causes of infections, and treatment are same; yet, their introductions are unique.

2. Microorganisms cause the disease only when people ignore the laws of nature and follows bad things such as eating fast food, improper diet, etc.

3. Nature itself is a remedy for all sicknesses. We simply need to pursue some normal laws to re-establish well-being conditions.

4. Naturopathic frameworks treat the entire body and psyche, not the manifestations. In this manner, it is having a total solution for medical problems.

5. An incessant malady grows gradually and treatment for them takes additional time. A few specialists state that if an individual experiences a sickness from half a year, treatment may take additional time close to around a quarter of a year.

6. Modern sciences smother sicknesses, which come to the surface amid the beginning of common fix treatment remains after that individual gets perpetual fix.

7. Ayurveda is the mother of regular fix; yet, one contrast is that common fix does exclude medication.

12.5 MECHANISM OF ACTION

Naturopathic medication is not characterized by its strategies. Professionals use a wide assortment of treatments and procedures, from the "nature fix" conventional (e.g., well-being through eating regimen, water treatment, change in way of life, expel lethal components), to the utilization of healthful enhancements and natural concentrate so as to control or impact the body's organic chemistry and physiology. Some naturopaths are professionals with extra information and who have undertaken additional studies of homeopathy, needle therapy, Ayurveda. Most of naturopaths, be that as it may, utilize a disconnected methodology (Bradley et al., 1999). Likewise, some state laws have arrangement that enable naturopaths to straightforwardly offering vaccinations, perform minor mobile medical procedure, and recommend a few pharmaceuticals.

12.6 METHODS OF NATUROPATHY

The naturopathy framework goes for the correction of human framework from undesirable to solid conditions, and works and acknowledges strategies for fix which are in resemblance with the helpful based standards of nature.

TABLE 12.1   Advantages and Disadvantages of Holistic/Natural Medicine and Allopathic/Conventional Medicine/Western Medicine

Best Used for	Holistic/Natural Medicine	Allopathic Conventional/Western Medicine
	Nonemergency, acute, and/or chronic illness or conditions	Emergency, major injuries, or serious illness and ointments
Advantages	• Focused on the person with strong belief between patient and doctor • Try to meet the need of the whole body: physical, mental, and soul • To encourage prevention through a wellness approach so the body can heal itself • Side effect is rare • Gives individuals a better sense of control over their own health • Less expensive	• Highly trained professionals/practitioners required • Strict government guidelines/regulations • Quality control procedures • Without hesitation accepted by health insurance companies • Almost based on scientific research and evidence
Disadvantages	• Hesitate to accept by health insurance companies • Not always based on scientific evidence • Lack of government regulations • Alternative practicioners canot always diagnose serious disease • Some herbs can react badly to those on certain medications	• Serves to treat via allopathic drugs and use invasive methods (surgery) • More often treats the symptoms and not root of a conditions • Prescriptive treatments often lead to side –effects and more drugs • Patients are not always encouraged to make lifestyle changes to combat illness or dysfunction • Expensive

Such techniques expel from the aggregation of horrible issue and toxic substances without activity or medical procedure of the significant pieces of the body. They additionally urge the organs to dispose of and refine to better functioning of body.

To fix maladies, the first and foremost prerequisite is to modify the eating regimen. To expel gathered dreary issues and re-establish the balance condition of the framework, it is necessary to totally bar corrosive shaping nourishments, including proteins, starches, and fats, for possibly more than seven days and to limit the eating regimen to crisp the natural product which will purify the gastrointestinal tract.

Systems and benefits of different methods of naturopathy

1. Diet therapy
2. Fasting therapy
3. Mud therapy
4. Hydrotherapy
5. Massotherapy
6. Acupressure
7. Acupuncture
8. Chromo therapy
9. Air therapy
10. Magnet therapy

12.6.1 DIET THERAPY

As indicated by this treatment, nourishment must be taken in a regular interval. Crisp occasional natural products, new green verdant vegetables, and bean sprouts are phenomenal for weight control plans. Diets are comprehensively arranged into the following three kinds:

1. Eliminative regimen, for example, fluid lemon, citric juices, tender coconut water, vegetables soup, butter milk, and white grass juices.
2. Calming diet, for example, natural products, salads, boiled/steamed vegetables, bean grows, and vegetable chutney.
3. Productive diet, for example, healthy flour, unpolished rice, little heartbeats, grows, and curd.

Being essential, these weight control plans help in improving well-being, detoxifying the body, and rendering it insusceptible to illness. To this end, an appropriate blend of nourishment is fundamental. Our eating routine should comprise 20% acidic and 80% essential sustenance for well-being. Reasonable nourishment is an absolute necessity for any unique individual attempting to make great well-being (Murray and Pizzorno, 1998). Nourishment is considered as prescription in naturopathy.

Nutraceutical is a piece of naturopathy that assumes a significant job in dietary enhancements. The people think that quality of life in terms of income, excellent, and lifestyle has enhanced with economic developments (Lutz and Przytulski, 2006). Nonetheless, it has likewise contorted up a noteworthy test as "way of life sickness.." The primary casualty of this way of life is changing of nourishment propensities. Eating of shoddy nourishment has led to complexities, which has prompted various maladies identified by wholesome inadequacy. Nutraceuticals can assume a significant job in controlling sicknesses. There is no big surprise that an ever-increasing number of individuals are going to nutraceuticals (Bronner, 2005).

Characterization of nutraceuticals ought to be considered in two different ways: one is potential nutraceuticals and another is set-up nutraceuticals. A potential nutraceutical is one that ties a guarantee of a specific well-being or health advantages. Potential nutraceuticals just turned out to be a steady one after there are adequate clinical information to show such an advantage. It is disillusioning to take note of that the greater part of nutraceutical items is in the "potential class," holding on to wind up the built-up. The foodstuffs that are used as nutraceuticals are classified on the basis of its chemical constituents.

1. Nutrients
2. Herbal
3. Phytochemical
4. Probiotic miniaturized scale life form
5. Nutraceutical compounds.

Substances such as minerals, nutrients, proteins, and unsaturated fats are called as built-up or customary supplements. Most vegetables, wholegrain oats, dairy items, and foods grown from the ground items, for example, meat are wellsprings of nutrient. These help in treatment of heart infections, stroke, waterfalls, osteoporosis, diabetes, and malignancy. Minerals found in plant, creature, and dairy items are helpful in treatment of issues related to bones, iron deficiency, teeth-related problems, and muscle issues; they help improve nerve motivations and heart beats. Salmon fish contains unsaturated fat, for example, omega-3 polyunsaturated unsaturated fats and, therefore, is considered a great controller of the irritation, maintains mind capacity, and declines cholesterol statement in a human body.

Nutraceuticals extraordinarily guarantee to improve well-being and counteract unending maladies with the assistance of home-grown prescription (Baer, 1992; Pandey et al., 2010; Singh and Sinha, 2012).

12.6.2 *FASTING THERAPY*

Fasting, an outstanding demonstration of enthusiastically keeping away from a few or all nourishment, drink, or both, for a timeframe, is accounted for in old Indian writing. The word is derived from the early English, "Feastan" which signifies to quick, watch, be exacting. In Sanskrit "Vrath" signifies "assurance" and "Upavasa" signifies "close to God." Fasting is termed as complete or partial fast on the basis of time of prolongation. Fasting is a significant treatment strategy for well-being protection. Mental readiness is a fundamental precondition in fasting. Nonstop drawn out fasting ought to be done uniquely under the supervision of a capable naturopath (Bolar, 1999).

The durable time of the quick relies on the age of the patient, the kind of the sickness, and the sum and sort of medications utilized beforehand. Sometimes, it is suggested to take series of fasts of two or three days which will be increased by a day or days decided on the basis of patients health. No detriment is accurate to fasting patient provided the proper rest and care should be taken (Michalsen and Li, 2013).

In fasting, one only consumes drinking water, juices, or crude vegetable juices. The best, most secure, and better strategy is lime-juice fasting. The body consumes and discharges immense measure of collected waste amid fasting conditions. We can likewise help body purify by drinking basic juices. Sugar in juices will give vitality or solidarity to the heart. In this way, juice fasting is the best type of fasting. All juices ought to be set up from new natural product preceding drinking. Safeguarded or solidified juices ought not be utilized. A careful step, which must be seen in all instances of fasting, is the finished exhausting of the insides toward the start of the quick by purification with the goal that the patient is not troubled by gas or breaking down issue framed from the defecation staying in the body. Douches ought to be utilized no less than each substitute day amid the fasting time frame. The all-out fluid admission ought to be approximately 6-8 glasses. A great deal of vitality is spent amid the quick during the time spent wiping out bleak and dangerous waste materials. It is along these lines, of most extreme significant in the quick that the patient gets however much physical rest and mental unwinding as could be expected amid the quick.

The accomplishment of the quick depends essentially on how it is broken. The principal rules for break in the quick are: do not eat excessively, eat gradually, and bite your sustenance complete and make a few days for the stride by step change in the typical eating routine.

12.6.2.1   ADVANTAGES AND PHYSIOLOGICAL IMPACTS OF FASTING

Doctors of most societies, since forever, have suggested long-last fasting as treatment for the past condition from old to current. In spite of the fact that previous perceptions were contemplated without a logical technique or an understanding despite everything they point to use of fasting as a remedial methodology. Prior perceptions depended on creature conduct; yet today they depend on creature physiology. In this chapter, we will attempt to perceive how best fasting could be helpful in advancing ones well-being through the audit of writing that depicts physiological and metabolic advantages.

Noticeable among the physiological impacts looked at by fasting (calorie restriction and intermittent fasting) are: expanded insulin affectability that reduces plasma glucose and insulin, focuses on upgraded glucose resilience, tumbles down the dimensions of oxidative worry as demonstrated by diminished oxidative harm to proteins, lipids and DNA, and expands protection from different kinds of stresses including heat, oxidative, metabolic anxieties, and increment resistant working.

Both whole and cell physiology are profoundly influenced by caloric limitation (CR) or irregular fasting (IF) routines. Fasting supports the cardiovascular system and minimizes the chance of myocardial infarction as it reduces body fat and mass (Shetty et al., 2015). Notwithstanding heart assurance, a more prominent continuance to stress is actuated in the liver—the supplement center of *Homo sapiens*. The closeness of elective vitality stores, for example, ketone bodies (e.g., β-hydroxybutyrate) empowers *Homo sapiens* to live in extra worries of life. Intemperate and harmful blood glucose is decreasing by an improved affectability to insulin and glucose and its use as a vitality source (Nakamura et al., 2014).

12.6.3   MUD THERAPY

Mud treatment is straightforward and successful treatment systems. The mud utilized for this ought to be spotless and taken from 3 to 4 ft. profundity from the outside of the ground. There ought to be no pollution of stone pieces or substance composts and so forth in the mud (Rastogi, 2012).

Mud is one of the five components of the nature with the gigantic effect on the body well-being and affliction. Focal points of utilizing mud are as follows:

1. Its dark shade assimilates every one of the shades of the sun and replaces them to the body.
2. Its shape and consistency can be changed effectively by use of water.
3. Mud holds dampness for quite a while, when connected over the body part, it produces cooling.
4. It is modest and effectively accessible.
5. Mud ought to be dried type of powdered and sieved to separate stones, grass particles, and different pollutions before utilizing.

12.6.3.1 MUD PACK-LOCAL APPLICATION

Keeping absorbed mud dainty, wet muslin fabric, transform it into a slight level block, contingent upon the extent of the patient's guts, and then apply it. The residency of the mud pack application is 2030 minutes. Mud to be applied in cold weather conditions and after application mud pack is covered completely using blanket. It is one of the best natural therapies for the treatment of rheumatoid arthritis (Bolar 1999; Codish et.al., 2005; Rastogi, 2012).

Benefits:

1. When applied to the abdomen, it gets relief from the indigestion. It is effective in decreasing intestinal heat and contraction of the digestive tract.
2. A thick mud pack applied to the head in congestive headache relieves the pain immediately. Hence, this is recommended whenever there is necessity for a prolonged cold application.
3. Application of the pack over the eyes is useful in cases of conjunctivitis, hemorrhages in the eyeball, itching, allergy, errors of refraction like short sight and long sight, and especially in glaucoma where it helps reduce eyeball tension.

Mud Pack for Face:

Mud paste is applied on the face and allowed to dry for 30 minutes. This is helpful in improving the complexion of the skin and removing pimples and open skin pores which in turn facilitate elimination. This is also helpful in eliminating dark circles around the eyes. After 30 minutes, face should be washed thoroughly with cold water (Bolar, 1999; Rastogi, 2012).

Mud Bath:

Mud may be applied to the patient in sitting or lying position. This helps improve the skin condition by increasing the circulation and energizing the skin tissues. Care should be taken to avoid catching cold during the bath. Afterward, the patient must be washed thoroughly with cold water jet spray. If the patient feels chilling effect, warm water should be used. The patient is then dried quickly and transferred to a warm bed. The duration of a mud bath may be 45-60 minutes (Bolar,1999; Rastogi, 2012).

Benefits:

1. The effects of mud are refreshing, invigorating, and vitalizing.
2. For wounds and skin diseases, application of mud is the only true bandage.
3. Mud therapy is used for giving coolness to body.
4. It dilutes and absorbs the toxic substances of the body and ultimately eliminates them from the body.
5. Mud is used successfully in different diseases such as constipation, headache due to tension, high blood pressure, and skin diseases.

Luis Espejo Antunez et al. (2013) evaluated the effects of mud therapy on perceived pain and quality of life related to health in patients with knee osteoarthritis. Treatment with mud compresses relieves pain, affecting the hands, and reduces the number of swollen and tender joints in the hands of patients suffering from rheumatoid arthritis. This treatment can augment conventional medical therapy in these patients (Codish et al., 2005).

12.6.4 HYDROTHERAPY

Hydrotherapy is a branch of nature cure. It is treatment of disorder using different types of water applications. The different types of water applications are in practice since ancient times. Hydrothermal therapy uses the effect of temperature, as in hot and cold baths, saunas, wraps, etc., and all its forms, such as solid, liquid, vapor, ice, and steam, either internally or externally. Water is, without doubt, the most ancient of all remedial agent for disease.

This great healing property has now been systematized and made into a special type of science. Hydrotherapeutic applications are generally used in different forms of temperature. The temperature forms of the applications are given in Table 12.2.

TABLE 12.2 Hydrotherapeutic Applications Used in Different Forms of Temperature

Sr. No.	Temperature	°F	°C
1.	Very cold (ice application)	30–55	−1–13
2.	Cold	55–65	13–18
3.	Cool	65–80	18–27
4.	Tepid	80–92	27–33
5.	Warm (neutral)	92–98 (92–95)	33–37 (33–35)
6.	Hot	98–104	37–40
7.	Very hot	>104	>40

Impacts and use of water:

1. Taking bath appropriately with virus-free water is a decent type of hydrotherapy. Such sorts of showers open up all the skin pores and makes the body crisp and light. By taking such a shower, blood flow in all body framework and muscle improves. Bathing in rivers, natural ponds, or under waterfall is one type of natural and traditional hydrotherapy.

2. It can be connected to a restricted region or to the entire body surface on account of the most adaptable vehicle for delivering the ideal thermic and mechanical impacts.

3. Water productively retaining warmth and furthermore forgets heat with availability. It can, in this way, be utilized either for expelling heat from the body or engrossing warmth to it. Despite the fact that cool water is for the most part utilized, the reason for existing is not to remove or lower the body heat, thereby expanding the essential vitality to produce more warmth than what is lost.

4. Water is used as solvent and called as universal solvent; used internally as an enema or colonic irrigation and more water uptake helps to remove various toxic components from body such as urea, uric acid, etc.

It ought to likewise be note that the successful utilization of these strategies needs a specific dimension of imperative vitality. Where that vitality is excessively low, these are pointless. As in intense conditions there is a higher level of essential vitality, and thus there is an increasingly shot of imperative response. Under constant conditions, where the essential power is lower, these showers are less helpful; however, in such a case, the packs are valuable since they are milder in their applications.

Water is utilized in numerous forms in treatment. The different types of treatment techniques are compress and fomentation.

1. Cold pack: stomach cold pack
2. Heating pack: throat pack, wet support pack, chest pack, stomach pack, knee pack, and full wet sheet pack
3. Hot and cold pack: head, lung, gastro-hepatic, kidney, pelvic and stomach, and hot and cold pack
4. Fermentation
5. Showers: different sorts of shower are
    *   Hip bath—cold, neutral, hot, stiz bath, and elective hip shower
    *   Spinal bath and spinal spray—cold, neutral, and hot
    *   Foot and arm bath: cold, hot foot bath, arm bath, foot and arm shower consolidated, contract arm bath, and complexity foot bath
6. Sauna bath
7. Sponge bath
8. Jet–spraymassages: a kind of shower—cold, neutral, hot, alternate, circular stream kneaded by splashing water
9. Affusion baths: cold, neutral, hot, and hot and cold
10. Cold shower
11. Trauma
12. Submersion bath: cold submersion shower, cold inundation with grinding, unbiased drenching shower, drenching impartial half-shower, graduated submersion shower with epson salt, asthma shower, whirlpool shower, and submerged back rub
13. Bowel purge: graduated enema, vaginal irrigation, cold irrigation, neutral irrigation, and hot irrigation.

One of the systems of hydrotherapy is colon treatment.

12.6.4.1　*COLON HYDROTHERAPY*

It is a technique to expel and flush out poisons from the colon or internal organ. This procedure is like a douche, yet lengthier. To wash out or detoxify a stagnated fecal issue from the colon by utilizing clean, sifted water under delicate weight, the quantity of session is changed in a person. The vast majority of individuals require a progression of 3-6 treatment to purging of the colon. Colonic hydrotherapy may cause the dispersal and assimilation of poisons and microorganisms into the body. This is on the ground that the fecal issue is strong close to the rectum, and hydrotherapy, which separates this issue into a suspension, may encourage the ingestion of microscopic organisms and poisons into the foundational dissemination (Seow-Choen, 2009).

12.6.4.2 ADVANTAGES AND PHYSIOLOGICAL IMPACTS OF HYDROTHERAPY

- The slow reclamation and recuperating properties of hydrotherapy depend on its mechanical as well as warm impacts. It abuses the body's response to hot and cold upgrades, to the expand utilization of warmth, to weight applied by the water and to the sensation it gives. The nerves convey driving forces felt at the skin further into the body, wherever they are instrumental in invigorating the resistant framework, affecting the get together of pressure hormones, empowering the course and absorption, empowering blood stream, and alteration of torment affectability. For the most part, heat calms and alleviates the body, hindering the movement the action of interior organs. Cold, interestingly, animates and strengthens, expanding interior movement.
- Its mechanical power following up on weight happens amid the shower 50 to 90 % body weight decline when submerged in a shower, a pool, or a whirlpool, feel like weightlessness, body is calmed from the consistent draw of gravity; water likewise has a hydrostatic impact. It has feel like back rub as the water tenderly manipulates your body. Water in movement, invigorates receptors on the skin, boosting blood course and discharging tight muscles.

Cold application is a diagram on the ground that with the helpful use of any substance, the body expels heat, thereby prompting debilitated tissue temperature. Cold application declines the tissue blood stream rate by causing vasoconstriction and lessens tissue digestion, oxygen usage, and irritation to muscles. Topical cold treatment tumbles down the temperature of the skin and basic tissues to a profundity of 2–4 cm, diminishing the enactment limit of tissue nociceptors and the conduction speed or speed of the nerve signals. This outcome in a nearby analgesic impact called neuropraxia. Thermotherapy is the restorative utilization of any substances to the body that adds warmth to the body bringing about expanded tissue temperature.

Warmth treatment is conveyed by three systems: conduction, convection and change. It helps develop damaged tissues and treats bloodstream infections by providing protein, supplements, and oxygen. With the help of this treatment, the tissue digestion rate is also expanded up to 10%–15% by increasing the tissue temperature by 1 °C. This expansion in digestion helps in mending process by expanding both catabolism and gives the medium to tissue fix.

12.6.5 *MASSO THERAPY*

Back rub is a superb type of aloof exercise. The word is taken from the Greek word "massier" that intends to work, from French "erosion of plying," or from Arabic "massa" which means "to contact, feel, or handle" from Latin "massa" meaning "mass, mixture."

Back rub is the physical control routine with regards to delicate tissue with physical (anatomical), useful (physiological), and now and again mental reason and objectives. In the event that effectively done on a stripped body, it very well may be profoundly animating and strengthening. Back rub is the basic strategies finished with hands, fingers, lower arms, knees, elbows, feet, or a by utilizing gadget. The reason for back rub is for the most part for mitigated body pressure or agony. Back rub is additionally a technique for Naturopathy and fundamental for looking after well-being. Target tissues could humiliate muscles, ligaments, tendons, skin, joints, or other connective tissue, just as body liquid vessels. There are more than 80 distinctive perceived back rub techniques. It intends to expand blood course and reinforcing body parts. In winter season wash up in the wake of rubbing the entire body is outstanding routine with regards to keeping up well-being and quality. To give consolidated advantages of back rub and sun beams treatment. The filling of activity can be gotten from back rub. The back rub is a substitution of activity for the individuals who cannot do the activity (Ichiman, 2016).

Different oils are utilized in the back rub as greases such as mustard oil, sesame oil, coconut oil, olive oil, and smell oils which additionally have helpful impact.

There are seven major methods of controls in back rub and these are: (1) touch, (2) effleurage (stocking), (3) friction (scouring), (4) petrissage (working), (5) tapotement (percussion), (6) vibration (shaking or trembling), and (7) joint development.

Back rub developments are shifted by infection condition and parts connected. Other types of back rubs supportive in most illness are the vibratory back rub, powder rub, water rub, and dry back rub. Powder of neem leaves and flower petals of neem are likewise utilized as greases for back rub.

12.6.5.1 PHYSIOLOGICAL EFFECTS OF MASSAGE

1. Reflex impacts (reaction intervened by the sensory system)
2. Vasodilation of supply routes
3. Stimulation of peristalsis (helps in digestion)
4. Increase or decline in muscle tone
5. Increases action of the organs in the stomach pit
6. Triggers the unwinding reaction
7. Soothing or invigorating impact on muscles
8. Stimulates the heart, increment quality and rate of constriction
9. Increase effectiveness of the invulnerable framework
10. Mechanical impacts (reaction coming about because of legitimately connected manual weight)
11. Increase venous return
12. Increase lymphatic stream, lymphatic seepage
13. Circulatory effectiveness
14. Loosening of mucous (respiratory framework)
15. Breakdown of fibrosis/bonds
16. Stretch to abbreviated muscles/extricates muscle strands
17. Increase muscle temperature
18. Increase metabolic rate locally and vaporous trade
19. Stretches scar tissue
20. Decrease muscle tone/increment muscle tone
21. Increase scope of movement
22. Restoration of legitimate joint mechanics/biomechanics
23. Elimination of muscle irregular characteristics
24. Strengthen debilitated muscles.

12.6.5.2 ADVANTAGES OF BACK RUB

In general, back rub, which manages almost all organs of a body, is useful from various perspectives. It solidifies the muscle in the sensory system, influences breath, and speeds up the end of poisonous and waste material from the body through the different eliminative organs, for example, the lungs, kidney, skin, and entrails. It additionally lifts blood course and metabolic procedures (Ichiman, 2016). A back rub evacuates facial wrinkles, rounds out hole on cheeks and neck, and facilitates unbending nature, sore muscles, and skin deadness.

Restorative research has demonstrated that the back rub incorporates relief from discomfort, diminishes uneasiness and discouragement, and briefly lowers circulatory strain, pulse, and state nervousness. Hypotheses behind what the back rub may do incorporate blocking vibe of agony (entryway control hypothesis), enacting the parasympathetic sensory system which may animate the arrival of synapses, counteracting fibrosis or scar tissue, expanding the progression of fluid body substance, and improving rest; such impacts are yet to be upheld by well-exquisite clinical examinations.

12.6.6 ACUPRESSURE

Pressure point massage is an antiquated recuperating workmanship that uses the fingers or any blunted articles to press key focuses called "Acu Points," for example. Vitality put away focuses on the skin surface musically to empower the body's regular self-healing capacities. At the point when these focuses are squeezed, they discharge strong pressure and advance the dissemination of blood and body's life power to help recuperating.

Needle therapy and pressure point massage utilize similar focuses, whereas pressure point massage utilizes the delicate yet firm weight of hands or any needle point. Although needle therapy uses needles, pressure point massage has been treated as a mending craftsmanship in any event for large number of times in a year (Lee, 2010). More than 3000 kinds of recorded conditions were accounted for to finish well-being treatment. Presently, acupoint therapy is treated using intact skin electric nerve sensory input and the use of specific wavelength laser light which shows better and lasting effect.

Pressure point massage foremost and acupoint tactile information depends on similar standards of needle therapy. By applying weight, electric tangible information or laser light of needles, it attempts to invigorate explicit reflex focuses situated along the lines of vitality which keep running in the body, called meridians. There are crucial energies that can move through the meridians in a reasonable and even way, the outcome is great well-being. When you experience agony or disease it means that there is square or break in the vitality stream inside your body.

To locate the proper point, delicately test the territory it is discovered that point which gives an "interesting bone" feeling or is touchy, delicate or sore. At that point press adequately difficult to make the point agonizing. Invigorate by applying steady weight for five seconds on and five seconds off turning weight. Normally one moment is adequate for every treatment session.

Pressure point massage might be powerful to acquiring ease from cerebral pains, eye fatigue, sinus issue, neck torment, spinal pains joint pain, muscle hurts, and strain because of stress, ulcer torment, menstrual issues, lower spinal pains, obstruction, and tension, sleep deprivation.

There is incredible focal points to utilize pressure point massage as an approach to keep up the body's well-being. The mending contact of pressure point massage declines strain, builds flow, and empowers the body to unwind profoundly. By alleviating pressure, acupressure therapies make solid protection from illness and improve well-being. The pressure point massage procedure can be used as a self-treatment and is simple to learn. The self-treatment pressure point massage is connected to utilizing the finger to apply weight to pressure point and weight focus (Abu et al., 2015).

12.6.7 ACUPUNCTURE

Needle therapy is the method of embedding and controlling fine documenting structure needle into explicit areas on the body for restorative purposes.

The word needle therapy originates from the Latin "acus," meaning "needle," "plunger," "to prick."

As indicated by customary Chinese therapeutic hypothesis, a needle therapy focuses are arranged on meridians along which Qi, the crucial vitality, streams. There is obscure anatomical or histological reason for the presence of needle therapy focuses.

In China, the act of needle therapy can maybe be hustled as far back as the Stone Age, with the Bian shi, or honed stones. Needle therapy's roots in China are dubious. The most punctual Chinese restorative content that initially portrays needle therapy is the Yellow Emperor's great of inside Medicine (History of Accupucture) Huangdi Neijing, which was composing around 305-204 B.C.

A few hieroglyphics (a composition arrangement of old Egypt) have been discovered going back to 1000 B.C. that may show an early utilization of needle therapy. As per one legend, needle therapy began in China when a few fighters who were injured by bolts in fight encountered a help of agony different pieces of the body, and thusly individuals began exploring different avenues regarding bolts and later use needles as treatment. Needle therapy spread from China to Korea, Japan, Vietnam, and somewhere else in East Asia. Portuguese preachers in the 16th century were among the ones who carries techniques of needle therapy toward the West.

12.6.7.1　*CONVENTIONAL THEORIES OF ACUPUNCTURE*

In conventional Chinese medication, "Well-being" is viewed as a state of parity of components of regular world (Yin and Yang) inside the body. Some have contrasted Yin and Yang with the dynamic and quiet. Especially, vital in needle therapy is the free progression of Qi, a hard-to-interpret idea that overruns Chinese logic and is usually deciphered as "indispensable activity." Qi is insignificant and subsequently is Yang and Yin, the material partner of blood (promoted to separate it from physiological blood and incredibly generally revere it). Needle therapy treatment keeps up the progression of Qi and blood, tonifying where there is lack, channelling out where there is abundance, and upgrading free stream where there is stagnation. A major guideline of the medicinal writing of needle therapy is "no agony, no blockage; no blockage, no torment." (Thambirajah, 2011)

Conventional Chinese medicine regards the constitution as a whole that includes many "frameworks of capacity" for the most part named after anatomical organs anyway in a roundabout way identified with them. The Chinese expression for this framework is Zang Fu (i.e., organs of the body are partitioned into sets comprising of one organ with a Yin nature and one with Yang nature, so as to have separate frameworks to perform various tasks from physical organs; Zang Fu are promoted in English, in this manner Lung, Heart, Kidney, and so on.)

Malady is involved as lost parity of Yin, Yang, Qi and blood (which looks somewhat like homeostasis). Treatment of illness is endeavoring to adjust the action of at least one frameworks of capacity through the movement of needles, weight, heat, and so on touchy segments of the collection of small volume customarily called "needle therapy focuses" in English, or "XUE" in Chinese. This is alluded to in traditional Chinese medicine as treating "examples of disharmony".

The greater part of the primary treatment focuses square measure found in the "12 fundamental meridians" and two of the "eight additional meridians" (Du Mai and Ren Mai) a sum of "14 channels," that square measure speaks to in established as pathways through which Qi and "blood " stream. Different focuses (known as "ashi focuses") may likewise require as they are accepted to be in the place where inertia has accumulated (Thambirajah, 2011).

Arrangement of maladies, side effects, or conditions for which needle therapy has been appeared as a compelling treatment in allergic rhinitis, depression, headache, nausea, and heaving, including morning infection, pain in the epigastria, face, tennis elbow, lower back, knee, amid dentistry and after task, primary dysmenorrheal, rheumatoid joint inflammation, sciatica, cervical and lumbar spondylosis, bronchial asthma, and insomnia.

12.6.8 CHROMO THERAPY

Seven shades of sun beams have various therapeutics impacts, for example, yellow, violet, indigo, blue, green, orange, red. These seven hues work adequately for being sound and compelling in treatment of various sicknesses.

Chromotherapy depends on the impact of hued light with various frequencies on human neurohormonal pathways, accurately on melatonin and serotonin pathways in cerebrum. There is a proof that unmistakable electromagnetic range of light we see as hues can have sway on human well-being, Circadian beat or natural clock is intricate basic physiological and natural cycle in human living being. The fundamental speculation of chromotherapy is that particular shades of the unmistakable range are activators or inhibitors of complex physiological, organic and biochemical procedures in human cerebrum, for example, blend of different neurohormones (Radeljak, et al., 2008).

In chromo treatment, diverse turmoil were treating by uncovering water and oil to sun for determined period in shaded jugs and hued glasses, are utilized as gadgets.

12.6.9 AIR THERAPY

Outside air is most fundamental for good well-being. The upsides of air treatment can be accomplished by methods of air shower. Everybody should clean themselves up for 20 minutes or over 20 minute every day, if conceivable. Mix of virus rub at morning and exercise is generally profitable. In this treatment, one should walk each day without garments or wearing light garments at a disengaged clean spot where sufficient natural air is accessible. Another substitute strategy is to build the room without rooftop and encompass by screen like dividers in order to pursue the free section of air, yet anticipating any view inside the room (Kulkarni et al.,2015; Sharma, 2002).

Madanmohan et al. (2013) assessed that the eight-week yoga treatment program produces noteworthy improvement in anthropometric and cardio-vascular parameters and lipid profile in patients with basic hypertension. It is likewise presumed that an extensive yoga treatment program can possibly improve the gainful impact of standard therapeutic administration of basic hypertension and can be utilized as a viable reciprocal or integrative treatment program. This investigation gives a logical premise to additionally connected research on the impact of yoga treatment in hypertensives.

System: So as to cold impact of water or cold air, the operational hubs, which control the flow, exchange the blood to the surface in huge sums, flushing the skin with warm, red, blood vessel blood. The progression of the blood is extraordinarily quickened and end of unfortunate issue from the outside of the body is correspondingly expanded.

Advantages: Air shower has been helping and impacting upon a large number of nerve endings on the outside of the body. Air treatment has great outcomes in the case of the occurrence of anxiety, neurasthenia, stiffness, skin, mental, and other perpetual issues.

12.6.10 MAGNET THERAPY

Magnet treatment is a clinical procedure wherein human body parts are dealt with and restored by the use of magnet to the body of the patients. It is the most straightforward, low and totally easy procedure of treatment with no reaction or delayed consequences. Only a single material is utilized in magnet treatment (Vergari, 2002).

Magnet is connected legitimately to the body parts by the restorative treatment. Magnet is accessible in various forces for the general treatment of the body. Attractive magnetic belts are additionally accessible for various body parts, for example, belly, knee, wrist; and magnetic jewelry, glasses, and wrist trinkets are additionally utilized for treatment (Mundell, 2013).

Advantages:

1. Magnet therapy helps in adjusting the vitality.
2. It improves flow to the connected region.
3. It increases the glowness in the body.

12.7   CONCLUSION

The modern medical system treats the symptoms and only suppresses the illness; however, it does not find the important cause behind the illness. Toxic medicines which can suppress or relieve some ailments usually have harmful side –effects. Drugs usually delay the self-healing efforts of the body and make recovery more difficult. Drugs also affect dietary deficiencies by destroying nutrients, utilizing them, and preventing from the absorption. Moreover, the toxicity they produce at a time when the body is least capable of coping with it. The power to restore health thus exists not in drugs, but within nature. The modern systems approach is more on combative lines after the disease has set in, whereas nature cure system lays greater emphasis on preventive method and adopts measures to attain and maintain health and prevent disease. The modern medical system treats each disease as a separate entity, requiring specific drugs for its cure, whereas the nature cure system treats the organism as a whole body and search to restore harmony to the whole of the patient's being.

Naturopathy plays the important role in the today's lifestyle; it cures the various diseases from their root by removal of toxic or morbid matters from the body.

KEYWORDS

- **naturopathy**
- **acupuncture**
- **Vedas**
- **yoga**
- **immune system**

REFERENCES

Abu, N. A.; Norasikin, M. A.; Maksom, Z. An interactive 3D acupressure model for self-treatment in reducing pain. *J Teknol*. 2015, 77(19), 97–105.

Antunez, L. E.; Puértolas, B. C.; Burgos, B. I.; Porto Payán, J. M.; Torres Piles, S. T. Effects of mud therapy on perceived pain and quality of life related to health in patients with knee osteoarthritis. *Reumatol Clin*. 2013, 9(3), 156–160.

Baer, H. A. The potential rejuvenation, as consequence of the holistic movements. *Med Anthropol.* 1992, 13(4), 369–383.

Bolar, P. K. *A Complete Handbook of Nature Cure.* 1999, healthlibrary.com (accessed January 10, 2019).

Bradley, R. S. Philosophy of naturopathic medicine. In: Murray MT, Pizzorno JE, eds., *Textbook of Natural Medicine*, 2nd ed. New York, NY: Churchill Livingstone, 1999, pp. 41–49.

Bronner, F., ed. *Nutritional and Clinical Management of Chronic Conditions and Diseases.* Boca Raton, FL: CRC Press, 2005.

Codish, S.; Abu-Shakra, M.; Flusser, D.; Friger, M.; Sukenik, S. Mud compress therapy for the hands of patients with rheumatoid arthritis. *Rheumatol Int.* 2005, 25(1), 49–54.

Downey, C. Naturopathic medicine. In: Novey, D. W., ed. *Clinician Complete Reference to Complementary/Alternative Medicine.* Philadelphia, PA: Mosbey, 2000, pp. 274–282.

Ernst, E.; Lee, M. S. Acupressure: An overview of systematic reviews. *J Pain Symptom Manage.* 2010, 40(4), e3–e7.

Hamidli, S.; Yetkin, A.; Yetkin, Y. The meaning of life: Health, disease and the naturopathy. *Int J Psychol Couns.* 2010, 2(1), 9–16.

Hufford, D. J. Evaluating complementary and alternative medicine: The limits of science and of scientists. *J Law Med Ethics.* 2003, 31(2), 198–212.

Ichiman, Y. *Standard Textbook for Medical Massage (Anma) for Visual Impaired People in Asia.* Ibaraki, Japan: National University Corporation, Tsukuba University of Technology, 2016.

Kulkarni, D. D.; Doddoli, S. G.; Bhogal, R. S. Effect of yoga training on bio-energy dynamics with reference to bioelectrical impedance and Tridosha in dominant nostril breathing types—a pilot study. *Int J Innov Appl Res.* 2015, 3(10), 47–58.

Lutz, C. A.; Przytulski, K. R. *Nutrition and Diet Therapy*, 4th ed. Philadelphia, PA: F. A. Davis, 2006.

Madanmohan, A. B.; Bhavanani, Z. S.; Vithiyalakshmi, L.; Dayanidy, G. Effects of eight week yoga therapy program on cardiovascular health in hypertensives. *Indian J Tradit Knowledge.* 2013, 12(3), 535–541.

Michalsen, A.; Li, C. Fasting therapy for treating and preventing disease—current state of evidence. *Forsch Komplementmed.* 2013, 20(6), 444–453.

Mioara, D.; Constantin, M.; Horia, L. Hidrotherapy. *Balneo Res J.* 2012, 3(1), 23–27.

Mundell, W. C. Static magnet therapy for pain relief: A critical review. *OA Altern Med.* 2013, 1(2), 19.

Murray Michael, N. D.; Pizzorno Joseph, N. D. *Encyclopedia of Nature Medicine*, 2nd ed. Roseville, CA: Prima Publishing, 1998.

Nakamura, S.; Hisamura, R.; Shimoda, S.; Shibuya, I.; Tsubota, K. Fasting mitigates immediate hypersensitivity: A pivotal role of endogenous D-beta-hydroxybutyrate. *Nutr Metab (Lond).* 2014, 11: 40.

Pandey, M.; Verma, R. K.; Saraf, S. A. Nutraceuticals: New era of medicine and health. *Asian J Pharm Clin Res.* 2010, 3(1), 11–15.

Radeljak, S.; Zarkovic-Palijan, T.; Kovacevic, D.; Kovac, M. Chromotherapy in the regulation of neurohormonal balance in human brain-complementary application in modern psychiatric treatment. *Coll Antropol.* 2008, 32(Suppl. 2): 185–188.

Rastogi, R. Current approaches of research in naturopathy: How far is its evidence base? *J Homeop Ayurv Med.* 2012a, 1–2.

Rastogi, R. Therapeutic uses of mud therapy in naturopathy. *Indian J Tradit Knowledge*. 2012b, 11(3), 256–259.

Seow-Choen, F. The physiology of colonic hydrotherapy. *Colorectal Dis*. 2009, 11, 686–688.

Sharma, K. D. Yoga and naturopathy: The true Science of healing, *Indian J Tradit Knowledge*. 2002, 1(1), 22–24.

Shetty, B.; Shetty, G. B.; Shetty, P.; Shantaram, M. Effect of naturopathic fasting therapy on serum lipid profile and haematological indices in healthy individuals. *Res J Pharma Biol Chem Sci*. 2015, 6(2), 1295–1299.

Singh, J.; Sinha, S. Classification, regulatory acts and applications of nutraceuticals for health. *Int J Pharma Biol Sci*. 2012, 3(1),177–187.

Thambirajah, R. *Energetics in Acupuncture: Five Element Acupuncture Made Easy*. Edinburgh, NY: Churchil Livingstone/Elsevier, 2011, pp. 1–481.

Vergari, G. *Magnet Therapy: Discover the Powerful New Force in Health and Recovery*. Norstar Publishing, United Kingdom, 2002, p. 132.

Index

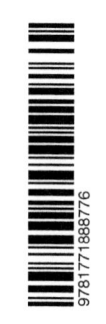